Internet of Things

Technology, Communications and Computing

Series Editors
Giancarlo Fortino, Rende (CS), Italy
Antonio Liotta, Edinburgh Napier University, School of Computing, Edinburgh, UK

The series Internet of Things - Technologies, Communications and Computing publishes new developments and advances in the various areas of the different facets of the Internet of Things. The intent is to cover technology (smart devices, wireless sensors, systems), communications (networks and protocols) and computing (theory, middleware and applications) of the Internet of Things, as embedded in the fields of engineering, computer science, life sciences, as well as the methodologies behind them. The series contains monographs, lecture notes and edited volumes in the Internet of Things research and development area, spanning the areas of wireless sensor networks, autonomic networking, network protocol, agent-based computing, artificial intelligence, self organizing systems, multi-sensor data fusion, smart objects, and hybrid intelligent systems.

Indexing: *Internet of Things* is covered by Scopus and Ei-Compendex **

More information about this series at https://link.springer.com/bookseries/11636

Yassine Maleh • Lo'ai Tawalbeh • Saad Motahhir
Abdelhakim Senhaji Hafid
Editors

Advances in Blockchain Technology for Cyber Physical Systems

Springer

Editors
Yassine Maleh
ENSA Khouribga
Sultan Moulay Slimane University
Beni Mellal, Morocco

Lo'ai Tawalbeh
Computing and Cyber Security
Texas A&M University
San Antonio, TX, USA

Saad Motahhir
ENSA
Sidi Mohamed Ben Abdellah University
Fez, Morocco

Abdelhakim Senhaji Hafid
Department of Computer Science
University of Montreal
Montreal, Quebec, QC, Canada

ISSN 2199-1073 ISSN 2199-1081 (electronic)
Internet of Things
ISBN 978-3-030-93645-7 ISBN 978-3-030-93646-4 (eBook)
https://doi.org/10.1007/978-3-030-93646-4

This Springer imprint is published by the registered company Springer Nature Switzerland AG
The registered company address is: Gewerbestrasse 11, 6330 Cham, Switzerland

Contents

About the Editors

Yassine Maleh is a cybersecurity professor and practitioner with industry and academic experience. He has a Ph.D. degree in Computer Sciences. Since 2019, he is working as a professor of cybersecurity at Sultan Moulay Slimane University, Morocco. He worked for the National Ports Agency (ANP) in Morocco as a senior security analyst from 2012 to 2019. He is the founding chair of IEEE Consultant Network Morocco and founding president of the African Research Center of Information Technology and Cybersecurity. He is a senior member of IEEE and member of the International Association of Engineers and the Machine Intelligence Research Labs. Dr. Maleh has made contributions in the fields of information security and privacy, Internet of Things security, and wireless and constrained networks security. His research interests include information security and privacy, Internet of Things, networks security, information system, and IT governance. He has published over 70 papers (book chapters, international journals, conferences/workshops), 12 edited books, and 3 authored books. He is the editor-in-chief of the *International Journal of Information Security and Privacy* (IJISP) and the *International Journal of Smart Security Technologies* (IJSST). He serves as an associate editor for *IEEE Access* (2019 Impact Factor 4.098), the *International Journal of Digital Crime and Forensics*, and the *International Journal of Information Security and Privacy*. He was also a guest editor of a special issue on "Recent Advances on Cyber Security and Privacy for Cloud-of-Things" of the *International Journal of Digital Crime and Forensics*, Volume 10, Issue 3, July–September 2019. He has served and continues to serve on executive and technical program committees and as a reviewer of numerous international conferences and journals such as Elsevier *Ad Hoc Networks*, *IEEE Network Magazine*, *IEEE Sensor Journal*, *ICT Express*, and Springer *Cluster Computing*. He was the publicity chair of BCCA 2019 and the general chair of the MLBDACP 19 symposium and ICI2C'21 Conference.

Lo'ai Tawalbeh completed his Ph.D. degree in Electrical and Computer Engineering from Oregon State University in 2004, and M.Sc. in 2002 from the same university with GPA 4/4. Dr. Tawalbeh is currently an associate professor in the

Department of Computing and Cyber Security at Texas A&M University-San Antonio. Before that he was a visiting researcher at the University of California-Santa Barbra. Since 2005 he taught/developed more than 25 courses in different disciplines of computer engineering and science with a focus on cybersecurity for the undergraduate/graduate programs at New York Institute of Technology (NYIT), DePaul's University, and Jordan University of Science and Technology. Dr. Tawalbeh won many research grants and awards with over than 2 million USD. He has over 80 research publications in refereed international journals and conferences.

Saad Motahhir, Eng., Ph.D., IEEE Senior Member has previous expertise acting in industry as Embedded System Engineer at Zodiac Aerospace morocco from 2014 to 2019, and more recently became a professor at ENSA, SMBA University, Fez, Morocco, since 2019. He received the engineering degree in embedded system from ENSA Fez in 2014. He received his Ph.D. Degree in Electrical Engineering from SMBA University in 2018. He has published a good number of papers in journals and conferences in the last few years, most of which are related to photovoltaic (PV) solar energy and embedded systems. He published a number of patents in Morocco patent office. He edited a Springer book and acted as a guest editor of different special issues and topical collections. He is a reviewer and is on the editorial board of different journals. He was associated with more than 30 international conferences as a program committee/advisory board/review board member.

Abdelhakim Senhaji Hafid is Full Professor at the University of Montreal. He is the founding director of Network Research Lab and Montreal Blockchain Lab. He is a research fellow at CIRRELT, Montreal, Canada. Dr. Hafid published over 240 journal and conference papers; he also holds three US patents. He supervised the graduation over 50 graduate and postgraduate students. Prior to joining the University of Montreal, he spent several years, as senior research scientist, at Bell Communications Research (Bellcore), NJ, USA, working in the context of major research projects on the management of next-generation networks. Dr. Hafid was also assistant professor at Western University (WU), Canada; research director of Advance Communication Engineering Center (venture established by WU, Bell Canada and Bay Networks), Canada; researcher at CRIM, Canada; visiting scientist at GMD-Fokus, Germany; and visiting professor at the University of Evry, France. Dr. Hafid consulted for a number of telecommunication companies and startups in North America; he also gave talks/keynotes in a number of international conferences. He co-founded Tipot Technologies Inc. (research and development Platform for IoT). Dr. Hafid has extensive academic and industrial research experience in the area of the management and design of next-generation networks. His current research interests include IoT, Fog/edge computing, Blockchain, and intelligent transport systems

Introduction

Yassine Maleh

The fourth industrial revolution, or Industry 4.0, brings together digital and physical technologies to create responsive and interconnected operations. Companies use AI, robotics, edge computing, and the cloud to make informed and timely decisions from the supply chain to the smart factory [1]. Solutions designed for the Industrial Internet of Things (IIoT) use connected sensors and edge devices to help improve product quality and operational efficiency in factories in real time. From workshops and worksites to decision-making stations, tools and equipment are now networked to become interconnected. Through sensors, detectors, and other monitoring devices, machines and operators are collecting increasingly huge amounts of data, the management, analysis, and exploitation of which constitute enormous challenges for companies [2]. The Internet of Things brings a wealth of promise in productivity, quality, performance, and security [3].

This development helps reduce nonconformances since connected sensors and detectors record failure data more accurately, making it easier to identify the source of nonconformance and then remedy it. Implementing Internet of Things tools contributes to quality improvement through better supervision and monitoring of the various production stages.

However, rapid advancements in enabling technologies have also exposed such systems to severe and profound risks. If such risks are not managed, the benefits provided by them will soon be lost. Advanced computing and Blockchain have great potential to create new foundations for most distributed systems by efficiently establishing trust among nodes [4]. It is a fundamental technology to enable decentralization and play an essential role in critical Industrial IoT applications [5].

The Blockchain has become famous for its ability to secure bitcoin transactions. It could very quickly prove to be indispensable for the deployment of all connected

Y. Maleh (✉)
ENSA Khouribga, Sultan Moulay Slimane University, Beni Mellal, Morocco
e-mail: y.maleh@usms.ma

Y. Maleh et al. (eds.), *Advances in Blockchain Technology for Cyber Physical Systems*, Internet of Things, https://doi.org/10.1007/978-3-030-93646-4_1

objects, by adding a layer of security via a chain associated with the identity of the connected object, making it resistant to hacking attempts [6]. This identification chain, characteristic of the Blockchain, allows connected objects to receive commands and securely communicate with each other. Indeed, the Blockchain will enable objects to interact with each other without communicating via a third party, thus limiting the outflow of information or potential attacks from the outside [7].

Blockchain provides a practical solution to enable a secure, decentralized public ledger that offers a huge plethora of new and exciting technology applications in several areas, such as the Internet of Things (IoT), cyber-physical systems (CPS), manufacturing and supply chain.

The new generation of the smart industry is based on automation, hyper-connectivity through cyber-physical systems, Industrial Internet of Things (IIoT), and Big Data. In terms of growth, IIoT is an accurate indicator of the speed of adopting this cutting-edge technology.

Blockchain technology has infiltrated all areas of our lives, from manufacturing to healthcare and beyond. CPS is a field that has been significantly affected by this technology and maybe more so in the future.

In this context, this book will go in-depth, showing how Blockchain can be used for CPS applications. It comprises a good balance between theoretical and practical issues, covering case studies, experience, evaluation approaches, and best practices, explicitly considering many aspects of cyber-physical system applications, including, Internet of Things, Intelligent Transportation, Supply Chain, Smart Grids, etc. The book will explain relevant concepts, review the state of the art, present representative Blockchain-based solutions that have been proposed, and discuss open challenges.

1 Research Challenges in Blockchain Applications for Cyber-Physical System

Although Blockchain is still mostly in experimental form, companies worldwide are eager to use this technology to help them reduce costs, improve processes, improve tracking and security of product data, and reduce fraud and counterfeiting.

Coupled with new technologies (AI, connected objects, industrial infrastructure, digitalization), Blockchain technology can propel companies toward Industry 4.0: connected and interoperable machines, information transparency, connected supply chains, etc.

In the industrial field, this technique should stand out. A field that requires a significant flow of information (documents to be validated, payments to be made), many stakeholders, and a need for control and transparency on the entire production chain.

- *Blockchain and intermodality*: Smart cities include important innovations in their development concerning transportation. First of all, for the problems of relieving congestion on certain roads, in cities for example. Innovative autonomous vehicles are raising high hopes in this respect. Combined with Blockchain, autonomous transport could manage traffic with maximum efficiency. It would be achieved by connecting vehicles to each other and traffic lights, toll booths, etc. The gain in terms of road congestion would be enormous. In addition to securing all exchanges between vehicles and connected objects related to traffic, the Blockchain also allows combined processing of all processed data. This could also help optimize autonomous vehicles' charging times and ensure the consumer receives an optimal price.

- *Saving energy with Blockchain*: Blockchain can become indispensable to the smart city by enabling it to achieve one of its stated goals: energy efficiency. Indeed, enabling interoperability between connected objects ensures optimized operation of objects and, therefore, automatically saves energy [8]. The same is true for transport: optimizing journeys, vehicle consumption, and recharging will optimize the electricity consumption of the entire road sector. Moreover, Blockchain is increasingly used to save energy: it manages electricity distribution on the networks, according to supply and demand. As a result, the distribution adjusted, but intermittent storage makes sense, thanks to this new efficiency [8].
- *Product management*: Blockchain technology can contribute to supply chain product management in multiple ways by providing product transparency, traceability, and security [9]. Authors in [10] propose a product ownership management system (POMS) that empowers customers to reject counterfeit products that might have cloned genuine RFID tags. They implement two smart contracts in this system, one for managing the manufacturers' information and the other for managing the products' information, and together, they verify the possession of products.
- *Power industry*: The power industry has undergone significant transformations over the past several years, with utilities embracing newer technologies and newer power generation sources. With a deluge of smart and IoT-enabled devices (ranging from smartphones to smart meters to electric vehicles) having variable power demands and mushrooming of many power generation schemes, the power grids are becoming very complex to handle. Blockchain as a tool can accelerate this global energy transformation by lowering the transaction costs and operating the grid in a more efficient manner [11].
- *Manufacturing industry*: In the manufacturing industry, manufacturers have to publish technical manuals of their products, distributed in the repair and maintenance departments. These technical records have to be released and timely updated, which is a tedious process and involves tons of paperwork. By using Blockchain technology, the technical publication can rest on this framework and is accessible to Blockchain users without worrying about the version changes or losing the latest publication [9]. Blockchain has significantly enhanced the working efficiency in the manufacturing industry by uploading the data on the shared ledger. For example, in the automobile manufacturing industry, tracking of the spare parts becomes vital, because the availability of the parts in real time is not known [12].

- *Security and regulations*: Security also remains a concern. Organizations are not interested in an open identity model. Banks and regulators want to have tight control. The development of a single digital identity passport will be a critical step [13].

  Regulation is also critical to creating an open digital environment for commerce and financial transactions [14]. Current physical certificates must be digitized to take full advantage of a fully electronic system [15]. In addition to the concerns already expressed, we need to consider other obstacles on the road to full adoption of these technologies and find the right answers to questions like:

  - Who will take primary responsibility for maintaining and managing the Blockchain?
  - What about admitting new participants to the Blockchain?
  - In terms of transactions, who do we turn to for validation and who determines the visibility of these?

2 Research Solutions

Chapter "Blockchain for Cyber-physical Systems: Challenges and Applications" of this book elucidates a deep analysis and review of various CPS applications where Blockchain has been used. Healthcare systems, transportation, and cybersecurity are many applications that can benefit from Blockchain technology and discussed in this chapter.

Chapter "Blockchain-Based Medical Records System" proposes a system that allows the interoperable exchange of medical records with proper authorization. This system ensures that a user of the system must ask for a patient's permission to view and upload their medical records [16]. To achieve this, smart contracts on the Blockchain network were deployed. Through the use of the deployed smart contract, a patient can grant read and write permissions to different entities namely hospitals and insurance agencies. The use of Blockchain ensures that the authorization mechanism is tamper-proof.

Chapter "Security of IoT Based e-Healthcare Applications Using Blockchain" summarizes the potential attacks on IoT sensor-based healthcare systems along with potential security provisions to such applications by using Blockchain technology using a deep literature survey. It then proposes a device-independent integrated approach for establishing a trust-based immutable safety model in IoT-based healthcare applications by securing patient data using Blockchain.

Chapter "Privacy-Preserving k–Means Clustering over Blockchain-Based Encrypted IoMT Data" proposes privacy-preserving $k-$means based on Paillier. All transactions are recorded in a distributed, immutable ledger for the participants' authenticity and secure data sharing [17]. Three medical datasets are used, and performance analysis exhibits that secure $k-$means achieves accuracies of 94.95%, 81.88%, and 78.10% on BCWD, HDD, and DD dataset, where standard techniques provide 96.60%, 81.00%, and 77.00%, respectively. On the other hand, secure $k-$means takes 2200 seconds, 1500 seconds, and 2605 seconds, where the standard method takes 3357 seconds, 2534 seconds, and 3709 seconds on BCWD, HDD, and DD datasets, respectively. Therefore, secure $k-$means can protect the privacy of the data owners, achieves almost comparable accuracy to the conventional methods, and outperforms them in time consumption.

Chapter "Blockchain for Smart Transport Applications" focuses on security aspects of smart transportation and how Blockchain can be used as a security solution for smart transportation systems [18]. It briefly explains Blockchain and its types and its advantages. The chapter outlines few schemes that provide security solutions for the Blockchain-based smart transportation system. A comparative analysis is also provided to study the effectiveness of the schemes.

Chapter "Blockchain-Based CPS and IoT in the Automotive Supply Chain" analyzes the application of Blockchain technology in different sectors, and then discusses its implementation in the automotive supply chain. This chapter highlights the opportunities cyber-physical production systems and Industrial IoT bring to automotive supply-chain management [19]. A Blockchain-based Cyber-Physical Production Systems (CPPS) and IIoT model is suggested to enhance the SCM efficiency. An implementation of this model in a car manufacturing factory is presented with a focus on its advantages and limitations. The chapter unravels the Blockchain system, understands the CPS and CPPS with Industry 4.0, discusses CPPS in automotive supply chain and its challenges, and presents how Blockchain makes the whole automotive industry smarter and more efficient.

Internet of Vehicles (IoV) over Named Data Networking (NDN) has recently emerged as a new model to enable vehicular communications and improve road safety [20]. Nevertheless, a malicious vehicle can disseminate fake content to other vehicles in the network, affect driving decisions, and result in traffic congestion or even accidents. Blockchain technology has brought believable achievements in every research field such as academia, healthcare, genetic engineering, and transportation management, where preserving security is the primary priority. This chapter proposes a new system that brings Blockchain to NDN-based IoV namely BIoVN. In addition, a novel bioinspired algorithm of name HoneyGuide is introduced in the data forwarding process that is used in BIoVN. This chapter aims to secure vehicular communications over NDN. These previously mentioned propositions, alongside the ease-of-use of machine learning techniques, inspire the authors of chapter "BIoVN: A Novel Blockchain-Based System for Securing Internet of Vehicles Over NDN Using Bio-inspired HoneyGuide" to address the issue of the security of the cached and forwarded content.

Blockchain technology can safely and reliably track the creation process of digital twins. To ensure data security in the case of multiple untrusted parties sharing data and improve users' satisfaction with the use of digital assets in daily transactions [21]. Chapter "Blockchain-Based Communication for Digital Twins" combines the digital twins technology with Blockchain technology, and for multiple scenarios in Blockchain networks, a credible and efficient edge computing resource allocation method is proposed based on deep reinforcement learning (DRL) theory.

The log storage system of the Blockchain is mainly used as the interface for writing and reading data operation logs, and a hybrid storage strategy is put forward for the log storage system. As for the trusted resource allocation under the decentralized model of the Blockchain network, to prevent the system from directly offloading the computing tasks submitted by the user each time to the edge server, first, the task is submitted to the system. Then a more reasonable resource allocation strategy is implemented with user satisfaction as the standard. The simulation experiment results show that the improved environment after introducing the log storage system reduces the storage overhead by about 75% compared with the Blockchain benchmark environment. When the number of servers is greater than 5, compared with other resource allocation algorithms, user satisfaction of the resource allocation algorithm based on DRL theory has been significantly improved compared with Q-learning algorithm. Its user satisfaction has increased by 15%. In conclusion, the digital twins use credible source data as input. In this process, the Blockchain ensures data management security, and the data analysis is performed to predict events and evaluate related factors. The resource allocation method based on DRL realizes credible resource utilization based on recorded data on the Blockchain.

Chapter "The Role of Blockchain Technology in Enhancing Security Management in the Supply Chain" analyzes the role of Blockchain technology in enhancing security management in supply chain management. The information was collected from various sources, organized into two privacy and security cases variables, and analyzed using EViews software. Data analysis was done by performing three tests: correlation, cross-correlation, and Granger's causality test. The findings revealed a correlation between the use of Blockchain technology in the supply chain and security and privacy complaints among organizations and individuals [17]. It also reveals a unidirectional causality between security concerns and the degree of application of Blockchain technology in the supply chain. The chapter represents one of the very few empirical studies that have been done about the subject. It's expected to change the public's understanding of Blockchain technology related to security and privacy, two of the essential aspects of supply chain management in contemporary corporate society. The main limitation of this study comes in the form of a lack of readily available data. Several data sources, including existing literature and statistics from reputed sites like Statista, had to be used to ensure sufficient information was collected to facilitate the study's key objectives successfully.

Chapter "Using Hyperledger Fabric Blockchain to Improve Information Assurance of IoT Devices for AI Model Development" shows that using Hyperledger Fabric Blockchain for Navy logistics assets can be applied to various be applied to data supporting artificial intelligence (AI) and software development in terms of system safety and the timely acquisition of data. Data-driven AI/machine learning (ML) requires trusted data for their use in AI functions [22]. It requires significant amounts of training data from diverse sources, including Internet of Things (IoT) devices/sensors. Unauthorized alterations to data supporting AI/ML could go unnoticed within the AI function build process but surface during operation in hazards affecting unwanted human death or resource destruction. AI/ML controlling hardware usually falls into the two highest software control categories: Levels 1 and

2, risk of death, disability, or resource destruction. The chapter shows how trust can be implemented through distributed consensus to ensure that only authorized people can modify data and that the modification is traceable and transparent. Distributed ledgers provide system safety through BC provenance, immutability, and policy enforcement through smart contracts.

The decentralized, open and unmodifiable nature of the Blockchain makes it a transparent, publicly verifiable system. In addition, since records are replicated to many distributed nodes, the Blockchain architecture allows you to eliminate the problem of a single point of failure [23]. Combining these properties will enable us to consider Blockchain technology as the basis for many applications. Such applications can be solutions for the Internet of Things (IoT) and Cyber-physical Systems (CPS). The possibility of interaction between a huge number of heterogeneous devices is required. In this regard, chapter "Developing Instrument for Investigation of Blockchain Technology" presents an approach to researching several aspects of Blockchain formation. The main goal of the work is to create a research tool for Blockchain technology and analyze its capabilities. Such an analysis allows us to determine the scope of the possible use of the Blockchain, depending on the underlying construction methods, in various fields of application, such as the Internet of Things (IoT) and Cyber-physical Systems (CPS) with specific application orientation. To perform the analysis, those aspects that explain the basic principles of Blockchain formation are selected. Some minor tasks are especially simplified so as not to distract attention from the main ones. The chapter also demonstrates the main possibilities of placing information in a distributed registry. A study of the specific implementation of the Blockchain formation system with the previously indicated features has been carried out [24]. A model of a multinode system is constructed, and an analysis of the functioning of a number of software solutions in the field of research modeling of Blockchain construction is carried out. The time characteristics of the system depending on the specified parameters are investigated.

Chapter "Trust Models for Blockchain-Based Self-Sovereign Identity Management: A Survey and Research Directions" provides the first thorough review in the literature addressing trust management for Blockchain-based self-sovereign identity. A formal and comprehensive trust model proposed for Blockchain-based self-sovereign IDM will be explored. Besides reviewing trust requirements, the chapter also surveys the state-of-the-art Blockchain technology for self-sovereignty in identity management [25]. This survey provides a critical analysis of existing research that sheds light on various opportunities to enhance the security and privacy of Blockchain-based self-sovereign identity management and improve trust management. The chapter concludes by presenting research gaps and suggestions for future work in the area.

Blockchain technology is the fastest developing technology in recent years, and it has had a significant impact on a wide range of industries and companies in the industry 4.0 era [9]. Privacy, security, and trust are three of the most pressing issues facing digital marketing today. In chapter "Blockchain-Enabled Trust Management for Digital Marketing in the Industry 4.0 Era," the authors strive to investigate security, privacy, and trust challenges in digital marketing, and how Blockchain technology can be used to influence digital marketing to increase consumer trust and security. It was done via a systematic review of the literature published between 2017 and 2021. The review identified that the use of Blockchain technology in digital marketing and marketing management would continue to grow. It has proven effective in providing solutions to both existing and upcoming company challenges and situations in Industry 4.0. Blockchain can influence digital marketing by removing intermediaries and delivering trusted cybersecurity services with a high level of transparency and accountability. Also discussed are the possible problems and limitations of Blockchain-enabled digital marketing.

The last chapter "Applying Advanced Wireless Network Cluster-Tree Topology to Optimize COVID-19 Sanitary Passport Blockchain-Based Security in a Constrained IoT Platform" proposes a smart, synchronized, and secure medical IoT platform that monitors a public area using a set of tests. The people in place must present their vaccination QR code. A first test is performed to read this QR code. If the code is validated, it conducts a second test to validate the person's identity corresponding to the vaccination code using the facial recognition algorithm. In the positive case, the person will be able to access the public area. Citizens who are still not vaccinated must present a negative PCR test not exceeding 48 hours. Then, two verification tests were performed, one test to read the barcode of the PCR test and a second test for facial recognition of its carrier. In the situation where a person is presented with neither their vaccination certificate nor a PCR test, we develop a strategy of three tests with three IoT nodes.

3 Conclusion

This book is intended for researchers, practitioners in the field, engineers, and scientists involved in designing and developing protocols and Blockchain applications for cyber-physical systems. It can also be used as the recommended textbook for undergraduate or graduate courses. The intended audience includes college students, researchers, scientists, and engineers, to advance the missions of anticipating, prohibiting, preventing, preparing, and responding to internal security. The book covers a wide range of CPS applications and scenarios, where Blockchain technology can be applied. The material covered is readable and a solid base for penetration into the comprehensive reference material on advanced communication concepts in the related research field. It is also a reference for selection by the audience with different but close to the field backgrounds. The wide variety of topics it presents offers readers of this book multiple perspectives on various Blockchain techniques and applications for cyber-physical systems.

References

1. S. Mumtaz, A. Alsohaily, Z. Pang, A. Rayes, K.F. Tsang, J. Rodriguez, Massive internet of things for industrial applications: addressing wireless IIoT connectivity challenges and ecosystem fragmentation. IEEE Ind. Electron. Mag. **11**(1), 28–33 (2017). https://doi.org/10.1109/MIE.2016.2618724
2. L. Zhou, D. Wu, J. Chen, Z. Dong, When computation hugs intelligence: content-aware data processing for industrial IoT. IEEE Internet Things J. **5**(3), 1657–1666 (2018). https://doi.org/10.1109/JIOT.2017.2785624
3. A. Karati, S.H. Islam, M. Karuppiah, Provably secure and lightweight certificateless signature scheme for IIoT environments. IEEE Trans. Ind. Informatics **14**(8), 3701–3711 (2018). https://doi.org/10.1109/TII.2018.2794991
4. P. Zhuang, T. Zamir, H. Liang, Blockchain for cybersecurity in smart grid: a comprehensive survey. IEEE Trans. Ind. Informatics **17**(1), 3–19 (2021). https://doi.org/10.1109/TII.2020.2998479
5. Y. Maleh, M. Shojafar, M. Alazab, I. Romdhani, *Blockchain for Cybersecurity and Privacy: Architectures, Challenges, and Applications* (CRC Press, USA, 2020)
6. W. Zhao, C. Jiang, H. Gao, S. Yang, X. Luo, Blockchain-enabled cyber–physical systems: a review. IEEE Internet Things J. **8**(6), 4023–4034 (2021). https://doi.org/10.1109/JIOT.2020.3014864
7. Z. Zheng, S. Xie, H. Dai, X. Chen, H. Wang, An overview of Blockchain technology: architecture, consensus, and future trends, in *2017 IEEE International Congress on Big Data (BigData Congress)*, (2017), pp. 557–564. https://doi.org/10.1109/BigDataCongress.2017.85
8. T.M. Fernández-Caramès, P. Fraga-Lamas, Towards post-quantum Blockchain: a review on Blockchain cryptography resistant to quantum computing attacks. IEEE Access **8**, 21091–21116 (2020). https://doi.org/10.1109/ACCESS.2020.2968985
9. T. Alladi, V. Chamola, R.M. Parizi, K.R. Choo, Blockchain applications for industry 4.0 and industrial IoT: a review. IEEE Access **7**, 176935–176951 (2019). https://doi.org/10.1109/ACCESS.2019.2956748
10. K. Toyoda, P.T. Mathiopoulos, I. Sasase, T. Ohtsuki, A novel Blockchain-based Product Ownership Management System (POMS) for anti-counterfeits in the post supply chain. IEEE Access **5**, 17465–17477 (2017). https://doi.org/10.1109/ACCESS.2017.2720760
11. E. Mengelkamp, B. Notheisen, C. Beer, D. Dauer, C. Weinhardt, A blockchain-based smart grid: towards sustainable local energy markets. Comput. Sci. Res. Dev. **33**(1), 207–214 (2018). https://doi.org/10.1007/s00450-017-0360-9
12. M. Bhatia, S.K. Sood, Quantum computing-inspired network optimization for IoT applications. IEEE Internet Things J. **7**(6), 5590–5598 (2020). https://doi.org/10.1109/JIOT.2020.2979887
13. T.M. Fernández-Caramés, From pre-quantum to post-quantum IoT security: a survey on quantum-resistant cryptosystems for the internet of things. IEEE Internet Things J. **7**(7), 6457–6480 (2020). https://doi.org/10.1109/JIOT.2019.2958788
14. W. Charles, N. Marler, L. Long, S. Manion, Blockchain compliance by design: regulatory considerations for Blockchain in clinical research. Front. Blockchain **2**, 18 (2019). https://doi.org/10.3389/fbloc.2019.00018
15. B. Bhushan, A. Khamparia, K.M. Sagayam, S.K. Sharma, M.A. Ahad, N.C. Debnath, Blockchain for smart cities: a review of architectures, integration trends and future research directions. Sustain. Cities Soc. **61**, 102360 (2020). https://doi.org/10.1016/j.scs.2020.102360
16. What are the differences between electronic medical records, electronic health records, and personal health records?, (2015). https://www.healthit.gov/providers-professionals/faqs/what-are-differences-between-electronic-medical-records-electronic
17. M.A. Khan, K. Salah, IoT security: Review, blockchain solutions, and open challenges. Futur. Gener. Comput. Syst. **82**, 395–411 (2018). https://doi.org/10.1016/j.future.2017.11.022
18. H. Rathore, A. Mohamed, M. Guizani, A survey of blockchain enabled cyber-physical systems. Sensors (Switzerland) **20**(1), 1–28 (2020). https://doi.org/10.3390/s20010282

19. J. Chi et al., A secure and efficient data sharing scheme based on blockchain in industrial Internet of Things. J. Netw. Comput. Appl. **167**, 102710 (2020). https://doi.org/10.1016/j.jnca.2020.102710
20. M. Singh, S. Kim, Branch based blockchain technology in intelligent vehicle. Comput. Netw. **145**, 219–231 (2018). https://doi.org/10.1016/j.comnet.2018.08.016
21. I. Yaqoob, K. Salah, M. Uddin, R. Jayaraman, M. Omar, M. Imran, Blockchain for digital twins: recent advances and future research challenges. IEEE Netw. **34**(5), 290–298 (2020). https://doi.org/10.1109/MNET.001.1900661
22. Y. Maleh, Y. Baddi, M. Alazab, L. Tawalbeh, I. Romdhani, *Artificial Intelligence and Blockchain for Future Cybersecurity Applications* (Springer, Switzerland, 2021)
23. H. Rathore, A. Mohamed, M. Guizani, A survey of Blockchain enabled cyber-physical systems. Sensors **20**(1) (2020). https://doi.org/10.3390/s20010282
24. J. Rosales, S. Deshpande, S. Anand, IIoT based augmented reality for factory data collection and visualization. Proc. Manuf. **53**, 618–627 (2021). https://doi.org/10.1016/j.promfg.2021.06.062
25. S. Shetty, X. Liang, D. Bowden, J. Zhao, L. Zhang, Blockchain-based decentralized accountability and self-sovereignty in healthcare systems, in *Business Transformation through Blockchain: Volume II*, ed. by H. Treiblmaier, R. Beck, (Springer International Publishing, Cham, 2019), pp. 119–149

Blockchain for Cyber-Physical Systems: Challenges and Applications

Yassine Maleh, Swathi Lakkineni, Lo'ai Tawalbeh, and Ahmed A. Abd El-Latif

1 Introduction

The concept of "blockchain" became widely known in 2009 with the emergence of the cryptocurrency bitcoin, and relatively recently, the prospects of blockchain application in other areas began to be discussed. Blockchain is a technology of storing and processing data in a chain of blocks in computer networks, which does not refer to any particular domain [1]. Each block in the chain can contain arbitrary data, including production processes, which allows describing the possibilities of using this technology in production systems [2].

The growth of cyber-physical systems (CPS) and the Industrial Internet of Things necessitate the resolution of several data interchange and processing issues, including storage, access, security, and so on. Furthermore, there is a contemporary trend toward developing distributed systems rather than centralized ones. One of the most essential characteristics of the Internet of Things is its nodes' autonomy and capacity to interact with one another [3]. This is a service-based interaction in which

Y. Maleh (✉)
ENSA Khouribga, Sultan Moulay Slimane University, Beni Mellal, Morocco
e-mail: y.maleh@usms.ma

S. Lakkineni
University of the Cumberlands, Williamsburg, KY, USA
e-mail: s.Lakkineni@ucumberlands.edu

L. Tawalbeh
Computing and Cyber Security, Texas A&M University, San Antonio, TX, USA
e-mail: Tawalbeh@tamusa.edu

A. A. Abd El-Latif
Menoufia University, Al Minufiyah, Egypt
e-mail: ahmedabdellatif@ieee.org

Y. Maleh et al. (eds.), *Advances in Blockchain Technology for Cyber Physical Systems*, Internet of Things, https://doi.org/10.1007/978-3-030-93646-4_2

specialized nodes deliver services to other nodes in the network. Some blockchain implementations use a smart contract mechanism to enable such interaction.

A smart contract is a self-executing script that is stored with other data in the blockchain. Each smart contract has its algorithm written in a specific programming language and automatically conducts any activities without the involvement of other parties. A smart contract monitors the fulfillment of specified circumstances and, using the given algorithm, takes choices based on them. Because every network member may sign a contract, this mode of engagement extends to Internet of Things nodes. This technique creates a dependable environment for transferring network nodes and makes services visible and uniform. Furthermore, because all contracts are already maintained in blockchain, there is no need to construct a separate service registry [4].

Blockchain technology is highly general; many of its applications are now employed in various domains of human activity [5]. To effectively use all of the benefits of blockchain technology for developing CPS and the Industrial Internet of Things, it is required to design the ideal blockchain network topology based on the tasks to be done and select the most relevant tools (software and hardware).

A cyber-physical system (CPS) results from the integration of computation with physical processes. On the other hand, some argue that it is a system that combines environmental elements with the computational part. Data acquired from the environment and actions in the environment correspond to the environmental elements. From the moment there is a translation of data from the environment into the digital world, it is the responsibility of computing to handle this data. CPSs monitor and control the physical world, with the possibility of having sensor networks, as well as associated actuators [6]. Thus, this type of systems depends on the synergy between physical and computational components. On the other hand, and unlike traditional embedded systems, the CPSs emphasize a holistic view of the system, that is, it is seen as a whole, and not only as several isolated modules. Figure 1 shows a basic architecture of a CPS.

CPSs have applications in a wide variety of areas, including high-reliability medical and life-support systems and devices, traffic control and safety, advanced automotive systems, process control, energy conservation, environmental control, aviation, instrumentation, distributed robotics (telepresence, telemedicine), defense systems, manufacturing, smart structures, and control of critical infrastructures (e.g., power grids, water resources, and communication systems) [7].

So, after a detailed consideration of the concepts of "Industrial Internet of Things" and "cyber-physical production systems," we can go directly to implementing blockchain technology in their structure.

There are two types of blockchain networks: global and private. The first is the most advanced and is typically employed to tackle global challenges. Global peer-to-peer (P2P) networks are extremely stable due to many members, but they are inappropriate for building corporate networks comparable to the industrial networks mentioned above. The fundamental drawback is that all data exchange activities are rigidly bound to the cryptocurrency utilized in one or more global blockchains. Changes in exchange prices in the cryptocurrency market are nearly hard to foresee,

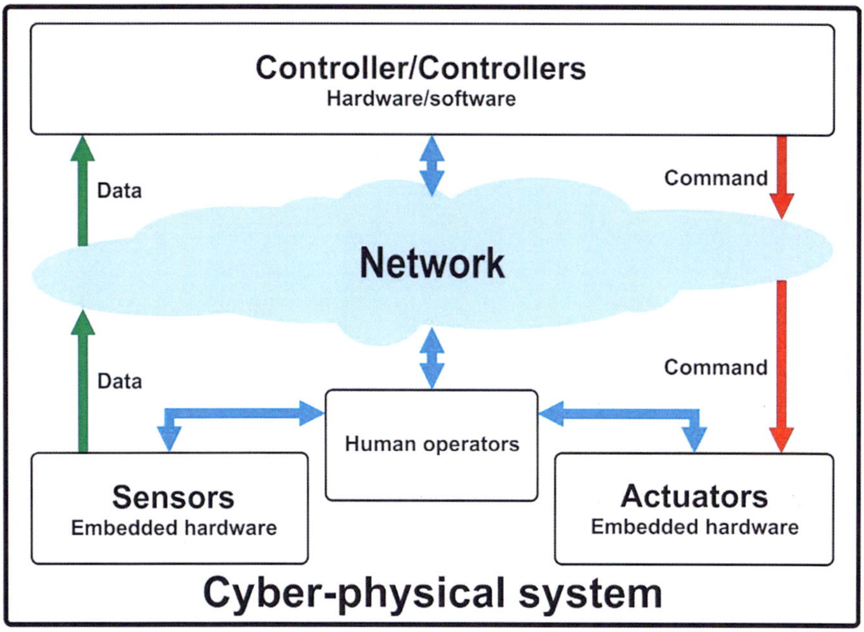

Fig. 1 Cyber-physical system architecture

making the cost of ownership of the planned CPS challenging to estimate. As a result, the architecture now under construction will be built on a private blockchain. The following blockchain functionalities should be implemented in the CPS:

- Organizing a single information space for inter-machine interaction within the CPS
- Ensuring the cybersecurity of the CPS
- Ensuring easy scaling and restructuring of the CPS
- Provision of redundancy of equipment and communication channels
- Organization of a single data storage facility
- Implementation of "digital twins" technology through the use of smart contracts
- Ensuring performance of common tasks for the CPS through the use of smart contracts

Blockchain technology was primarily used to protect storage systems, smart contracts, financial transactions, and notaries. Other applications, like healthcare, supply chain, transportation, and cybersecurity, swiftly recognized its benefits, as the sector realized it could increase its efficiency by implementing blockchain. This has resulted in an active field of inquiry, with researchers and scientists currently looking at various uses for this technology. Among the most commonly mentioned uses are healthcare, transportation, and cybersecurity. The main contributions of this chapter are as follows:

- Provide a detailed and in-depth analysis of the applications in CPS systems where blockchain is implemented.
- Identify the various challenges and limitations of blockchain applications.

This chapter is organized as follows: Section 2 introduces the core concepts of blockchain technology. Blockchain applications in CPS systems including healthcare, transportation, and cybersecurity are discussed in Sect. 3. Section 4 discusses the limits of the blockchain and offers suggestions for the future. Section 5 concludes this work.

2 Blockchain Technology

It all started on November 1, 2008, when an anonymous article titled "Bitcoin: A Peer-to-Peer Electronic Cash System," signed by the pseudonym Satoshi Nakamoto, was published. It described the theoretical foundations for creating a new generation of electronic currency: decentralized, transparent, and independent of central banks and regulators [8]. However, it was not widespread and in the first months it was discussed in academic circles – among cryptographers, mathematicians, and programmers. Bitcoin, the world's first blockchain, which is the embodiment of the concept of this article, was launched on January 3, 2009, and has been successfully functioning for almost 10 years. Several thousand blockchains have emerged during this time, replicating bitcoin with minor variations and bearing little resemblance to its progenitor. Satoshi Nakamoto's identity is still unknown, as he stepped away from bitcoin development in 2010 and has never revealed his name or even the country where he lives. Researchers and journalists have put forward many theories about Satoshi, but none have been confirmed. There have also been many imposters who have claimed to be Satoshi Nakamoto, but not one of them has been able to provide sufficient evidence to back up their claims. To date, the public is likely to accept only one way to confirm Satoshi's identity: ownership of bitcoins he mined in 2009–2010. Satoshi is credited with more than a million bitcoins, which have never come to fruition except for a few test transactions sent to prove the blockchain is working. In particular, Satoshi sent the first-ever 10 BTC blockchain transaction to the famous cryptographer Harold (Hal) Finney, who was actively involved in the discussion of the theoretical foundations of bitcoin. However, while all the glory of creating bitcoin as the world's first workable blockchain undoubtedly belongs to Satoshi Nakamoto, the blockchain did not emerge as an isolated discovery that appeared out of nowhere, from nowhere. In fact, blockchain is the result of the synthesis of several trends in information and financial technology, united by the insights of Satoshi Nakamoto, whoever he may be. Among the technologies and solutions from which bitcoin and blockchain emerged are commonly cited:

1. BitGold, a virtual monetary system created in theory by cryptographer Nick Szabo back in 1998, more than 10 years before bitcoin appeared. BitGold was never put into practice, but its concept is almost identical to bitcoin in some

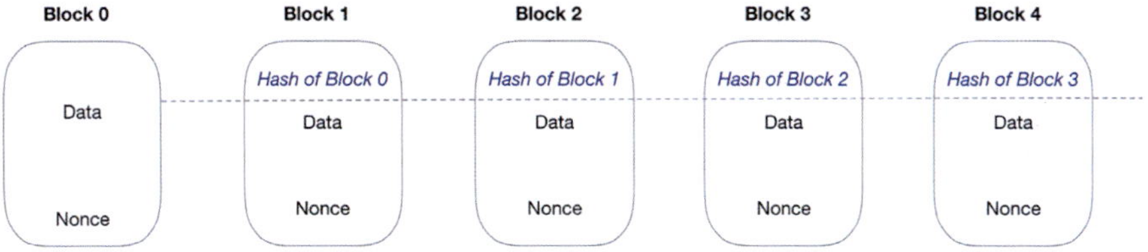

Fig. 2 Transaction records of blockchain

aspects of a decentralized payment network. Nick Szabo has been repeatedly "put on a pedestal" by declaring that he is Satoshi Nakamoto, but Szabo himself denies it. He is also the author of the term "smart contract." The smart contract was realized with cryptocurrencies and will be met many times in this book.

2. Proof-of-Work method, created by cryptographer Adam Beck in 2003 to protect against spam in the e-mail service HashCash. In the HashCash system, a user had to perform a certain amount of computation on their computer to send an email. This spared the system from mass mailings, which were most often commercial or malicious spam. The Proof-of-Work method was used in the bitcoin blockchain to confirm blocks of transactions while ensuring the issuance of new coins. Figure 2 shows the illustration of transaction records of blockchain.
3. Public key cryptography emerged in the last century to ensure the security of electronic communications, including financial transactions. Bitcoin uses elliptic curve cryptography (ECDSA), sends transactions, and creates addresses using a classical pair of private and public keys. Like any other blockchain token, owning bitcoins is analogous to owning the private key needed to send it to another network participant.
4. Using a specific algorithm, hacking technology, that is, obtaining a unique "fingerprint" of the original character set. It is theoretically impossible to get the same hash for two different character sets (so-called collision) or the original character set from the hash. The bitcoin blockchain uses the widely used hashing standard SHA2-256, while other blockchains often use other hashing algorithms. The hash tree is used to form a block header, and calculating a hash of the required complexity is a computational task that must be performed to create a new block and generate bitcoins (mining).
5. BitTorrent technology of a peer-to-peer distributed file storage and transmission network. The block distribution method in the bitcoin network is much the same as the distribution of files using torrents. In addition, peer-to-peer (P2P) file exchanges do not have a single control center, except for the source content and torrent files.

The blockchain industry becomes more mature every year, and many new projects are created with the identified exploitation problems of pioneers such as bitcoin and Ethereum in mind. In addition to the term "blockchain," the phrase "distributed ledger" is also often used. There is some conceptual difference between the two, because a distributed ledger is a broader concept. We can even say that blockchain is a particular case of a distributed ledger. Government and corporate projects often create distributed registries with a hierarchical rather than peer-to-peer structure, where some nodes have a higher level of authority and can influence the entire network and make decisions without the support of the majority.

Blockchain Categories

We distinguish two approaches, the decentralized public approach and the centralized private approach. The concept of blockchain appeared thanks to the emergence of cryptocurrencies and in particular of bitcoin.

Cryptography is used to validate transactions and produce new money in the bitcoin electronic currency system, which is peer-to-peer or decentralized. There will be no need for a trusted third party using the protocol's decentralized fiduciary mechanism. Decentralization dictates that anybody can contribute to the code's development (although an entry fee must be obtained). If a transaction has been approved, it is added to the bitcoin blockchain; otherwise, it is removed. To maintain transaction security and integrity, a variety of cryptographic techniques are employed. At the moment, bitcoin is the most popular public blockchain system on the Internet. Bitcoins are created in exchange for the processing of each transaction, according to the software source code. Users (miners) use their computing power to verify, save, and guarantee transactions and burn them into the blockchain. This work done by miners is called Proof of Work (PoW) and consists of solving algorithmic problems that are part of the bitcoin protocol. Once the transaction is validated, it is time-stamped, added to the blockchain, and is then visible to the recipient and all members of the network. The blockchain described above and used to generate bitcoins is a public blockchain.

A blockchain is considered private if the consensus principle is validated by a small and specified number of participants instead of a public blockchain. The ability to participate in transactions and the verification tasks are both determined by an organization. There is a private network with a predetermined number of nodes, such as a blockchain of location, described in the literature. A cryptography-based technique is not required in a private blockchain. There are no miners, no Proof of Work, and no monetary reward on a private blockchain. These are the primary differences between public and private data storage and transmission systems (public key infrastructure). Thus, a blockchain of any sort is a low-cost, decentralized, and fully secure storage and data transmission mechanism.

Currently, private blockchain applications may be divided into two broad categories:

Applications for asset transfer (monetary use, but not only: securities, votes, industrial patents, connected objects, diploma security, stocks, bonds, etc.).

Applications of blockchain as a registry: Assets and products will be easier to find as a consequence. In a smart contract, the terms and conditions of a contract are

Table 1 Different types of blockchain

Type of blockchain	Name	The register record	Realization of a transaction	Validation	Example
Public	Blockchain without permission	Open to everyone	Anyone	Anyone, as long as they make a significant investment in computing power (Proof of Work) or in holding cryptocurrency (proof of stake)	Bitcoin, Ethereum
Private	Blockchain public permission	Restricted to authorized participants	Participants authorized	All or part of the authorized participants	Banks operating a shared registry
	Consortium	Restricted to authorized participants	Authorized participants	All or part of the authorized participants	Banks operating a shared registry
	Permitted private (enterprise blockchain)	Totally private or limited to a set of authorized nodes	Limited to the network operator	Limited to the operator of the network	Internal bank register shared between subsidiaries

automatically implemented by autonomous programs without the need for human interaction.

The choice of the consensus protocol is a crucial element of blockchains. Behind the technical dimension of this question lies a strong issue of security. However, this classification into a public, permissioned, and private blockchain are reductive, given the many characteristics that can be played on [9]. Table 1 shows two examples of classifications. In practice, these classifications are always imperfect. With the open-source software used in blockchains, creating many variants and playing with multiple parameters is possible, depending on the intended use. Some of these parameters are technical, while others relate to the governance of the system.

Blockchain Explanation

Classic blockchain is much like existing electronic payment systems (EPS) and interbank financial messaging networks (such as SWIFT), but has a number of differences in how it is communicated and managed. Blockchain nodes, called wallets, are analogous to bank accounts, just as a bitcoin address is analogous to a customer's bank account number or SWIFT bank identifier [10]. A blockchain wallet is an instance of blockchain access and transaction software. The wallet can run on almost any electronic device with an operating system, including a server, PC, laptop, or smartphone. A blockchain wallet has similarities to online banking, which provides access to money in a bank account. Still, the blockchain user has sole and complete control over their money and can independently start any number of wallets without providing personal information or documents to any organization. At the same time, the user is solely responsible for all actions with the wallet, and all technical and legal problems he will have to solve. The blockchain circulates virtual units of account, which can be used as money or perform certain technical functions. These units have the same name: bitcoin (bitcoin, BTC, from "bit," a minimal unit of information, and coin, a coin) in the bitcoin system. Because bitcoin was conceived as the electronic equivalent of gold, cryptocurrency monetary units are commonly referred to as coins. At the same time, the broader term "token," long used in IT systems and games, is now used for nonfinancial blockchains. As blockchain systems became more complex and tiered networks emerged, more or less established terminology emerged:

- Units of account that circulate directly on the blockchain are still called coins.
- Derivative units transmitted within the transactions of the main blockchain, that is, using it as a transport medium, are called tokens.
- In generalizations, tokens can be all virtual units of account circulating in the blockchain, regardless of the levels at which they are applied.

Each wallet has one or many identifiers, to which coins (tokens) can be sent. Each address is unique and the probability of creating two identical addresses in different wallets is almost zero. The movement of coins (tokens) between wallets in a blockchain is certified by a user's unique private key, which he uses to make a cryptographic signature of the transaction, thus certifying his authority as a wallet owner. A wallet's private key is the only proof of token ownership, and anyone who receives a copy of that key will have the same power in the blockchain as the owner of the original wallet. Therefore, the security of private keys requires the highest level of security possible. Hacking into the bitcoin network from the outside is now virtually out of the question, as its reliability has been confirmed by many years of operation. However, individual hacking wallets or centralized services that handle cryptocurrencies and tokens cannot be ruled out. A wallet can also be lost after a hardware failure or natural disaster. The wallet or private keys can be stored in any number of copies, as long as you manage to keep them secure. If all wallet copies are lost, all bitcoins associated with it will forever remain immovable in the blockchain, since the private key is the only guarantor of their transferability. Therefore, the node owner (wallet) must be fully responsible for the safety of its assets.

To transfer coins (tokens) in a blockchain, so-called transactions are made – debits from one address and credits to another in financial blockchains or the transfer of information messages with different content in other types of blockchains. Each transaction is a financial message composed according to the established rules and signed with the sender's cryptographic key. The transaction contains the amount of coins (tokens) to be transferred, the sender's signature, and the recipient's address based on his public key. To use the coins transmitted in a transaction, a private key must be paired with the public key specified in the transaction. Once transmitted to the network, the transaction must be validated, written into a blockchain block, and distributed to all nodes in the bitcoin peer-to-peer network. The block contains a header for transmitting technical information and a list of transactions in which user data – payment or any other transaction – is transmitted. A blockchain consists of blocks connected in series. The previous block's hash is supplied in each new block's header. In this way, an unbreakable chain is formed. It can only be broken or changed by recalculating all block headers and reassembling the chain from the break point. This requires using computational resources equivalent to or greater than those expended in assembling the original chain. This means that the long-term security of a classic blockchain depends on total computational power. The most trusted blockchains are those that require a resource expenditure that is incomparable to the benefit gained.

Benefits of Blockchain

The blockchain, a distributed ledger technology, became known after bitcoin appeared in 2009 [8]. At first, the cryptocurrency was worthless, until it was used to pay for the first purchase – two Papa John's pizzas. It was blockchain technology that allowed the seller to transfer the money. How does it work? Every transaction on the blockchain network, such as a transfer of funds from one person to another, is translated as a block and added to the chain of other blocks in the chain. Such transfers are secure because they are effectively an encrypted message that only the recipient can open and use the contents. All transactions are undisputed and recorded in a blockchain that resembles a large ledger. With a blockchain, participants in a process store information about transactions and transactions between them. This data is visible to all of them and cannot be deleted or changed retroactively. This technology property opens up great opportunities for businesses and consumers: many processes can be simplified, paperwork and intermediaries can be eliminated, and all transactions can be controlled in real-time. Companies can significantly reduce business costs, and their customers can get the final product at a lower price. Perhaps the only drawback of the technology to date is the small number of parties connected to it. In a blockchain network, all parties, including the manufacturer and often even the customs authorities, must be involved to maximize the efficiency of the commercial cycle, from the production of the product to its delivery to the customer.

The benefits of the blockchain include the following: eliminating the need for transactions, transparency and immutability via the shared register and the fact that transactions cannot be deleted nor altered or altered, as well as the high quality of the blockchain's data due to its completeness, consistency, date, and wide availability. The use of a common transaction register reduces the likelihood of data loss or unavailability due to any malfunction. There are various new concepts governed by the blockchain, which will usher in significant change. The blockchain is a massive database. Take a look at some of the most frequent features; however, keep in mind that they can change depending on the intended use:

A Decentralized System
The blockchain differs from conventional digital platforms in that it is a decentralized system, with a copy of the vast record maintained by each member. To avoid forgeries and other threats, there is no central server, only a collaborative administration system. As a result of this disintermediation, costs should be reduced.

A Transparent System
The system is also completely open: anybody with access to the Internet may look up the register and hence the history of transactions at any moment (or by all network members). As a result, with a blockchain, it is feasible to guarantee complete asset or product tracking. While one individual uses a pseudonym, his actions are fully tracked.

A Reliable System
The blockchain cannot be tampered with or altered in any way. Information that has been recorded in the blocks cannot be changed or erased after it has been placed there. With this new technology, an electronic document may be just as valuable as paper as a proof document. Because the copies are being multiplied, the decentralized approach provides some protection against piracy as well.

An Automated System
The blockchain offers complete independence, up to the point of flawless monitoring, without the need for a middleman. Transactions are carried out using computer programs. Self-executing "smart contracts" will be available.

An Efficient System
All of blockchain's advantages come together to guarantee maximum economic efficiency: time savings and lower costs due to the elimination of middlemen and automation and lower mistake rates and lawsuits. Such assets are understandably attractive when a lack of trust is frequently regarded as a significant impediment to growth. There are disadvantages to these benefits as well. The blockchain revolution must overcome several technological, organizational, and societal obstacles in order to succeed.

3 Blockchain (BC) for Cyber-Physical Systems (CPS) Applications

There are a lot of conversations going on in the world about blockchain technology. According to Gupta [11], the blockchain is an information recording system customized with security features that make it impossible to hack attacks or cheat in the system. Abadi and Brunnermeier [12] indicate a ledger system that decentralizes the records by distributing them across all the blockchain networks. Transactions in the blockchain system are distributed to all the system participants, making it hard to cheat or steal. They further indicate that technology's correctness, decentralization, and cost efficiency make an excellent record-keeping system. Their comparison of the system to the traditional centralized system highlights the tremendous revolution blockchain has brought to the record-keeping industry. One of the blockchain's essential features is the algorithms that permit record-keepers to rewind and undo false reports in the ledger's historical records. Besides the finance industry, the system is quite useful in procurement, Internet apps, among other industries where transparency is highly required.

Businesses succeed by improving interaction with their stakeholders. Blockchain technology achieves this by offering a distributed ledger. The strategy allows businesses to utilize a shared database of transactions. The technology applies encryption mechanisms that focus on authentication and authorization of transactions.

E-commerce is one of the major industries that the advancement of blockchain has been enormously beneficial to the continuous and overall growth of the industry. Blockchain innovation is ready to change the Internet business industry. It offers an unrivaled mix of security, straightforwardness, and cost-productivity. Entrepreneurs hoping to grow their endeavors should accept this turn of events and reclassify how they work. Blockchain technology has positively disrupted the finance and e-commerce sectors by offering new and effective payments, smart contracts, sufficient trading execution, and smart contracts.

Blockchain technology holds the key to unlocking new possibilities for organizations on a global scale to be more agile, efficient, and efficient while providing an attractive price, security, and security model. Blockchain is a technology enabling digital transactions between a human or an entity and an outside entity. The blockchain is a distributed database that stores information about everything. It is used to store a wealth of information about users, the networks that connect them, and any other connected devices. Each blockchain is called a "wallet" or "blockchain" because of the blockchain's blockchain. These wallets store information recorded by the network. This information is known as the "transaction." Blockchain provides a framework for decentralized application developers, allowing users to communicate directly and secure their assets using cryptography. It will enable users to store digital signatures for goods and services securely. They can also send messages to each other. Some of these messages can be used to make payments, but this is still more than an intermediary service. It is the application-level layer where the application can perform operations. The application can be any operating system or application program with a blockchain backend [13].

It is considered that blockchain can create a new way to control transactions and data flow on multiple layers without needing to have an exchange infrastructure and data centers. It allows the developers to add features to it as an efficient, trusted, and efficient solution. It is interesting to understand why and how blockchain works. The concept of blockchain differs from traditional banks as they are created by the people and are made to transact and manage their assets by themselves to avoid risk and make money. Blockchain is a platform that helps to create a central resource that will automatically act as a ledger where these resources and the records could be verified to validate the information.

Blockchain is undoubtedly the major innovation. There has been a massive growth in the application of blockchain to the industry. Not only has it led to the transformation of companies like Airbnb and Dropbox, and many others. This change in the way businesses operate will help drive growth and improve the business's efficiency. It will also help the government to get more people employed, the better quality of services. It has played an essential role for businesses by providing a cost-effective and efficient alternative to cash payments. Blockchain technology allows the payment system to function. It is also a useful feature in several applications, including digital identity, e-commerce, insurance, property management, e-payments, and crowdfunding. Some of the uses of this technology are as follows: It facilitates the flow of data and information; the ability to verify every transaction made on a network; the ability to do things that are impossible in conventional financial institutions and businesses can achieve in less than 2 s; and the speed at which companies and industries can create new products and services.

Blockchain technology is all about consensus, which is where the blockchain works. Blockchain has made a significant impact on businesses and industries in several industries. For example, the largest IT company in the USA, Accenture, said they are exploring a blockchain implementation to build blockchain-based applications. As with any technology, there are several uses for blockchain – for example, to enable automated, low-latency bidding through automated contract systems or systems of record for large, complex businesses or to make certain forms of transactions more secure and efficient.

This chapter focuses primarily on emerging blockchain applications for cyber-physical systems, namely, medical records, transportation, e-commerce, finance, and cyber security. Table 2 lists the various systems covered in the study, along with their respective application domains.

Applications in Transport

All tracking and tracing data must be collected, consolidated, and archived for traceability to be achieved. Using tracking information, I am able to keep track of where my product is at all times. The history of all the steps my thing has taken

Table 2 CPS application domains

Systems	Applications	Benefits
Transportation	Automobile electronics, rail systems, road networks, aviation, and airspace management	Facilitation of complex flow and equipment compliance management Simplification of payment procedures Traceability of flows Reverse logistics
e-commerce	Monitoring and tracking of the supply chain to ensure openness in the market, redesign of the payments system, secure e-commerce platform, product testimonials for the real deal	Alternative payment methods Better order processing Enhanced payment security Faster transactions
Healthcare	Automobile electronics, rail systems, road networks, aviation, and airspace management	Medical data management Clinical trial optimization Drug traceability and anti-counterfeiting
Finance	Fraud prevention, financial inclusion, money laundry prevention, trade finance, smart assets, and smart contracts	Uberization of banking services Facilitation of fund transfers More secure and efficient transactions
Cybersecurity	Keyless signature infrastructure, user anonymity, validate transactions in cyber-physical systems, data authentication	Permanent data security Decentralization on a blockchain could replace certification authorities Advanced authentication

is referred to as tracking data. According to Wattanakul et al. [14], monitoring data comes in three forms: setting data, transport-related conditions, and business-related transactions. A frequency (real-time, event-driven, batch, or offline) and a collection method are often specified for each of these kinds of data. Wattanakul et al. [14] state that this traceability data may be divided into three categories based on their characteristics: (1) master (for non-changing data, such as IDs); (2) transactional (e.g., departure and arrival times); (3) status conditions (e.g., temperature, humidity). This data may also be utilized to make tactical and strategic choices after being collected, cleansed, aggregated, and archived. When it comes to supply chain traceability, the traditional design relies on information gathering mostly through EDI exchanges, with some calls to web services thrown in. In some instances, the carriers' information systems are the only source of all traceability information. Therefore, these calls and exchanges are necessary. It is possible to get this information by using email, phone chats, or even text messaging. These systems have the following drawbacks.

- The impossibility of having traceability information in real-time because it is necessary each time that the information is collected, seized, and made available by the information system of the carrier, so that it can be recovered by the other participants of the logistic chain.

- The carrier's centralized storage of traceability information poses a serious threat to the information's availability and dependability. The way a carrier collects and transmits information affects the quality of the data significantly.
- Since the information is delayed, any flaws in the supply chain are left uncorrected. However, even if web services are available, the logistics operator's information system must travel to receive the information, which results in delays before events or issues in the transportation chain are reported. The following obstacles must be solved in order to develop supply chain traceability systems that are dependable and responsive.
- The sharing of traceability, that is, the availability of the data to all supply chain stakeholders, will bring transparency to the operations and processing within the supply chain.
- The possibility of having responsive systems can receive notifications of events occurring in the supply chains and adapt to these events. Events can be, for example, changes in customer needs or an unforeseen event that occurs in the supply chain that will require a quick decision to be made so that there is no delay in delivery, or at least if there is a delay, to minimize it as much as possible. This would require implementing actions that would be reactively triggered by events occurring in the transport chain.
- Verification of the reliability and quality of the information transmitted by the system.

The US Department of Transportation's national ITS architecture may be seen in Fig. 3.

Integrated sensor elements in connected automobiles provide them a 360-degree picture of their surroundings and allow them to monitor them. For example, navigation systems, cameras, proximity sensors, light sensors, and radiofrequency

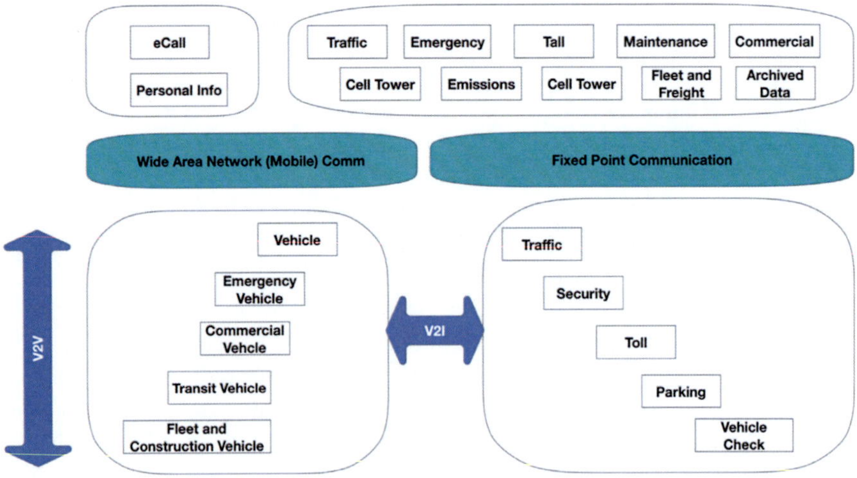

Fig. 3 Intelligent transport systems (US DOT)

sensors are just a few examples of what they can do. To keep vehicles and infrastructure informed in the case of an accident, sensor fusion synchronizes input from several sensors with real-time data from the road. It is because of this that driver assistance technologies like lane departure warnings and accident prevention techniques are becoming more commonplace. All of these aspects have resulted in modern automobiles being equipped with a wide range of communication devices and protocols that make data sharing between vehicles much easier. Dedicated Short Range Communications (DSRC) is the only protocol currently approved for use in 5.9 GHz intelligent transportation systems (ITS) [15]. The IEEE 1609.2-4 messaging protocol and safety services make it feasible to use V2V communication scenarios such as electronic emergency brake lights, forward collision warning, and blind-spot detection. Basic Safety Requirements Per SAE J2735 Transmission of information between vehicles on the road is accomplished through the usage of messages (BSM). BSMs tell other cars of their location, size, and speed, so they are always aware of their surroundings. PKI-based certificates are used to encrypt the data, ensuring the security message remains intact.

Decentralized sensor and communication channel protection was suggested by Rathore et al. [16]. Using a blockchain-based architecture, they safely exchange communications between linked automobiles. Smart vehicle communication is implemented using the blockchain method and a reward-based system by Trust bit [17]. It rewards effective communication by exchanging trust bits. These trust bits were traded in the vehicle cloud using blockchain technology, which recorded and preserved historical documentation of those transactions. This makes it possible for cars to safely access all information contained in the trust bits, no matter how much time or space they have available. Chaudhary et al. [18] introduced a branch-based blockchain system that explores the concepts of a local dynamic blockchain and a master blockchain. Smart vehicle trust points are cryptographic identifiers that assure the safety and security of automobiles by ensuring that no two are the same. Vehicles communicate with one another by using the local dynamic blockchain to verify IDs. Calvo et al. [19] introduced a brand-new, secure linked car communication system built on the blockchain. A ring signature-based system is used to verify new cars' identities before allowing them to join the network.

In order to share the information provided by smart multi-party contracts, secure communication channels are used to reach agreement across cars utilizing a blockchain-based method. There was a proposal in [20] for a multi-signature blockchain. New automotive services like remote software upgrades are provided without disclosing any personal information about the automobiles they are used in. According to Yuan et al. [21], a blockchain-based seven-layer conceptual paradigm for intelligent transportation has been developed, enabling a safe and dependable decentralized environment. Ethereum-based smart contracts were integrated with car network technologies by Leiding et al. [22]. When cars grow more and more software-dependent, a critical issue arises: how will software upgrades be handled as new functionality is added?

As shown by Steger et al. [23], Overlay technology may be employed for this purpose. A cloud overlay network is used in this strategy to transport data between software providers, cloud storage systems, and car interfaces. The blockchain technology and software distribution mechanisms both employ these messages. Aguirre et al. [24] presented a vehicle system with a CPU set to execute data-driven functions, such as creating reports. The cryptocurrency protocol is used to create a vehicle-specific digital currency record for these activities. Each time you make a purchase, the value of this digital cash changes. It is kept in memory and sent across the network when you are making a transaction. Rowan et al. [25] presented a novel blockchain method to secure visible light and auditory side-channel communication in automobiles. It is proposed in [26] to use blockchain to secure the execution of energy refilling for self-driving electric cars. Chaudhary et al. [18] presented a blockchain-based reward-based intelligent vehicle-to-vehicle communication system. It enhances security and privacy while allowing for quick and safe communication between automobiles on the road. Table 3 lists transportation-related uses for the blockchain.

Blockchain in E-Commerce Industry

Innovation has consistently been developing, and entrepreneurs have made openings, and comfort has been rethought for shoppers. Versatile applications for Internet business organizations began to arise until, in the end, financial innovation and portable wallets turned into the standard. Over the past decade, blockchain has proven to foster and benefit various industries worldwide with its numerous advantages, for example, decentralization and transparency of the blockchain. Today, blockchain innovation is observing an ever-increasing number of exchanges [32]. In addition to being faster and significantly less expensive, a portion of the upper hands this innovation offers remember absolute control for the production network for improved effectiveness and amped-up evaluating for expanded benefit. The most recent information recommends that almost 350,000 bitcoin exchanges are affirmed each day. This advancement is above and beyond into increasing the security of exchanges to clear a path for better encounters for customers everywhere in the world. It is set to change the worldwide monetary scene, and online business organizations, particularly, will profit from the ascent of blockchain [33].

Electronic commerce also e-commerce is a business strategy that allows companies and people to buy and sell items and products over the Internet. Thus, there are various remarkable and promising attributes in applying blockchain in the electronic commerce sectors. Some of the advancement of blockchain in e-commerce industry includes transforming and upgrading financial transaction, trade finance, cross-border payments, and capital markets [34]. E-commerce helped firms develop a wider market for their products by offering more efficient, cheaper, and effective distribution channels. Nevertheless, blockchain technology has assured the growth of e-commerce by giving fast and safe alternatives to traditional payment methods.

Table 3 Blockchain application in transportation

Application domain	Applications	Contributions
Intelligent transport systems [21]	Intelligent transport systems	Consider new business models and practical application situations, as well as the reasons behind them
Charge it up [26]	Smart mobility systems delay, latency, security, and cost state channels	State channels can be used for control logs and connections in smart mobility systems
TangleCV [27]	Vehicular edge computing (VEC)	Update authentication procedures to address issues with communication between many cars and a single trustworthy edge computer node
Trustbit [17]	Intelligent vehicle communication	Identify and resolve issues relating to the trustworthiness and accuracy of data received and disseminated over the communication channel
Intelligent vehicle trust point [28]	Intelligent vehicle (IV)	Focus on IV communication's dependability, correctness, and security in the communication route, which are key challenges
CUBE [20]	Autonomous Car Network	Prevent harmful assaults by utilizing artificial intelligence (AI)
Fast and secure multihop algorithm for IVC [29]	Intervehicular communication (IVC)	Intervehicle wireless information injection resulting in life and money losses or any other type of hostile selfishness is an issue to address (e.g., traffic redirection for the adversarial benefit)
BEST [18]	Intelligent transportation system	Creating a safe energy trading environment for smart grid charging and discharging is a good place to start
Privacy-preserving blockchain-based electric vehicle charging [30]	Electric vehicle charging	Customers should look for charging stations in their area to find the best deal while maintaining their privacy
Blockchain for ITS [31]	Intelligent transport system	Bring up the issue of retrieving relevant data and deleting irrelevant facts and statistics while describing particular scenarios, like an accident.

Some of the significant advancements of blockchain in the e-commerce industry include the following.

More Effective Supply Chain Tracking and Monitoring
The production network is perhaps the most basic part of an online business. Online entrepreneurs regularly experience issues observing items, overseeing supplies, and incorporating the data set in any event. Blockchain innovation makes these undertakings simpler to accomplish. Blockchain permits Internet business organizations to deal with their stock all the more productively. The new technology eliminates the requirement for businesses to allocate resources to various assets in order to watch and monitor stock prices. One such following arrangement supplier is WOWTRACE. They are a group of blockchain engineers, scientists, advertisers, and vital organizers that gives constant item data to buyers on all means in the store network measure. In Vietnam, they help make the following of horticultural products more straightforward for the manageability of their agribusiness.

When a manager is taking a gander at an inventory network, one of the principal concerns is getting hacked or succumbing to false practices. Notwithstanding, by dispensing with the agent from the inventory network, blockchain innovation disposes of these dangers. Keeping a record and the following provenance is simpler since the item data can be obtained through radio-recurrence ID labels and implanted sensors. Monitoring an item from its origin to where it is at present can be followed with blockchain innovation. The expulsion of the broker likewise prompts a decrease in the general expenses.

Advancement of Transparency in the Marketplace
Straightforwardness is the commercial center that gives purchasers a conviction that all is good. Previously, one of the primary worries about web-based business organizations was the absence of straightforwardness. This issue was tended to with the improvement of the distribution chain. Because innovation makes it possible for people to stay abreast of even the tiniest changes in a transaction, everyone becomes more alert and feels more secure. It would create a decentralized business atmosphere in which any improper action by a shipper may be noticed on the web. Retail monsters, for example, Walmart and Unilever, have as of late pronounced blockchain projects, indicating their aim to acquire traction in blockchain-based contributions.

A More Efficient Makeover for Payments
Individuals have continuously begun utilizing digital money as an option compared to customary cash throughout the long term. A huge move toward virtual cash is getting increasingly evident, as seen with the ascent of bitcoin and other blockchain-based monetary forms. One of the essential reasons individuals are inclined toward utilizing digital money is the decentralized idea of blockchain innovation. There is no focal position, implying that solitary individuals engaged in an exchange can handle the activities. The estimation of blockchain likewise does not rely upon factors like legislative issues or a country's economy. Also, blockchain-based monetary forms are substantially more agreeable to utilize. They dispose of the

need to visit an administrative power to record since everything should be possible at home. Also, there are no charges to opening a record as a virtual cash wallet is free [35].

A Secure Platform for E-Trade Business
Each type of security is essential in the web-based business industry. The blockchain-based web-based business stage gives exhaustive security, including information and wallet insurance. Information security is particularly fundamental since specific organizations keep up client data, for example, their location, telephone number, and different subtleties. Blockchain-based monetary standards do expect buyers to uncover delicate information. The only piece of information attached to every client's wallet is a haphazardly produced novel identifier.

Genuine Item Reviews
Counterfeit item surveys are pervasive in a lot of online organizations. Numerous customers will, in general, succumb to these bogus audits and wind up being disillusioned with the buy. Notwithstanding, this is not the situation with blockchain [36]. This innovation confirms surveys and restricts entrepreneurs from deleting history without telling clients. Blockchain keeps information in squares that are added to a chain of comparable data blocks. Each square requires a check from across an organization of PCs before it is added to the chain, making it exceptionally difficult to adjust.

Decreased Costs for Retailers and Consumers
Shippers make a critical bit of their benefits by cutting the absolute installment in the web-based business. The expense to the shopper increments when more players are associated with the installment organization. With blockchain, there is immediate contact among purchasers and merchants since the agent is killed. It likewise enables client outreach and eventually decreases the expense for buyers. The autonomy from go-betweens likewise benefits retailers because of reducing the number of expenses charged to extra gatherings. As of now, this innovation is considered one of the primary drivers to accomplish a generous expense saving. As per a Santander FinTech study, circulated record innovation could lessen yearly monetary administration foundation costs between $15 billion and $20 billion by 2022 [32].

Despite the benefits of blockchain in e-commerce systems, there are various challenges, for instance, technological and regulatory issues, that need to be addressed to ensure blockchain's full potential and benefits in the e-commerce industry.

Blockchain in Healthcare Industry

With its mechanism for stabilizing and safeguarding the data set with which users can interact through various types of transactions, blockchain technology has

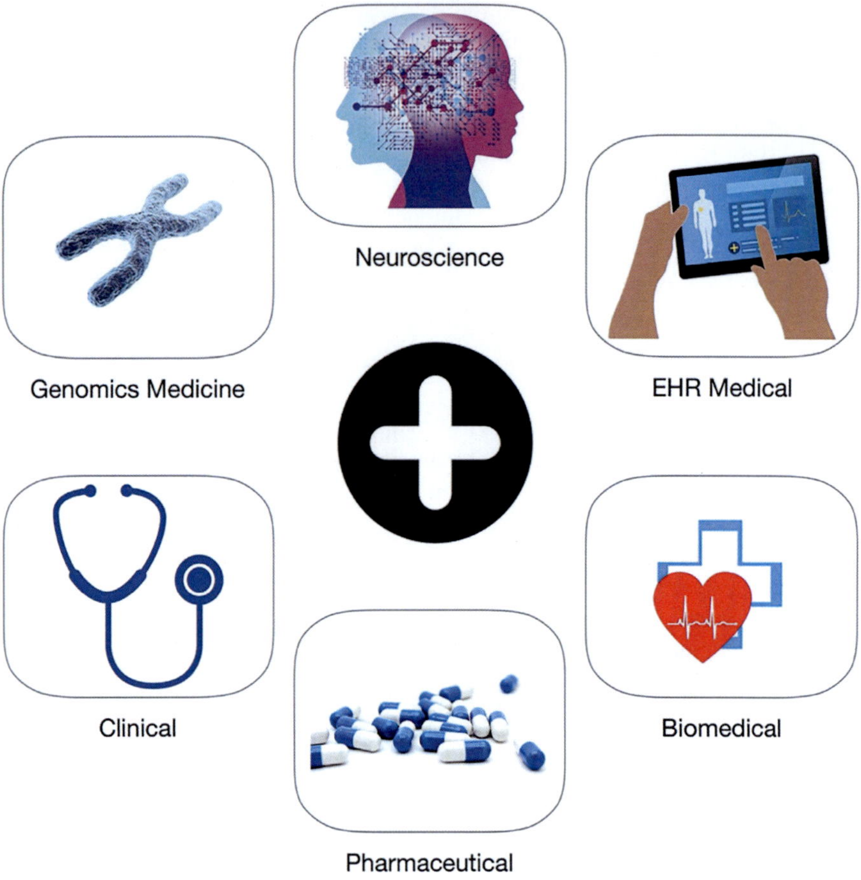

Fig. 4 Blockchain in healthcare industry

enormous possibilities for biomedical, genomic, telemedicine, remote monitoring, eHealth, neuroscience, and personalized healthcare applications in general (see Fig. 4).

Following are a few health and medical fields where blockchain technology has enormous promise. Biometrics, blockchain, and EHR (EHR) Doctors, hospitals, and medical devices have all pushed for the digitization of medical records in the last decade since digitizing this data make it easier to access and share and provides a foundation for better and faster decision-making. One of the most common applications of blockchain technology in healthcare is electronic medical records.

Patients who later split from the data of a healthcare provider may easily lose access to past data if they use electronic health records (EHRs), which are not intended to keep track of the records of several medical facilities and patients. Many academics have considered the use of blockchain technology for EHR maintenance since it is vital to discover an innovative approach to administer EHRs in order to encourage patients to access their current and historical health data. A "MedRec" prototype utilizes specific benefits for handling authentication, secrecy, integrity, and easy data exchange, all in one package. They offer patients an immutable, entire history and simple access to their healthcare information from different physicians or treatment facilities through a decentralized record management system [37]. "MedRec" does not keep any records of your medical care and does not need to be adjusted. It alerts the patient, who is in charge of the record's location, and records a mark on a blockchain. The mark serves as a guarantee that you have purchased a replica of the document. Similarly, it gives patients more control over their health by shifting the organization's burden of care to them. Administrator associations perform the function of patient agent for patients who do not want their information processed. Today, users' patient entries are complex in design, require more labor, and have different user interfaces depending on the foundation. An interface is included in the MedRec architecture to help maintain connections with healthcare records when they travel between organizations (https://medrec.media.mit.edu/). With EHR implementation, medical data sharing typically confronts significant restrictions such as loss of controlled data and dependability and verification and safe backup of medical information. Medical Data Sharing MeDShare, developed by Xia et al. [38], is a safe system that ensures the transmission of medical data between parties who are not trusted. MeDShare may be used by cloud service providers, hospitals, and health research to share medical data and manage electronic health records. More substantial data sources, personalized audit control, and minimum dangers to privacy and safety are all characteristics of good governance organizations (GROs).

Typically, electronic health records (EHRs) include highly confidential patient data that must be shared among physicians, radiologists, nurses, pharmacists, researchers, and others in the healthcare industry to provide accurate diagnosis and treatment. As long as this extremely sensitive patient information is stored, transmitted, and disseminated among many organizations, it poses a significant danger to patients' health and the preservation of patients' medical records. The incidence of these hazards can become bigger than the history of previous and subsequent treatment, follow-up, and rehabilitation procedures in patients with chronic illnesses (such as cancer or HIV). In order to provide successful therapy, it is critical to maintain the patient's current medical history. Blockchain-based architecture for organizing, storing, and sharing cancer patients' electro-medical information has been developed by Dubovitskaya et al. [39] to get around these restrictions. To access, manage, and store encrypted patient data, they turned to permissioned blockchain technology. These frameworks may be used to put blockchain technology to practical use in healthcare activities, such as accessing and controlling patient data privacy and security.

Another historical example is the Estonian medical records blockchain initiative. This year, when it planned to retain millions of private medical records and make them readily available for healthcare professionals and insurance firms, Estonia returned to the forefront of core technologies [40]. It is possible that the increasing usage of blockchain technology in medicine throughout the world is a good sign for patients who want their medical information to be as accurate and unaltered as possible. Any effort to gain access or make adjustments is easily identified and tracked over the blockchain. Not only does this assist in maintaining the integrity of the patient data, but it also identifies any illegal activity such as blockchain fraud or record fabrication. It will be much easier to share and evaluate approved medical care records now that they have been standardized. Most of the patient's providers had already seen it by the time of the appointment. Patient management algorithms that take into account prescription interactions, hypersensitivities, and pharmacological solutions make it possible to process all of these things across all blockchain records quickly. Consequently, blockchain innovation will benefit from medical record management [41], faster validation of clinical information (clinical data), enhanced security, and better care organization.

Blockchains in Clinical Research

Challenges in clinical research range from patient confidentiality and safety to privacy and integrity of healthcare data to record-keeping to clinical trial enrolment. Blockchain, the Internet's next-generation, has the potential to solve these issues. Healthcare experts are using blockchain technology to try to find solutions to these issues. Blockchain, artificial intelligence (AI), and machine learning will soon be taking over the healthcare industry. The Ethereum-permissioned protocol described by Nugent et al. [42] provides smart contract capabilities on the blockchain, which is employed in clinics together with data administration systems. The study's primary goal was to find a solution to the challenging challenge of patient recruitment. As a result of the study's findings, it was recommended that Ethereum smart contracts be used in clinical trials for data management system transparency in order to make transactions more efficient. As a result, one of the current uses of blockchain technology in clinical research is patient enrollment. Benchoufi et al. [43] devised a procedure that allows for the collecting of patients' informed consent, which is bound to protocol changes, the storage and tracking of the consent in a secure, unfalsifiable, and publically verifiable manner, and the real-time exchange of this information. The management of the medication supply chain is a major use of blockchain in the medical business. In many sectors, supply management is vital, but it is even more so in healthcare due to the industry's rising complexity. Any lapse in the healthcare supply chain directly impacts the health of the patients [44]. Because of the numerous moving elements and individuals involved, blockchains are extremely insecure and include several security gaps that hackers may exploit. Due to greater data openness and improved product traceability, blockchains offer a secure platform for eliminating this problem and preventing fraud in some circumstances. Manipulating the blockchain is difficult since the string in the blockchain cannot be confirmed or modified from the smart contract.

Blockchains in the Pharmaceutical Industry

Healthcare delivery's fastest-growing and most important areas are pharmaceuticals and biotechnology. This industry helps bring novel and potentially life-saving pharmaceuticals to market. Medical items and pharmaceuticals supplied to the general public benefit from this by being safer and more effective. As a result, the pharmaceutical industry helps speed up patient recovery by evaluating and processing safe medications [45]. Many times, pharmaceutical firms are confronted with counterfeiters trying to undermine production or infiltrate the system with substandard products at the wrong moment, posing a significant danger. As a result, the creation and distribution of fake drugs have become one of the significant health dangers across the world, particularly in impoverished nations. Blockchain might be an ideal solution for evaluating, monitoring, and guaranteeing the production processes of new pharmaceuticals during R&D. To fight the creation of counterfeit medications, Hyperledger recently announced a counterfeit drug initiative that uses blockchain technology.

Because of this, there is a pressing need in developing nations to keep track of, analyze, and assure the whole process of generating new drugs and dispensing them to patients by utilizing digital technology. Digital drug control systems (DDCS) are a potential long-term answer to the problem of counterfeit pharmaceuticals. Sanofi, Pfizer, and Amgen have started a cooperative pilot initiative to inspect and assess novel medications using a blockchain-based DDCS. A blockchain-based strategy might help track medication manufacture and location while also improving traceability. This technique would also safeguard the drug supply chain and verify the quality of pharmaceuticals delivered to customers or end-users [46].

Blockchain technology makes it possible to track healthcare expenses across multiple sites in real-time. It gives a complete record of the healthcare costs. It can help them to get cheaper health insurance or insurance policies. It reduces the cost of each server, and it makes healthcare more affordable [47]. It is an innovation that has allowed people to exchange their physical files for digital resources. Blockchain technology allows creating digital certificates in one-way communication and recording data to retrieve the data by blockchain. A surgeon can create a software program that can process the scans and deliver digital certifications. It represents a new way to track and manage health data about health services and medical services and products. It can also be applied in the digital healthcare industry.

A blockchain is a system in which individuals will store information in an immutable format. When needed, they will access the data and perform various actions such as transferring the data. It also includes a contract in which individuals agree to provide the data, and if there is any breach of the agreement, individuals will not be able to access the data. It is also possible for an individual to create a service where the data can be collected, recorded, and shared. Blockchain is a distributed ledger technology that enables instant, decentralized record-keeping that allows an organization, a patient, or a patient's organization to create a trustable account on the blockchain with the patient's permission securely and efficiently. It provides the ability to control access to confidential information from a patient's history and from other organizations, doctors, patients, and medical professionals through blockchain-enabled apps, including those accessible online. The applications may be used in medical care and other specialized fields in the healthcare industry, allowing access to more critical data. It reduces fraud, tamper-

profits, and transaction data from getting into the hands of hackers, identity thieves, or malicious entities.

Blockchain is also beneficial in the verification of identity by adding security, transparency, and accountability. Blockchain makes it easier for consumers to obtain the information they need without any centralized systems. It helps with payments, medical records, and healthcare records by adding security [48]. Blockchain is also a method of ensuring accountability and confidentiality. So, it is not related to the healthcare industry. However, it is an initiative that aims to develop blockchain technology for bright medical instruments and medical records by providing better service and transparency. Blockchain technology for medical records and intelligent medical devices is not related to each other. The same technology can provide an alternative to traditional medical records and electronic medical records. Technology revolutionizes the healthcare industry and changes how it does healthcare and is helping the people. The blockchain will get global attention and help people save in financial benefits. They can save money on health and better health, significantly impacting the quality of life.

These benefits and savings will come to the healthcare industry that can be utilized for people's health. It would make people save money and save on financial benefits. They can save money on fitness and better health with a significant impact on the quality of life. These benefits and savings will come to the healthcare industry that can be utilized for people's health. Blockchain technology will significantly impact people's lives with a good quality of life and make them brighter and happier. It is also giving health to the people. The term "blockchain," as applied to healthcare, is quite broad. In this context, blockchain can be understood as a decentralized protocol used to create and manage data on a centralized database. While the blockchain itself does not allow for much information on a patient's records, some banks can provide data and services. The health industry is slowly getting better at using blockchain to help people. So far, it is hard to say whether it is the right technology for the job.

A blockchain that allows for the whole patient record and thus provides a way for patients to pay for treatments is the future – not only for financial records but also for everything in the patient's office. Blockchain is used as an extension of e-commerce for healthcare – which could be a new way to improve care. With blockchain, it could be much cheaper and just as reliable to order drugs online. Health records like the patient's and medications could also be much easier to store.

The advancement of blockchain is of increasing importance for the healthcare industry, facing massive healthcare costs and globalization, and a digital divide. Blockchain is an easy technique that can be used and utilized for good and wrong purposes. The advancement of blockchain in the healthcare industry is a trend of the health industry, where the technology was adopted to benefit the patients. Since blockchain is a decentralized technology, it allows anyone to participate in the market. It is a technology where all kinds of transactions are handled electronically. It is also the one in which the transaction is transparent, and they do not have to provide any proof of transaction history. There are already a growing number of healthcare organizations offering a blockchain-based solution.

Several technological changes need to be adopted so that healthcare institutions can meet blockchain requirements without any delay. The blockchain and other digital technologies have been put in play for various purposes, such as tracking medical records and saving money for the patients, such as medical records linked to each patient and making a smart contract with an intelligent address. Some of these uses are done with healthcare companies. In this kind of usage, blockchain technology and other technology are made available for them, making it possible to provide these advantages to patients. It helps companies and businesses. Blockchain is the platform for transactions, which the health industry is doing [48].

Moreover, in that way, it will benefit them. All the patients have used blockchain. They are going to be able to see how the whole transaction is being processed. The entire healthcare will be digitally encrypted, so they will not see a record that may be there to see the patients. Blockchain has changed how many healthcare industries are managed and think outside the box to do business with a blockchain-enabled smart drug market. This blockchain's rise allows a significant benefit in the healthcare industry.

The first step is to develop a health management system (HMS) that can support multiple doctors and hospitals. The next step is developing the software to help the hospital system connect with the patients, interact with the doctors, and monitor the patients on an electronic healthcare platform. At the time, most blockchain systems used the eHealth system to manage the transactions on this platform to manage the patient to determine the status. Blockchain technology allows doctors to use the blockchain to exchange information and manage patients. The information communicated from the users to the doctors is not subject to data exchange or the doctors to pay the money; all is done in smart contracts that execute in real-time.

With blockchain in healthcare, only a medical specialist can manage the network. The doctor does not even need to implement it as only a person's medical status has to be displayed. Then the doctor can perform the services as prescribed. According to the company, blockchain technology has increased the availability and transparency of data in medical records. However, it has also led to new solutions where hospitals can use the latest technological advances in innovative medicine to improve care. As per the company, the blockchain-based solution allows hospitals to have an open and transparent system.

Furthermore, the blockchain's information is encrypted with the user's key, which the hospital then uses to identify patients. One of the main innovations of blockchain is the possibility of managing the data that is generated. With this technology, the data will be stored in the system, but it will be automatically transferred through blockchain. Moreover, this creates a significant opportunity for the healthcare industry since it is a way to make the system more transparent, both to the users and the healthcare providers, with ease and simplicity. Blockchain gives the healthcare industry access to data, usually held in various blockchains, such as the blockchain-based healthcare database containing information like hospitals' operations, billing, and treatment history.

Blockchain is the protocol of the Internet. It is the system of information, storage, sharing, and communications. The Ethereum blockchain system allows

Table 4 Blockchain applications in healthcare

Application domain	Applications	Contributions
MedRec [37]	Electronic health records	Give patients access to all of their medical records, make care auditable, and share their data
MIStore [49]	Sharing healthcare information for administrative or economic purposes	Develop a medical insurance data storage system based on the blockchain
GAA-FQ [50]	Granular access control to electronic medical records (EMR)	Authorize varied authorization granularity levels while keeping the underlying blockchain data structure compatible
BlockHie [51]	Interoperability of personal healthcare data and electronic medical records	Off-chain and on-chain privacy and security storage and verification
Analytical methods in healthcare [41]	Health data collection, storage, and sharing	Analytics in healthcare using blockchain and artificial intelligence (AI)
Blockchain and Internet of Things (IoT) powered [40]	Integration of large amounts of data into the mining process	Validating transactions requires a consensus mechanism, which reduces the computational cost of mining blocks
MedShare [38]	Shared cloud storage model for data	Reduce data processing and anonymization delay by reducing latency
MedBlock [52]	Information about healthcare that is shared for research and therapeutic purposes	It is challenging for pharmaceutical professionals to develop precise remedies using data collected under different rules since current EMR systems do not have a standard data management and sharing strategy in place. Take action right away to resolve this issue

anyone and everyone to transact on the Ethereum blockchain using a global, secure network of smart contracts. It is a decentralized solution. Its blockchain is on this decentralized platform that it can use and exchange without centralized intermediaries. A blockchain is a decentralized computer. When people can send transactions over the public Internet, they can publish those transaction blocks, meaning the public knows which blocks they sent transactions. The computers in these computers know who did which transactions and that transaction were sent. Blockchains are public ledger-based systems. They support decentralization and privacy. They also work as trusted computing and storage systems and as the virtual wallet of the people. Table 4 presents the various blockchain applications in healthcare.

Blockchain in Cryptocurrency and Finance Industry

Many financial networks exist, which include banks, brokers, bonds, and real estate. If they have a car or vehicle and buy fuel, this happens on a blockchain where they own the supply and use a blockchain. In cars with their drivers and their distributed ledgers, anyone on the network can verify anything that goes on there, and they would sell it and repurchase it at a fair price as if it is not there. It is a technology that takes the underlying information that has been used to create the payment system from the banking system and distributes that information to millions of potential customers without any intermediaries that the bank can hold the lead. They can do so using blockchain to create a consensus mechanism, this is what has made this whole technology, and it is a vast revolution. It allows anyone to transact anything to anybody in any block or chain using an Ethereum smart contract. So, in banking and finance, the payment of funds is possible. It may be the next thing to move banks and companies from analog to digital technology [53].

Banks are trying to provide customer services and financial information in digital format. Banks are trying to replace old banks and financial information with blockchain technology. The banking sector is looking for blockchain solutions to solve its economic problems. The banking and finance industry is about business and technology, about managing money. In the bank and finance industry, the term "blockchain" refers to dealing with information or a system of transactions that can move money. When they talk about blockchain technology, they can be sure of being part of the banking and finance industry. Blockchain technology creates a secure and reliable system of payments or electronic transactions. This technology is for money generation. It involves ledger-like records. One can read and see the information by using a device called a device with chips. There is no need for any financial information to be stored in a ledger using this technology. They can keep all of the transactions or the blockchain data, and it is not required to have any financial information in the catalog. This technology is related to the banking and finance business; information is simpler than a ledger system. The processes are done automatically on blockchain technology. The server carries out the network to verify all transactions and the data with a chip. It is not required to take the money received, and the information is kept in the blockchain data. As the blockchain data is transparent, there is no need to keep records in a ledger system [53].

Advancing blockchain is about increasing speed, safety, and availability for financial and banking users. Blockchain is the most scalable, tamper-proof, and secure way to transfer money anywhere. It has already revolutionized how our money is held in the cloud and has been the most disruptive technology for the digital economy. The latest industry is taking this future and applying it to the real world. As part of blockchain technology, it provides a framework that facilitates transactions to and from accounts. As a result, a person or entity of the financial institution may act as an intermediary between bank and customer at one point in the process. As a bank, in the event of a transaction completed by any person, it can act as a transaction broker at the same time. This way, the person, entity acts as a transactor, and the account's funds are released in one or two transactions that would otherwise be delayed. As an institution that wants to develop new technologies, it works to become a blockchain platform provider. It is a good business strategy.

The technology can facilitate the technology companies. It provides a framework for banking companies to create a blockchain and provide transparency and record-keeping. With these changes, banks need to adopt bitcoin. The advancement of blockchain in the banking and finance industry is very significant. Moreover, now everyone has the technology to build smart contracts that operate entirely on blockchain. The banks can now offer the customers financial products they can make without a central bank. So, this process is accelerating, the pace of changes in this process is improving every day.

What is more, a considerable sector has already started working on what blockchain can be and how to implement it in various industries, where the blockchain itself cannot be used for this purpose. For example, the banking industry, which has a huge problem, is working on smart contracts to manage credit and debt to give customers better banking. Both the credit and debit are stored in a specific virtual account.

The advancement of the blockchain in the banking and finance industry results in many reasons, including high transaction volume, efficient use of the network, rapid pace of payment settlement, and low costs. For the banking industry, there are several advantages like a single ledger of account details provides transparency and ease of access, automated settlement helps in making large transactions quickly and cost less, the rapid scale of payments in financial services makes transactions possible in less time than before, and the fast growth of business in financial services makes it easier for new ideas to advance. Each register is an encrypted archive stored in a distributed ledger called distributed ledger of payment details, which is also a data storage system. These two information systems are complementary. The fact that they are separate from each other is why they are complementary. However, one may prefer to use one over the other. It is a different phenomenon compared to the familiar concept of the ledger of account details.

They contain a copy of the ledger of account details, but these two copies exist on separate machines with their state. It is not easy to know how they are synchronized since they are not synchronized. One may prefer to use a ledger of payment information in the banking and finance industry, allowing transaction speed to be reduced and making transactions cheaper. The difference between banks and finance firms is noticeable. It is a distributed ledger technology that enables instant, decentralized record-keeping that allows an organization, a bank, and a medical practitioner. The applications may be used in the finance industry.

Blockchain is an online, distributed ledger that can record everything that ever happened to something. A blockchain ledger, or a distributed ledger, is like a digital currency. Instead of using physical money, one uses computers to create a virtual currency to pay for goods and services. A block is a file; one can write their file in the blockchain and read it later. The blockchain is a record of files, including all bitcoin network records, the personal bitcoin wallet, and any changes they have made to it, like transactions or mint coins. The blockchain is a platform that allows for a distributed, transparent, decentralized, and secure computing platform for managing all financial information. It offers a peer-to-peer platform for managing digital assets and transactions. It is a secure mechanism by which transactions are recorded, and it acts as a centralized trust to manage digital assets.

This industry deals with a lot of currency, and as a slip of a finger, there are millions of transactions happening worldwide. The blockchain advanced how the financial operations occur; most individuals needed cash, but it was no more complete. The blockchain has introduced more cryptocurrencies technologies, but the most known one is bitcoin [54]. Organizations and companies can invest as much as possible; there is no limit to how it happens in the traditional transaction. The traditional transaction has limitations and cannot allow the user or organization to have a certain amount of money in one transaction. However, the blockchain has changed this industry as they can transact as much as possible, and there are no limitations.

Peer-to-peer global financial transaction is made possible. The traditional transaction has a lot of limitations, and one of them concerns the boundaries. The transaction happening by a third party has limits to the boundary of the nation-state its origins. For example, the user cannot complete a transaction in the UK for a bank account in France or Spain [55]. The traditional transaction system works in the country, and in the case of advancement, it works for neighboring countries only. The blockchain has advanced the transitions from the sender to the receiver directly without the third party. It makes the transaction process easy and fast, reducing the time taken to wait for verification. It does not require any third party to validate the process hence confidentiality in the transaction.

Blockchain is useable for protecting market communications from dissimilar clients, strategy holders, and insurance companies. It is useful to exchange, purchase, and record insurance policies. Making a complete transaction is traditionally expensive; the sender must pay the sending fee and the receiver with a withdrawal fee. Blockchain has made the financial industry on another level. Making a transaction with the blockchain requires less transaction fee, and to some extent, there is no fee to complete a transaction. It has increased the adoption of cryptocurrency to compete for the transaction rather than the traditional way that is not reliable.

The stock of trading is another advancement in the financial industries as a result of blockchain technology. Different nation-states have different currency and are not a problem when it comes to the stock of exchange. The seller may sell the currency to the buyer at a higher price, but the buyer ends up selling it at a lower price due lack of currency standardization. Blockchain has transparency, and the currency remains the same globally. The price of bitcoin currency remains standard in all the nations; if it rises, the impacts apply when there is a drop. It has made the transaction permanent and reduced the cyber activities concerning traditional transactions.

Blockchain technology is essential to all industries and plays a critical role in stimulating productivity. This technology is highly secure when used through the use of cryptographic encryption mechanisms. Only those with perfect encryption keys will be able to decrypt the information. There is no limitation with blockchain technology as a way to complete your transaction. It is available all the time provided there is an Internet connection. Indeed, the financial industry has evolved due to the blockchain technology whack allowing you to save as much as you have. Blockchain indeed makes the financial industry successful due to the unlimited number of transactions with lower transaction fees. It also allows more customers to purchase the products in the whole world due to currency standardization.

There are a lot of conversations going on in the world about blockchain technology. According to [11], the blockchain is an information recording system customized with security features that make it impossible to hack attacks or cheat in the system. Abadi et al. [12] indicate a ledger system that decentralizes the records by distributing them across all the blockchain networks. Transactions in the blockchain system are distributed to all the system participants, making it hard to cheat or steal. They further indicate that technology's correctness, decentralization, and cost efficiency make an excellent record-keeping system. Their comparison of the system to the traditional centralized system highlights the tremendous revolution blockchain has brought to the record-keeping industry. One of the blockchain's essential features is the algorithms that permit record-keepers to rewind and undo false reports in the ledger's historical records. Besides the finance industry, the system is quite useful in procurement, Internet apps, among other industries where transparency is highly required. My focus will be on the finance industry, exploring the technology's advancements in the industry.

Fraud Prevention

The blockchain has contributed to the reduction of fraud in the finance industry tremendously. The financial organizations dealing with money and assets transactions are highly exposed and susceptible to experiencing losses brought about by fraud or crime. The financial sector has previously depended on a centralized system for record-keeping. Hackers and crime agents are well versed with this kind of design, and it is effortless for them to manipulate it as one access to such a system would give them the ultimate power to do as they please. Blockchain is a secure, non-corruptible technology that depends on a hard decentralized network for attackers to manipulate or penetrate. Each transaction is recorded and stored in the form of a cryptographic mechanism. The mechanism has an almost impossible way of being corrupted, and if corrupted, there are easy ways of tracing the attackers. The difficulty is linking all blocks so that if one breach is detected, they all detect and show the change. The linked blocks also reduce the time of tracing the breach, reducing the time for the attackers to conduct any illegal business in the system.

Financial Inclusion

The current banking regulations and restrictions highly prevent banks' use by many people who are left looking for an alternative solution. Financial inclusivity is the ability and opportunity for everyone to use a formal financial system for economic growth and development. The low cost associated with blockchain gives start-ups a chance to compete with central banks. The start-ups rely on the alternative that comes with digital identification and mobile devices to access financial services.

The hassle-free system has a competitive opportunity for innovators willing to serve small bankers hence achieving financial inclusivity.

Money Laundry Prevention

With the anti-money laundry regulations taking place in most developed and developing economies, knowing your customer policy has made the registration of a customer in a banking institution quite an expensive affair. It is estimated that financial institutions spend between USD.60 million and USD.500 million enrolling a customer to their records. They are required to conduct a background check or what is commonly known as customer due diligence. The process is undertaken to reduce or eliminate global money laundry and curtail criminal organizations such as terrorists and drug groups. Due diligence by one bank or institution on a customer makes the information about the customer access to other financial institutions in the blockchain in the blockchain system. The workload is reduced tremendously, and there is no repetition of efforts from the same industry. This advantage from the blockchain system highly motivates business leaders in the financial sector to acquire and join the blockchain system to reduce their operations costs, optimizing their organization's profits [11].

Digital Currency

As the blockchain system increases financial inclusivity and allowing innovators, the digital currency known as cryptocurrency is the new wave of financial assets. The cryptocurrency highly relies on the blockchain system to increase its credibility and security features. The currency is now used in different parts of the world as an alternative to traditional money. Although the cost of accessing digital currency is currently high, the business community is working to reduce the barrier by providing a continuous exchange of money.

Trade Finance

Trade finance has been made easier on the blockchain system. Transaction of complex trading in the traditional system is considered a long and tedious process that involves a lot of paperwork and can also be costly. In blockchain technology, trade finance is an essential application that eliminates lengthy processes and involves experts conversing with the system. The experts' role is to engage the traders involved in the complex transaction by signing them in the system, export and import needed ledgers. Once agreed upon, the transaction automatically completes the rest of the task in an impressively short time. All the parties are privy to the activities being conducted in the system. In a practical example, the Barclays Bank in Israel completed a transaction record of 4 h. In the traditional design, the transaction would have taken 7–10 days to complete.

Smart Assets and Smart Contracts

Smart assets and smart contracts are features in the blockchain system that are automatically executed. A smart asset application in the blockchain is used to store records of asset transactions and eliminate the long process of buying and selling paperwork documentation. Once the transfer of assets is done, the blockchain system holds up this information digitally, updating any information or activity conducted on the system. On the other hand, smart contracts are an application that facilitates the ease of agreements. It enables financial transactions by increasing the speed and simplifying the process to reach or complete a transaction [56]. The application ensures that the information transferred is accurate, and its approval is dependent on the written code. The errors and execution time of this application are favorably dropped at the extreme level, and all the parties involved are privy to the transaction.

The blockchain technology is quite a handful of tasks to understand but, once understood, reduces financial fraud in the finance industry, reduces activities by criminal organizations such as the terrorist, makes trading finance a light task, increases the cryptocurrency trading, allows financial inclusivity, and transfers assets and achievement of contracts an easy task. Although there are challenges associated with the system as it is based on peer-to-peer transactions where everyone in the network is privy to the information and allowed to add data, blockchain is a solution to most of the challenges experienced in the financial industry. It has experienced adoption across the globe, and its scalability is expected to open up more opportunities to innovators and financial consumers. The universal adoption of blockchain means that the system will open up cross-border money transfers and scale-up trade across the board.

Cryptocurrency is not one asset but many asset classes. Cryptocurrencies are in many ways like credit cards and stock markets, but they have different characteristics. They are different types of asset classes that provide services to specific groups. Like fiat currencies are used in many cases to control prices, but the money is not limited to that kind of role. Like stocks, crypto can become assets like a bank payment system. The blockchain is a network of digital records, immutable digital copies of all electronic transactions, including paper money. The history of all bitcoin transactions in the peer-to-peer networks is stored in a database called the blockchain. The blockchain records all of the transactions in bitcoin. To create a digital asset, one needs to hold bitcoins in an exchange, which is similar to a bank account, so that a user can convert those bitcoins into dollars. The user can transfer those dollars to the person.

It is worth mentioning that many people may not know about it due to the market's low visibility and its lack of development. Since there is nothing wrong with the idea of a blockchain platform, it will be a good idea to write a blog about it. It would help people understand the ecosystem better as it is very technical and might even make them rethink their views. They might start using the services that it provides. Crowdfunding has been a great way to fund projects, and this has helped people with cryptocurrency needs. Since we are beginning the coin economy based on the blockchain, our goal is to increase adoption and decrease entry barriers for anyone interested in cryptocurrencies and technology. We are already working on the coin economy to create the platform to allow anyone with a computer or smartphone to earn tokens from creating projects. Many people have launched their cryptocurrency projects, but with the current market structure, it takes them. Ripple, Bitcoin, and Bitcoin Cash are now the most popular cryptocurrencies in the market. They were even the most significant currencies in the world. Many people are interested in bitcoin, and it seems people are ready to make some progress in the future with it [57].

The industry of blockchain is rapidly growing because of its simplicity, ease of use, and flexibility. This technology could be applied to several sectors such as IT, manufacturing, finance, banking, real estate, insurance, education, real estate, healthcare, transportation, retail, and more. Blockchain technology enables smart contracts to be validated, making it easier to do business, track and manage the digital assets from start to finish. It also allows the transfer of digital assets through the Internet and creates a new era of investment in digital assets. Blockchain has enabled much innovation in the development of the cryptocurrency market. The community has gone so far as to put in the necessary resources to allow technology development.

Blockchain innovation has made the technology more efficient as its market expands to more prominent industries such as banking, insurance, healthcare, and the Internet of Things.

One exciting aspect of the blockchain is that it allows for a decentralized peer-to-peer financial system. Unlike traditional financial networks, where parties rely on their networks, no central authority provides them with services. As a result, the peer-to-peer network that creates a blockchain has no financial institutions. There have been few breakthroughs in cryptocurrencies since the beginning. Bitcoin has been in this space for quite a while before it caught onto bitcoin, before it caught onto the crash of the bitcoin bubble that went after its price, then finally it caught onto the Ethereum Bubble. Blockchain has already taken many coins in the space and has given them significant growth that will surely go bigger. The world over, blockchain solutions have been used to build real-time settlements, make real-time payments, provide the backbone for the banking system, verify digital identities, enable financial contracts, and allow digital transactions among people who know each other. With blockchain, it is increasingly apparent that real-time solutions are the currency's backbone. The technology itself provides more applications than just cryptocurrency [58].

There have been few breakthroughs in cryptocurrencies since the beginning. Bitcoin has been in this space for quite a while before it caught onto bitcoin, before it caught onto the crash of the bitcoin bubble that went after its price, then finally it caught onto the Ethereum Bubble. Blockchain has already taken many coins in the space and has given them significant growth that will surely go bigger. The world over, blockchain solutions have been used to build real-time settlements, make real-time payments, provide the backbone for the banking system, verify digital identities, enable financial contracts, and allow digital transactions among people who know each other. With blockchain, it is increasingly apparent that real-time solutions are the currency's backbone. The technology itself provides more applications than just cryptocurrency.

Blockchain Applications in Cybersecurity

The current techniques available in cybersecurity offer a centralized storage system to authorize access [59]. However, blockchain uses distributed ledger technology that gives it additional power of not getting compromised quickly. Blockchain achieves trust among the users by applying cryptographic and mathematical algorithms and does not depend on any third party. The characteristics of blockchain technology like authenticity, transparency, and immutability made it applicable to various other sectors. For example, it is applied in financial sector, medicine, IoT, education, and cybersecurity. Further discussed are some of the cybersecurity problems addressed by blockchain.

Secure Domain Name Service The centralized Domain Name Service (DNS) is susceptible to attacks as the core functions of resolving the domain name are located in a centralized location. A map can be established between DNS and hash using blockchain. Users can register, transmit, and revise domain names. Each block represents the public, private key of domain owners and resolved domain names. Since the information is distributed across the nodes, there is no centralized location to attack. Unlike a centralized DNS system, even if a node is attacked, there is no harm to other nodes in the network [60].

Keyless Signature Infrastructure Authentication schemes that rely on keys suffer from key distribution, key updating, and key revocation. Recent research in blockchain resolves this problem by using Key Signature Infrastructure (KSI). Each node in the blockchain stores the state of the data, network, and hash. KSI will constantly monitor the hash value with a timestamp. Any change in the data changes the hash value and helps to detect unauthorized access. There is no need to distribute, retain, or revoke keys when employing a timestamp-based monitoring system. In England, nuclear power plants and flood control systems both use KSI-based security protection system.

Secure Storage Information regarding finances and medical is usually stored in a centralized location, and unauthorized access brings various problems to the organizations and the users. By using the hash value concept of blockchain, the data can be stored efficiently. Apart from the areas discussed, there are other IoT equipment certification areas, cloud data desensitization, and secure data transmission. Though there are advantages of using blockchain in cybersecurity, there are gaps identified.

Gaps and Resolutions of Security Issues in Blockchain The frauds that happen in a cryptocurrency network are increasing. The increase in fraud each year is slowing down the cryptocurrency market. Weak security systems and lack of government regulations are blamed for it. Another gap is the increase of quantum power. An increase in quantum power will make hackers break the key used for encryption in the blockchain. It is therefore feared to be a cybersecurity threat.

Similarly, one more gap identified is inexperienced users in the blockchain networks. Users who are unaware of safe practices in the blockchain are prone to get attacked by scammers. Thereby they provide insecure access to the blockchain network. Further discussed are three solutions to handle the security gaps identified in the blockchain.

Quantum Computing The gap related to Quantum Computing can be overcome by using a key with a higher number of bits, because quantum computers can crack keys with lower number of bits quickly. Therefore, it is better to offer packages with 64-bit, 128-bit, and 256-bit cryptography so that users can choose depending on their requirements.

Dealing with Inexperienced Users Proper training has to be provided for inexperienced users not to give away their keys to the scammers. Similarly, it is better to add two or three layers of authentication for verification purposes. Another solution is to track transactions using network features, alerting users, and confirm access to their systems.

User Anonymity The user identity in the blockchain network is hidden. Due to this, scammers and hackers are taking advantage of it. When a public key gets flagged, there should be a possibility to track the user's identity. The tracking also should be enabled to government agencies that deal with cybersecurity. This feature would create fear among scammers, so the probability of fraud might be reduced. Although blockchain has many features to improve cybersecurity, some attacks happened in the blockchain.

Application of Blockchain Technology to Validate Transactions in Cyber-Physical Systems

A relatively new trend in cybersecurity is the development of protection mechanisms and systems based on blockchain technology.

The blockchain ensures transaction integrity in the absence of a reliable central hub. System users' tangible and intangible assets are subject to transactions specified as specific activities taken from a predetermined list. Blocks containing transaction information are linked together using hashing to build a chain. To make it more difficult for an attacker to undermine the blockchain, a specific method known as a consensus algorithm is employed to distribute identical copies of the blocks to all system members.

The main advantage of blockchain, which makes the technology attractive for various data protection applications, is the difficulty of violating the integrity of stored transactions. Any change to a single block might have disastrous consequences for the rest of the chain, and it will have to be rebuilt from scratch. However, the computational complexity of this task minimizes the probability of blockchain hacking [8].

At present, blockchain technology is actively used in cyber-physical systems for various purposes. As previously stated, the primary benefit of this technology is the ability to verify a variety of transactions that would otherwise be impossible in an untrusted environment. According to several studies, blockchain technology is crucial for the next fourth industrial revolution (Industry 4.0) [61, 62].

Furthermore, blockchain is being promoted alongside other promising technologies of our time as part of Industry 4.0 [63]. The Internet of Things [64], big data [65], fog computing [66], and augmented reality [61] are examples. In general, in the Industrial Internet of Things, blockchain is widely regarded as a key technology, helping transform traditional factories into modern smart factories that use the latest breakthroughs in digital technology.

Let us mention some examples of current research that offer specific scientific and technical solutions to applying blockchain technology to solve security problems in cyber-physical systems.

The secure management of diverse assets, including those in cyber-physical systems, is an important element of the known works. Blockchain technology was initially used in conjunction with bitcoin. Therefore, this is what happened. With the advancement of blockchain technology, the cryptocurrency industry grew and today plays an essential part in society's daily activities.

Over time, the number of applications of blockchain technology has expanded considerably. For example, a recent paper [67] analyzes the utility of blockchain in solving the security problems of the smart city, which is an example of a large-scale cyber-physical system. The authors consider such components of smart city functioning as transportation, healthcare, smart grids, financial systems, supply chain management, and data center networks; discuss blockchain technology capabilities in relation to these components; and suggest future research directions.

In general, blockchain technology research may be classified into numerous main categories.

The first group of studies is related to supply chain management using blockchain technology. This group is primarily general research, which does not focus on a specific area or a specific class of cyber-physical systems, but rather offers a general solution for secure blockchain-based supply chain management and discusses some aspects of the problem. In some cases, the proposed solutions are designed for use in cyber-physical systems for different purposes. In some cases, they are not explicitly specified in such a scope of use.

Thus, Saberi et al. [48] presented the classification of barriers that prevent blockchain technology implementation in supply chain management. Aceto et al. [63] discussed some of the challenges of overcoming these roadblocks. The precise asset is not provided in either scenario. A wide range of supply chain services and items are included in this set of research as well. Kshetri et al. [68] described real-world applications of blockchain for tracking raw materials, ingredients, or spare parts in various industries. In many cyber-physical systems, the emphasis is on leveraging blockchain technology combined with Internet of Things technologies.

If not to be classified in more detail, the first group includes studies devoted to related tasks arising in organization and management of production; for example, the work [69], which presents an architectural solution for data integrity protection in cyber-physical production systems used in co-production.

The second group of studies aims to tackle the problem of risk-free management of a certain asset or service, including supply chain management of the associated assets. Today, there are a plethora of such applications to choose from. Blockchain is used to control sales or distribution of electricity [70], fuel [71], computing resources [72], and software.

All of the previous studies have one thing in common: they all involve trading commodities for money. As a result, blockchain-based solutions borrow heavily from cryptocurrency concepts.

The next group includes researches devoted to the problem of the organization of trusted interaction between multiple devices. The specific tasks related to ensuring the integrity of some or other data operated by such devices may differ.

Many papers deal with the interaction of arbitrary IoT devices without reference to specific types of cyber-physical systems. Some examples of recent works in this direction are [73–75].

Most of these works focus on the energy efficiency of architectural solutions intended for use in IoT systems and propose various ways to achieve this property.

In terms of the tasks to be solved, the works under consideration can be divided into those that only ensure the integrity of transactions and those that, in addition, ensure the confidentiality of the data contained in transactions. For example, in a study [76], data on the location of Internet of Things devices are considered the object of protection. The authors point to the need to ensure the confidentiality of this data, so in the scheme they propose, blockchain is combined with encryption.

Turning from general solutions for the application of blockchain technology for data protection in cyber-physical systems, which are based on the Internet of Things technology, to particular cases, it is necessary to note such a class of cyber-physical systems as connected vehicles, including unmanned vehicles [77–79]. In 2019–2021, there is an "explosive" growth in the number of journal publications devoted to relevant research; hence, we can say that the security of this class of cyber-physical systems using blockchain technology is an example of a promising direction in the problem area under consideration.

Data Authentication in Cyber-Physical Systems

Digital evidence may be subject to an entire forensic process, encompassing the following stages: identification, collection, examination, analysis, documentation, and presentation [80]. Preservation of digital evidence is an essential principle that should be considered in all stages of this process. For this purpose, blockchain plays an important role in ensuring the integrity and proof of origin of the collected evidence. The complexity existing in the operation scenarios of cyber-physical systems imposes, to the security solutions and methods used, restrictive non-functional requirements concerning scalability, computational performance, use of the communication network, among others.

Several blockchain techniques have been proposed in the literature to provide data authentication and prevent cyber-physical attacks. Evsyutin et al. [81] provide an overview of strategies for embedding information into digital data in the Internet of Things applicable at the end of 2018. As a result, the focus of this review is on new research that has arisen in recent years. At the same time, we should stress that only digital watermark embedding methods will be addressed in the context of this study. In contrast, digital steganography methods are often unrelated to data integrity.

At the outset, it is vital to distinguish between several research projects focused on the development of methods and algorithms for concealing information in digital photographs (as well as other digital objects), to ensure the security of sensitive data in cyber-physical systems. Examples of works in this area are [79, 82]. Although their authors claim that their solutions aim to ensure data security in the Internet of Things, they do not present any examples of how their algorithms may be used in other domains. In many of these research, the authors are worried about the security flaws in telemedicine systems.

Because such studies are so prevalent, they should be classified as a different class. However, the works in this class do not extend beyond the boundaries of traditional embedding into multimedia data and will not be studied further.

The following set of works also includes traditional data embedding into multimedia products. The authors identify unique data transmission situations specific to such systems and explain the limits associated with them while stressing the applicability of their solutions in cyber-physical systems.

This group's works are not as extensively represented, but they should be separated from those in the first group.

A solution for secure picture transmission in telemedicine systems is also presented in [83]. Encrypted confidential pictures are placed in photographs with non-confidential material. In addition, the fingerprint (perceptual hash) of the secret picture is included in the image container for further authentication. The tracking of the picture transmission sequence is a distinguishing characteristic of this approach. To that end, the authors offer the concept of a picture fingerprint chain by analogy with blockchain technology.

The research of Zhang et al. [84] is fairly unique in that it entails embedding secret attachments in graphics used in printed items. On the other hand, this research explicitly describes the potential applications of the proposed approach in IoT systems and the associated situations. These scenarios include, for example, offering data authentication in order to protect products from being counterfeited. There should be an emphasis on the endurance of digital watermark embedding in the authors' discussion on steganographic embedding.

The following works are unrelated to multimedia and deal with inserting digital watermarks in data created and transferred through cyber-physical systems. A substantial portion of the work in this group is concerned with inserting digital watermarks in wireless sensor network data for integrity control.

A comparable approach is proposed, among other places, in one of the authors of this review's papers [80]. The ability to alter the degree of distortion generated by embedding is a distinguishing characteristic of this approach. As a result, it applies to sensor data of many physical types.

Hoang et al. [85] incorporate digital watermarks into wireless sensor network data to protect against clone sensor nodes attacks. The embedding is based on a gamming-like modification of the binary alphabet. It is argued that the algorithm's lightweight provides an advantage.

The algorithms embed the digital watermark components into the sensory data consistently and independently since they are not dependent on the values of these sensory data or some of their features. While traditional digital watermarking methods and algorithms and the problem of wireless sensor networks and the Internet of Things can create a digital watermark based on protected data, the notion is exceedingly wide.

In the simplest example, digital watermark components are created only based on the sensor data elements' values. The embedding approach provided in [86] is one example. This approach generates the digital watermark bit inserted in the next sensor value depending on previous sensor values.

Separate embedding of digital watermark components offers various advantages; nonetheless, this technique poses a challenge with timing. Think about what may happen if a message arrives with the data out of order. The digital watermark extraction will be hampered even if there is not an active intruder on the communication channel. Wang et al. [87] offer an answer to this dilemma. Sensor data is divided depending on the key into variable-length groups, as proposed by the authors in their paper. Digital watermark chains are produced and implanted for pairs of adjacent groupings. A series of digital watermarks are used to verify the sensor data. Separators and data synchronization are provided by the second digital watermark chain, which encodes group separators.

Creating a digital watermark that contains sensory quantity values and some of their characteristics is possible in a more complex scenario. Ferdowsi et al. [88] proposed a watermarking algorithm for dynamic authentication of IoT signals to detect cyber-attacks. It is possible for IoT devices (IoTDs) to extract a collection of stochastic characteristics from their produced signal and dynamically watermark these features into the signal using the suggested watermarking technique. In order to authenticate the signals collected by the IoT gateway, this approach allows the IoTDs to be authenticated. Hameed et al. [89] also addressed how to create a digital watermark using several aspects of the acquired data, such as data length, frequency of occurrence, and time of capture. Nguyen et al. [90] create a digital watermark based on CSMA/CA protocol collisions to deter sensor node clone attacks. Furthermore, the way the sensor data is portrayed is a distinguishing characteristic of the study. They are combined to make a matrix resembling a digital image. In general, such a system enables the adoption of methodologies that have proven successful when dealing with digital pictures with sensory data.

Using digital watermarks that represent binary sequences, all of the algorithms in the mentioned study work. Watermarking analog transmissions (particularly modulated signals) to solve signal source authentication is also a field of study. The answers discovered in those works are ideally comparable to those found in digital watermarking. The distinction is solely in the manner in which the signal is represented and, as a result, in the manner in which it is processed.

Sender authentication in systems matching to the NB-IoT (Narrow Band Internet of Things) standard focuses on research [91]. The notion of a radio-frequency watermark is used in the investigation. Rather than using binary sequences for future embedding, the watermark is first created as a digital one and then converted to modulated signals. The main benefit of the suggested technique is that it is more reliable since the useable signal and the watermark signal do not conflict with one another.

Watermarks can be used to deter certain sorts of assaults in some instances. Rubio-Hernan et al. [92] suggested an adaptive control-theoretic technique for detecting cyber replication attempts on networked control industrial systems. This refers to an intruder's effort to tamper with system control by replicating previously intercepted data sequences. The fundamental contribution of this work is not the embedding technique, which is borrowed from prior publications, but rather the approach for employing this algorithm to guard against an intruder.

Huang et al. [93] offer a technique for embedding reversible air signs in signals conveyed in "hard" real-time industrial control systems. The authors choose ship control systems as the most important field of application. A secret key must be delivered in advance through a secure communication channel before embedding can begin. The approach described here can detect attacks that aim to cause signal delay and distortion.

Finally, experiments integrating blockchain with digital watermark technology have begun to emerge as outlined in the preceding section. Different security concerns in cyber-physical systems are addressed by blockchain and digital watermarks. Their combined use may produce a better level of security than each of these methods alone. This concept has previously been explored in prior works, but mostly in one aspect, namely, the issue of digital rights management [79, 91]. The collaborative implementation of these technologies in other areas [7, 94, 95] is a promising research field whose advancement will benefit cybersecurity. Table 5 summarizes the many blockchain uses in cybersecurity.

4 Blockchain Limitations and Future Directions

Despite their wealth, blockchain promises continue to face many security issues, as seen by the numerous breaches and frauds that have been reported. This is a paradoxical snare to fall into for a system whose key touted attributes are dependability and inviolability. Computer assaults, on the other hand, are carried over to all types of interactions with other systems – primarily markets, which are vulnerable to the classic flaws of centralized systems, like banks [102].

In general, the speed, throughput, secrecy, scalability, and interoperability issues with blockchains have been shown and well documented. Mining issues, which are at the heart of the Proof of Work on which the bitcoin consensus is based, are also the subject of much debate. On the theoretical level, it is a matter of rigorously defining the conditions that will protect against malicious validation nodes. The level of 51% of the computing power held by a malicious entity is certainly considered the reference level. However, this value is the subject of

Table 5 Blockchain applications in cybersecurity

Application domain	Applications	Contributions
Quantum-Inspired Blockchain [96]	Smart edge utilities in IoT-based smart cities	Resist potential assaults from digital and quantum computers
Lightweight Blockchain-based Cybersecurity (LBC) [97]	IoT environments	Address intensive computational requirements and bandwidth consumption overhead
Blockchain Empowered Cooperative Authentication [97]	Vehicular edge computing	Protect and preserve the privacy and confidentiality of data while also ensuring mutual authentication is in place (e.g., reply attack)
BloCyNfo-Share [98]	Cybersecurity Information Exchange (CYBEX)	Describe how to share private information with other organizations or provide private information to other organizations access
BBDS [99]	Electronic medical records in cloud environments	Aim to disseminate medical data outside of the protected institutions' cloud
Ancile [100]	Electronic medical records	Preserve the privacy of patients' sensitive information
Secure and decentralized sharing [101]	Image sharing across domains. Personal health data may be processed in a batch using a Hyperledger fabric tree-based system	Allowing parties to come to an agreement without relying on a single authority
BlockChain [23]	Interconnected smart vehicles	Address the security and privacy threats that smart vehicles face, such as location tracking and remote vehicle hijacking

controversy in the research community. The number and distribution of nodes is also a sensitive issue. Economic issues are also beginning to mobilize researchers [10]. In addition to the monetary and financial aspects (competition between currencies, monetary systems), the themes concern the economics of mining, with the question of cooperative incentives for miners or the evolution of mining capacity. And the debates do not stop at bitcoin, since many depend on the consensus model chosen to replace the centralized decision-making system. Whatever the case may be, the technological environment for distributed registries will continue to evolve significantly.

The time it takes to complete a transaction on the blockchain is a major drawback. Due to the vastness of the bitcoin network, this process might take many hours to complete. Using a blockchain like Hyperledger Sawtooth helps reduce this latency since it allows the PoET (Proof of Elapsed Time) consensus process to be used, which is among the fastest and least resource-hungry in terms of reaction time, making it better suited to our current context, which includes networked object data and services [103]. Second, not everyone should have access to blockchain data. This challenge may be solved by utilizing private blockchains, which can govern blockchain access privileges and transaction execution rights. Second, the blockchain's consensus algorithms, notably the PoW (Proof of Work), are extremely energy-hungry in terms of calculations.

Additionally, the redundant data and redundant calculations necessary to decide whether or not a new block may be added to the blockchain are energy-hungry. The blockchain, in the end, represents a sea change in thinking. In other words, the network is becoming decentralized instead of centrally controlled. Customers may have difficulty adopting and integrating this technology into their existing ecosystems as a result of this.

Other limitations of blockchains are regularly invoked, particularly the tension between transparency and confidentiality, between anonymity and identification of stakeholders. Because the register is widely disseminated, stakeholders may easily access the plain language information it includes. When it comes to tracing transactions, this is a benefit, but when it comes to corporate confidentiality, such as in banking or healthcare, it is a redhibitory problem. The market looks for ways to reliably disguise information while engaging in activities that require the disclosure of some and the protection of others [58].

Another limitation is the high amount of energy required. Using blockchain necessitates a lot of power-hungry verification, validation, and cryptography procedures. If this technique is widely used, it might have significant negative environmental externalities.

Despite the widespread interest in blockchain, it must be recognized that distributed registries are not a panacea. Blockchain is not yet suitable for the fast processing of large amounts of data, especially video and audio, and use in a fast-changing environment for use in fast-changing environments. Blockchain is ideal for the long term and most as reliably as possible for storing information that changes infrequently. Therefore, the technology is promising for capturing customer data from banks, medical institutions, and insurance and logistics companies [104]. A distributed transaction registry will benefit patent offices and cadastral offices. Technology is suitable for law enforcement and tax authorities to record personal data. Brokerage and investment firms will benefit from blockchain as a registry of transactions. The technology's current capabilities are just an in-between. The continuous improvement of blockchain opens up prospects for its application in new industries. In its evolution, any technology must overcome resistance to change. Blockchain has already passed that stage, and, therefore, it will continue to evolve.

5 Conclusion

Rapid advances in computational and communication technologies drive the scientific community's interest and industry in cyber-physical systems. Using sensor, computational, and networking capabilities, cyber-physical systems contribute to a new generation of scientific and technical solutions that provide automatic decision-making processes in various fields, from automation of small domestic processes to transportation of materials, factories of the future, and mission-critical industries. Greater efficiency, dependability, and sustainability may be achieved by creating a smart infrastructure that integrates information technology approaches with physical systems like the power grid, transportation system, and supply chain.

An overview of various applications of blockchain in cyber-physical systems control protocols is presented. Although each industry has advantages in using blockchain, there are also challenges involved. However, blockchain is well known and adopted for its various benefits in various industries.

This chapter contributes to the thematic area by providing information on a poorly documented topic in the scientific literature. It became clear during the development of the work and deserved further theoretical investigation. The analysis of the results shows that the theme addressed has grown annually and has become of great relevance for the emergence and development of new applications using blockchain for cyber-physical systems. It is still too early to say whether blockchain technology is more appropriate in the context of cyber-physical system applications and to compare it with other technologies already used.

References

1. M. Swan, *Blockchain: Blueprint for a New Economy* (O'Reilly Media, 2015)
2. S. Porru, A. Pinna, M. Marchesi, R. Tonelli, Blockchain-oriented software engineering: Challenges and new directions, in *2017 IEEE/ACM 39th International Conference on Software Engineering Companion (ICSE-C)*, (2017), pp. 169–171. https://doi.org/10.1109/ICSE-C.2017.142
3. N. Teslya, I. Ryabchikov, Blockchain-based platform architecture for industrial IoT, in *2017 21st Conference of Open Innovations Association (FRUCT)*, (2017), pp. 321–329. https://doi.org/10.23919/FRUCT.2017.8250199
4. K. Christidis, M. Devetsikiotis, Blockchains and smart contracts for the internet of things. IEEE Access **4**, 2292–2303 (2016). https://doi.org/10.1109/ACCESS.2016.2566339
5. V. Daza, R. Di Pietro, I. Klimek, M. Signorini, CONNECT: CONtextual NamE disCovery for blockchain-based services in the IoT, in *2017 IEEE International Conference on Communications (ICC)*, (2017), pp. 1–6. https://doi.org/10.1109/ICC.2017.7996641
6. L. Monostori, Cyber-physical production systems: Roots, expectations and R&D challenges. Procedia CIRP **17**, 9–13 (2014). https://doi.org/10.1016/j.procir.2014.03.115
7. Y. Maleh, M. Shojafar, A. Darwish, A. Haqiq (eds.), *Cybersecurity and Privacy in Cyber-Physical Systems* (CRC Press, 2019)., [Online]. Available: https://www.crcpress.com/Cybersecurity-and-Privacy-in-Cyber-Physical-Systems/Maleh/p/book/9781138346673
8. S. Nakamoto, Bitcoin: A peer-to-peer electronic cash system. Decentralized Bus. Rev. **2008**, 21260 (2008)
9. S. Zhang, J.-H. Lee, Analysis of the main consensus protocols of blockchain. ICT Express **6**(2), 93–97 (2020). https://doi.org/10.1016/j.icte.2019.08.001
10. V. Chang, P. Baudier, H. Zhang, Q. Xu, J. Zhang, M. Arami, How blockchain can impact financial services – The overview, challenges and recommendations from expert interviewees. Technol. Forecast. Soc. Change **158**, 120166 (2020). https://doi.org/10.1016/j.techfore.2020.120166

11. S.S. Gupta, Blockchain. IBM Onlone (http//www. IBM. COM) (2017)
12. J. Abadi, M. Brunnermeier, Blockchain economics. Natl. Bur. Econ. Res. (2018)
13. T. Ahram, A. Sargolzaei, S. Sargolzaei, J. Daniels, B. Amaba, Blockchain technology innovations, in *2017 IEEE Technology & Engineering Management Conference (TEMSCON)*, (2017), pp. 137–141. https://doi.org/10.1109/TEMSCON.2017.7998367
14. S. Wattanakul, S. Henry, M.L. Bentaha, N. Reeveerakul, Y. Ouzrout, Improving risk management by using smart containers for real-time traceability. arXiv Prepr. arXiv1810.13332 (2018)
15. Y.L. Morgan, Notes on DSRC and WAVE standards suite: Its architecture, design, and characteristics. IEEE Commun. Surv. Tutorials **12**(4), 504–518 (2010). https://doi.org/10.1109/SURV.2010.033010.00024
16. H. Rathore, A. Samant, M. Jadliwala, TangleCV: A distributed ledger technique for secure message sharing in connected vehicles. ACM Trans. Cyber-Phys. Syst. **5**(1), Dec (2021). https://doi.org/10.1145/3404500
17. M. Singh, S. Kim, Trust bit: Reward-based intelligent vehicle commination using blockchain paper, in *2018 IEEE 4th World Forum on Internet of Things (WF-IoT)*, (2018), pp. 62–67. https://doi.org/10.1109/WF-IoT.2018.8355227
18. R. Chaudhary, A. Jindal, G.S. Aujla, S. Aggarwal, N. Kumar, K.-K.R. Choo, BEST: Blockchain-based secure energy trading in SDN-enabled intelligent transportation system. Comput. Secur. **85**, 288–299 (2019). https://doi.org/10.1016/j.cose.2019.05.006
19. J.A.L. Calvo, R. Mathar, Secure blockchain-based communication scheme for connected vehicles, in *2018 European Conference on Networks and Communications (EuCNC)*, (2018), pp. 347–351. https://doi.org/10.1109/EuCNC.2018.8442848
20. CUBE, Autonomous car network security platform based on blockchain, White Pap. Available online https://cubeint.io/wp-content/uploads/2019/10/Cube-Whitepaper-Centered-v2-3.pdf. Accesses 17 Jul 2017 (2017)
21. Y. Yuan, F. Wang, Towards blockchain-based intelligent transportation systems, in *2016 IEEE 19th International Conference on Intelligent Transportation Systems (ITSC)*, (2016), pp. 2663–2668. https://doi.org/10.1109/ITSC.2016.7795984
22. B. Leiding, P. Memarmoshrefi, D. Hogrefe, Self-managed and blockchain-based vehicular ad-hoc networks, in *Proceedings of the 2016 ACM International Joint Conference on Pervasive and Ubiquitous Computing: Adjunct*, (2016), pp. 137–140. https://doi.org/10.1145/2968219.2971409
23. A. Dorri, M. Steger, S.S. Kanhere, R. Jurdak, BlockChain: A distributed solution to automotive security and privacy. IEEE Commun. Mag. **55**(12), 119–125 (2017). https://doi.org/10.1109/MCOM.2017.1700879
24. J. Aguirre, J.P. Davis, F. Cenciarelli, U.S. Patent Application No. 14/887,594 (2017)
25. S. Rowan, M. Clear, M. Gerla, M. Huggard, C.M. Goldrick, Securing vehicle to vehicle communications using blockchain through visible light and acoustic side-channels. arXiv Prepr. arXiv1704.02553 (2017)., [Online]. Available: http://arxiv.org/abs/1704.02553
26. A.R. Pedrosa, G. Pau, ChargeltUp: On blockchain-based technologies for autonomous vehicles, in *Proceedings of the 1st Workshop on Cryptocurrencies and Blockchains for Distributed Systems*, (2018), pp. 87–92. https://doi.org/10.1145/3211933.3211949
27. H. Rathore, A. Samant, M. Jadliwala, A. Mohamed, TangleCV: Decentralized technique for secure message sharing in connected vehicles, in *Proceedings of the ACM Workshop on Automotive Cybersecurity*, (2019), pp. 45–48. https://doi.org/10.1145/3309171.3309177
28. M. Singh, S. Kim, Branch based blockchain technology in intelligent vehicle. Comput. Netw. **145**, 219–231 (2018). https://doi.org/10.1016/j.comnet.2018.08.016
29. W. Ben Jaballah, M. Conti, M. Mosbah, C.E. Palazzi, Fast and secure multihop broadcast solutions for intervehicular communication. IEEE Trans. Intell. Transp. Syst. **15**(1), 433–450 (2014). https://doi.org/10.1109/TITS.2013.2277890
30. F. Knirsch, A. Unterweger, D. Engel, Privacy-preserving blockchain-based electric vehicle charging with dynamic tariff decisions. Comput. Sci. Res. Dev. **33**(1), 71–79 (2018). https://doi.org/10.1007/s00450-017-0348-5

31. A. Balasubramaniam, M.J. Gul, V.G. Menon, A. Paul, Blockchain for intelligent transport system. IETE Tech. Rev. **38**(4), 438–449 (2021). https://doi.org/10.1080/02564602.2020.1766385
32. Q. Al-Maatouk, M.S. Othman, A. Aldraiweesh, U. Alturki, W.M. Al-Rahmi, A.A. Aljeraiwi, Task-technology fit and technology acceptance model application to structure and evaluate the adoption of social media in academia. IEEE Access **8**, 78427–78440 (2020). https://doi.org/10.1109/ACCESS.2020.2990420
33. A. Ghosh, S. Gupta, A. Dua, N. Kumar, Security of cryptocurrencies in blockchain technology: State-of-art, challenges and future prospects. J. Netw. Comput. Appl. **163**, 102635 (2020). https://doi.org/10.1016/j.jnca.2020.102635
34. Z. Liu, Z. Li, A blockchain-based framework of cross-border e-commerce supply chain. Int. J. Inf. Manag. **52**, 102059 (2020). https://doi.org/10.1016/j.ijinfomgt.2019.102059
35. T.M. Xuan, M.T. Alrashdan, Q. Al-Maatouk, M.T. Alrashdan, Blockchain technology in e-commerce platform. Int. J. Manag. **11**(10), 1688–1697 (2020)
36. M. Janssen, V. Weerakkody, E. Ismagilova, U. Sivarajah, Z. Irani, A framework for analysing blockchain technology adoption: Integrating institutional, market and technical factors. Int. J. Inf. Manag. **50**, 302–309 (2020). https://doi.org/10.1016/j.ijinfomgt.2019.08.012
37. A. Ekblaw, A. Azaria, J.D. Halamka, A. Lippman, A case study for blockchain in healthcare: 'MedRec' prototype for electronic health records and medical research data, in *Proc. IEEE open big data Conf.*, vol. 13, (2016), p. 13
38. Q. Xia, E.B. Sifah, K.O. Asamoah, J. Gao, X. Du, M. Guizani, MeDShare: Trust-less medical data sharing among cloud service providers via blockchain. IEEE Access **5**, 14757–14767 (2017). https://doi.org/10.1109/ACCESS.2017.2730843
39. A. Dubovitskaya, Z. Xu, S. Ryu, M. Schumacher, F. Wang, Secure and trustable electronic medical records sharing using blockchain, in *AMIA Annual Symposium Proceedings*, vol. 2017, (American Medical Informatics Association, 2018), pp. 650–659, [Online]. Available: https://pubmed.ncbi.nlm.nih.gov/29854130
40. E. Karafiloski, A. Mishev, Blockchain solutions for big data challenges: A literature review, in *IEEE EUROCON 2017-17th International Conference on Smart Technologies*, (2017), pp. 763–768. https://doi.org/10.1109/EUROCON.2017.8011213
41. T. Le Nguyen, Blockchain in healthcare: A new technology benefit for both patients and doctors, in *2018 Portland International Conference on Management of Engineering and Technology (PICMET)*, (2018), pp. 1–6. https://doi.org/10.23919/PICMET.2018.8481969
42. T. Nugent, D. Upton, M. Cimpoesu, Improving data transparency in clinical trials using blockchain smart contracts. F1000Res. **5**, 2541 (2016). https://doi.org/10.12688/f1000research.9756.1
43. M. Benchoufi, R. Porcher, P. Ravaud, Blockchain protocols in clinical trials: Transparency and traceability of consent. F1000Res. **6**, 66 (2017)
44. R. Klein, Assimilation of internet-based purchasing applications within medical practices. Inf. Manag. **49**(3), 135–141 (2012). https://doi.org/10.1016/j.im.2012.02.001
45. M. Uddin, Blockchain medledger: Hyperledger fabric enabled drug traceability system for counterfeit drugs in pharmaceutical industry. Int. J. Pharm. **597**, 120235 (2021). https://doi.org/10.1016/j.ijpharm.2021.120235
46. E. Fernando, Success factor of implementation blockchain technology in pharmaceutical industry: A literature review, in *2019 6th International Conference on Information Technology, Computer and Electrical Engineering (ICITACEE)*, (2019), pp. 1–5. https://doi.org/10.1109/ICITACEE.2019.8904335
47. W.J. Gordon, C. Catalini, Blockchain technology for healthcare: Facilitating the transition to patient-driven interoperability. Comput. Struct. Biotechnol. J. **16**, 224–230 (2018). https://doi.org/10.1016/j.csbj.2018.06.003
48. S. Saberi, M. Kouhizadeh, J. Sarkis, L. Shen, Blockchain technology and its relationships to sustainable supply chain management. Int. J. Prod. Res. **57**(7), 2117–2135 (2019). https://doi.org/10.1080/00207543.2018.1533261

49. L. Zhou, L. Wang, Y. Sun, MIStore: A blockchain-based medical insurance storage system. J. Med. Syst. **42**(8), 149 (2018). https://doi.org/10.1007/s10916-018-0996-4
50. X. Zhang, S. Poslad, Blockchain support for flexible queries with granular access control to electronic medical records (EMR), in *2018 IEEE International Conference on Communications (ICC)*, (2018), pp. 1–6. https://doi.org/10.1109/ICC.2018.8422883
51. S. Jiang, J. Cao, H. Wu, Y. Yang, M. Ma, J. He, BlocHIE: A BLOCkchain-based platform for healthcare information exchange, in *2018 IEEE International Conference on Smart Computing (SMARTCOMP)*, (2018), pp. 49–56. https://doi.org/10.1109/SMARTCOMP.2018.00073
52. K. Fan, S. Wang, Y. Ren, H. Li, Y. Yang, MedBlock: Efficient and secure medical data sharing via blockchain. J. Med. Syst. **42**(8), 136 (2018). https://doi.org/10.1007/s10916-018-0993-7
53. I. Eyal, Blockchain technology: Transforming libertarian cryptocurrency dreams to finance and banking realities. Computer (Long. Beach. Calif) **50**(9), 38–49 (2017). https://doi.org/10.1109/MC.2017.3571042
54. F. Calvão, Crypto-miners: Digital labor and the power of blockchain technology. Econ. Anthropol. **6**(1), 123–134 (2019). https://doi.org/10.1002/sea2.12136
55. D. Knezevic, Impact of blockchain technology platform in changing the financial sector and other industries. Montenegrin J. Econ. **14**(1), 109–120 (2018)
56. R.B. Sağlam, Ç.B. Aslan, S. Li, L. Dickson, G. Pogrebna, A data-driven analysis of blockchain systems' public online communications on GDPR, in *2020 IEEE International Conference on Decentralized Applications and Infrastructures (DAPPS)*, (2020), pp. 22–31. https://doi.org/10.1109/DAPPS49028.2020.00003
57. T. Aste, P. Tasca, T. Di Matteo, Blockchain technologies: The foreseeable impact on society and industry. Computer (Long. Beach. Calif) **50**(9), 18–28 (2017). https://doi.org/10.1109/MC.2017.3571064
58. M. Hashemi Joo, Y. Nishikawa, K. Dandapani, Cryptocurrency, a successful application of blockchain technology. Manag. Financ. **46**(6), 715–733 (Jan. 2020). https://doi.org/10.1108/MF-09-2018-0451
59. L. Koh, A. Dolgui, J. Sarkis, Blockchain in transport and logistics – Paradigms and transitions. Int. J. Prod. Res. **58**(7), 2054–2062 (2020). https://doi.org/10.1080/00207543.2020.1736428
60. J.M. Huang, S.B. Yang, C.L. Dai, An efficient key management scheme for data-centric storage wireless sensor networks. IERI Procedia **4**, 25–31 (2013). https://doi.org/10.1016/J.IERI.2013.11.005
61. T.M. Fernández-Caramés, P. Fraga-Lamas, A review on the application of blockchain to the next generation of cybersecure industry 4.0 smart factories. IEEE Access **7**, 45201–45218 (2019). https://doi.org/10.1109/ACCESS.2019.2908780
62. T. Alladi, V. Chamola, R.M. Parizi, K.R. Choo, Blockchain applications for industry 4.0 and industrial IoT: A review. IEEE Access **7**, 176935–176951 (2019). https://doi.org/10.1109/ACCESS.2019.2956748
63. G. Aceto, V. Persico, A. Pescapé, A survey on information and communication technologies for industry 4.0: State-of-the-art, taxonomies, perspectives, and challenges. IEEE Commun. Surv. Tutorials **21**(4), 3467–3501 (2019). https://doi.org/10.1109/COMST.2019.2938259
64. T.M. Fernández-Caramés, P. Fraga-Lamas, A review on the use of blockchain for the internet of things. IEEE Access **6**, 32979–33001 (2018). https://doi.org/10.1109/ACCESS.2018.2842685
65. M. Zhaofeng, W. Lingyun, W. Xiaochang, W. Zhen, Z. Weizhe, Blockchain-enabled decentralized trust management and secure usage control of IoT big data. IEEE Internet Things J. **7**(5), 4000–4015 (2020). https://doi.org/10.1109/JIOT.2019.2960526
66. H. Baniata, A. Kertesz, A survey on blockchain-fog integration approaches. IEEE Access **8**, 102657–102668 (2020). https://doi.org/10.1109/ACCESS.2020.2999213
67. B. Bhushan, A. Khamparia, K.M. Sagayam, S.K. Sharma, M.A. Ahad, N.C. Debnath, Blockchain for smart cities: A review of architectures, integration trends and future research directions. Sustain. Cities Soc. **61**, 102360 (2020). https://doi.org/10.1016/j.scs.2020.102360
68. N. Kshetri, Blockchain's roles in meeting key supply chain management objectives. Int. J. Inf. Manag. **39**, 80–89 (2018). https://doi.org/10.1016/j.ijinfomgt.2017.12.005

69. C. Yu, X. Jiang, S. Yu, C. Yang, Blockchain-based shared manufacturing in support of cyber physical systems: Concept, framework, and operation. Robot. Comput. Integr. Manuf. **64**, 101931 (2020). https://doi.org/10.1016/j.rcim.2019.101931
70. M. Li, D. Hu, C. Lal, M. Conti, Z. Zhang, Blockchain-enabled secure energy trading with verifiable fairness in industrial internet of things. IEEE Trans. Ind. Inf. **16**(10), 6564–6574 (2020). https://doi.org/10.1109/TII.2020.2974537
71. H. Lu, K. Huang, M. Azimi, L. Guo, Blockchain technology in the oil and gas industry: A review of applications, opportunities, challenges, and risks. IEEE Access **7**, 41426–41444 (2019). https://doi.org/10.1109/ACCESS.2019.2907695
72. A. Seitz, D. Henze, D. Miehle, B. Bruegge, J. Nickles, M. Sauer, Fog computing as enabler for blockchain-based IIoT app marketplaces – a case study, in *2018 Fifth International Conference on Internet of Things: Systems, Management and Security*, (2018), pp. 182–188. https://doi.org/10.1109/IoTSMS.2018.8554484
73. P. Koshy, S. Babu, B.S. Manoj, Sliding window blockchain architecture for internet of things. IEEE Internet Things J. **7**(4), 3338–3348 (2020). https://doi.org/10.1109/JIOT.2020.2967119
74. J. Luo, Q. Chen, F.R. Yu, L. Tang, Blockchain-enabled software-defined industrial internet of things with deep reinforcement learning. IEEE Internet Things J. **7**(6), 5466–5480 (2020). https://doi.org/10.1109/JIOT.2020.2978516
75. J. Chi et al., A secure and efficient data sharing scheme based on blockchain in industrial internet of things. J. Netw. Comput. Appl. **167**, 102710 (2020). https://doi.org/10.1016/j.jnca.2020.102710
76. D. Li, Y. Hu, M. Lan, IoT device location information storage system based on blockchain. Futur. Gener. Comput. Syst. **109**, 95–102 (2020). https://doi.org/10.1016/j.future.2020.03.025
77. M. Cebe, E. Erdin, K. Akkaya, H. Aksu, S. Uluagac, Block4Forensic: An integrated lightweight blockchain framework for forensics applications of connected vehicles. IEEE Commun. Mag. **56**(10), 50–57 (2018). https://doi.org/10.1109/MCOM.2018.1800137
78. G. Rathee, A. Sharma, R. Iqbal, M. Aloqaily, N. Jaglan, R. Kumar, A blockchain framework for securing connected and autonomous vehicles. Sensors **19**(14), 3165 (2019). https://doi.org/10.3390/s19143165
79. Y. Qian, Y. Jiang, L. Hu, M.S. Hossain, M. Alrashoud, M. Al-Hammadi, Blockchain-based privacy-aware content caching in cognitive internet of vehicles. IEEE Netw. **34**(2), 46–51 (2020). https://doi.org/10.1109/MNET.001.1900161
80. O. Evsutin, R. Meshcheryakov, V. Tolmachev, A. Iskhakov, A. Iskhakova, Algorithm for embedding digital watermarks in wireless sensor networks data with control of embedding distortions, in *Distributed Computer and Communication Networks*, (2019), pp. 574–585
81. O.O. Evsyutin, A.S. Kokurina, A review of methods of embedding information in digital objects for security in the internet of things. Comput. Opt. **43**(1), 137–154 (2019)
82. H. Prasetyo, C. Hsia, C. Liu, Vulnerability attacks of SVD-based video watermarking scheme in an IoT environment. IEEE Access **8**, 69919–69936 (2020). https://doi.org/10.1109/ACCESS.2020.2984180
83. H. Peng, B. Yang, L. Li, Y. Yang, Secure and traceable image transmission scheme based on semitensor product compressed sensing in telemedicine system. IEEE Internet Things J. **7**(3), 2432–2451 (2020). https://doi.org/10.1109/JIOT.2019.2957747
84. Y. Pu, N. Zhang, H. Wang, Fractional-order spatial steganography and blind steganalysis for printed matter: Anti-counterfeiting for product external packing in internet-of-things. IEEE Internet Things J. **6**(4), 6368–6383 (2019). https://doi.org/10.1109/JIOT.2018.2886996
85. T. Hoang, V. Bui, N. Vu, D. Hoang, A lightweight mixed secure scheme based on the watermarking technique for hierarchy wireless sensor networks, in *2020 International Conference on Information Networking (ICOIN)*, (2020), pp. 649–653. https://doi.org/10.1109/ICOIN48656.2020.9016541
86. Y. Xiao, G. Gao, Digital watermark-based independent individual certification scheme in WSNs. IEEE Access **7**, 145516–145523 (2019). https://doi.org/10.1109/ACCESS.2019.2945177

87. B. Wang, W. Kong, W. Li, N.N. Xiong, A dual-chaining watermark scheme for data integrity protection in internet of things. Comput. Mater. Continua **58**(3), 679–695 (2019). https://doi.org/10.32604/cmc.2019.06106
88. A. Ferdowsi, W. Saad, Deep learning for signal authentication and security in massive internet-of-things systems. IEEE Trans. Commun. **67**(2), 1371–1387 (2019). https://doi.org/10.1109/TCOMM.2018.2878025
89. K. Hameed, A. Khan, M. Ahmed, A. Goutham Reddy, M.M. Rathore, Towards a formally verified zero watermarking scheme for data integrity in the internet of things based-wireless sensor networks. Futur. Gener. Comput. Syst. **82**, 274–289 (2018). https://doi.org/10.1016/j.future.2017.12.009
90. V. Nguyen, T. Hoang, T. Duong, Q. Nguyen, V. Bui, A lightweight watermark scheme utilizing MAC layer behaviors for wireless sensor networks, in *2019 3rd International Conference on Recent Advances in Signal Processing, Telecommunications & Computing (SigTelCom)*, (2019), pp. 176–180. https://doi.org/10.1109/SIGTELCOM.2019.8696234
91. B. Zhao et al., Y-DWMS: A digital watermark management system based on smart contracts. Sensors **19**(14), 3091 (2019). https://doi.org/10.3390/s19143091
92. J. Rubio-Hernan, L. De Cicco, J. Garcia-Alfaro, Adaptive control-theoretic detection of integrity attacks against cyber-physical industrial systems. Trans. Emerg. Telecommun. Technol. **29**(7), e3209 (2018). https://doi.org/10.1002/ett.3209
93. H. Huang, L. Zhang, Reliable and secure constellation shifting aided differential radio frequency watermark design for NB-IoT systems. IEEE Commun. Lett. **23**(12), 2262–2265 (2019). https://doi.org/10.1109/LCOMM.2019.2944811
94. A. Sadeghi, C. Wachsmann, M. Waidner, Security and privacy challenges in industrial internet of things, in *2015 52nd ACM/EDAC/IEEE Design Automation Conference (DAC)*, (2015), pp. 1–6. https://doi.org/10.1145/2744769.2747942
95. A. Iskhakov, R. Meshcheryakov, Intelligent system of environment monitoring on the basis of a set of IOT-sensors, in *2019 International Siberian Conference on Control and Communications (SIBCON)*, (2019), pp. 1–5. https://doi.org/10.1109/SIBCON.2019.8729628
96. A.A. Abd El-Latif, B. Abd-El-Atty, I. Mehmood, K. Muhammad, S.E. Venegas-Andraca, J. Peng, Quantum-inspired blockchain-based cybersecurity: Securing smart edge utilities in IoT-based smart cities. Inf. Process. Manag. **58**(4), 102549 (2021). https://doi.org/10.1016/j.ipm.2021.102549
97. O. Abdulkader, A.M. Bamhdi, V. Thayananthan, F. Elbouraey, B. Al-Ghamdi, A lightweight blockchain based cybersecurity for IoT environments, in *2019 6th IEEE International Conference on Cyber Security and Cloud Computing (CSCloud)/2019 5th IEEE International Conference on Edge Computing and Scalable Cloud (EdgeCom)*, (2019), pp. 139–144. https://doi.org/10.1109/CSCloud/EdgeCom.2019.000-5
98. S. Badsha, I. Vakilinia, S. Sengupta, BloCyNfo-share: Blockchain based cybersecurity information sharing with fine grained access control, in *2020 10th Annual Computing and Communication Workshop and Conference (CCWC)*, (2020), pp. 317–323. https://doi.org/10.1109/CCWC47524.2020.9031164
99. Q. Xia, E.B. Sifah, A. Smahi, S. Amofa, X. Zhang, BBDS: Blockchain-based data sharing for electronic medical records in cloud environments. Information **8**(2), 44 (2017). https://doi.org/10.3390/info8020044
100. G.G. Dagher, J. Mohler, M. Milojkovic, P.B. Marella, Ancile: Privacy-preserving framework for access control and interoperability of electronic health records using blockchain technology. Sustain. Cities Soc. **39**, 283–297 (2018). https://doi.org/10.1016/j.scs.2018.02.014
101. V. Patel, A framework for secure and decentralized sharing of medical imaging data via blockchain consensus. Health Informatics J. **25**(4), 1398–1411 (2019). https://doi.org/10.1177/1460458218769699
102. P. Tasatanattakool, C. Techapanupreeda, Blockchain: Challenges and applications, in *2018 International Conference on Information Networking (ICOIN)*, (2018), pp. 473–475. https://doi.org/10.1109/ICOIN.2018.8343163

103. L. Da Xu, Y. Lu, L. Li, Embedding blockchain technology into IoT for security: A survey. IEEE Internet Things J. **8**(13), 10452–10473 (2021). https://doi.org/10.1109/JIOT.2021.3060508

104. J. Kolb, M. AbdelBaky, R.H. Katz, D.E. Culler, Core concepts, challenges, and future directions in blockchain: A centralized tutorial. ACM Comput. Surv. **53**(1), 1–39 (2020). https://doi.org/10.1145/3366370

Blockchain-Based Medical Records System

Nisarg Soni, Saurav Tayal, Tarun Kumar Singh, and Gourinath Banda

1 Introduction

Medical record management is a problem as old as medicine itself. Transfer and sharing of data between all the involved entities have always been the crucial components of the medical industry. Each entity involved in this process relies on non-tampered and complete data for the complete system to work. Along with interoperability, there is a confidentiality concern associated with protecting individual records from data leaks as health is a compassionate and private aspect of life.

The onset of technology brought electronic solutions to traditional paper-based systems. However, these systems are based on a centralized architecture, that is, client-server architecture. In such systems, data and functionality is served by a single server to multiple clients via the internet. All the client data and logic reside on the server. These systems have associated drawbacks like a single point of attack and failure where any attack or failure on the server would compromise the entire system. There is also an issue of centralization of power where the entity in control over the server controls all data and access to it. A better solution to the drawbacks would be decentralization. Decentralized systems are more robust against failures and attacks, without the centralization of power over the system. Blockchain is a distributed ledger. It is a decentralized system, and it is useful in targeting the problems associated with centralization.

This chapter proposes a system that allows the interoperable exchange of medical records with proper authorization. This system ensures that a user of the system must ask for a patient's permission to view and upload their medical records. To achieve

N. Soni · S. Tayal · T. K. Singh · G. Banda (✉)
Indian Institute of Technology Indore, Simrol, Madhya Pradesh, India
e-mail: gourinath@iiti.ac.in

Y. Maleh et al. (eds.), *Advances in Blockchain Technology for Cyber Physical Systems*, Internet of Things, https://doi.org/10.1007/978-3-030-93646-4_3

this, we deploy smart contracts on the blockchain network. Through the use of the deployed smart contract, a patient can grant read and write permissions to different entities, namely, hospitals and insurance agencies. The use of blockchain ensures that the authorization mechanism is tamper-proof.

This chapter is organized into eight sections. The first section describes the importance of medical records and how a blockchain-based decentralized approach is better than the current centralized ones for their storage. Section two explains the concepts used in the solution like blockchain, cryptographic techniques, Inter-Planetary File System (IPFS), and the patient lifecycle for visiting the hospital. Section three discusses some related works done in the field. Section four describes the proposed solution architecture. Section five explains some key implementation details, and section six details the various read/write workflows. Finally, sections seven and eight discuss our proposed solution analysis and the conclusion and the future work, respectively.

2 Related Works

Earlier attempts at electronic health records (EHRs) were based on centralized systems with a centralized access control mechanism where the patients had little to no say over access control of their data. Such systems first had an in-house centralized data storage like in the hospital or the clinic itself, which did not improve any security or privacy compared with earlier pen-and-paper systems. Over time, cloud-based storage took over and provided more security for the stored data. However, they still were centralized solutions and made the hospitals and the authorities the data owner instead of the patient, who should have been the actual owner of the data.

Ying et al. [1] proposed a policy preserving EHR systems based on CP-ABE to allow secure sharing of EHR data over the cloud. Though this allowed for more secure storage and health data sharing, it was still wholly a centralized solution. No kind of decentralization was explored in the solution.

Liang et al. [2] used blockchain to maintain the health data's integrity and allow secure access control. Their solution uses a cloud-based database to store health-related data, data requests from healthcare providers and insurance companies, data access records, and data access control policy. In their solution, the health data collected from the user's wearable devices and the data obtained from healthcare providers is processed to create a hashed data entry which is then uploaded to the blockchain network. Also, a decentralized permission management protocol is used to process each data access request, for the user's personal health data, from healthcare providers and health insurance companies. It is in the hands of the data owner to provide permission if they want. These access control policies ensure stability as they are stored on the blockchain in a decentralized manner. Besides that, every time an access request is made and any access activity is done, the blockchain, a distributed ledger, is used to record that activity. This record can be used for further auditing or investigation. In this case, though the blockchain provides security and integrity, the actual data is stored in a centralized database in the cloud, which brings a lot of problems, as already discussed.

Vijayakumar et al. [3] used hyperledger fabric blockchain to store all the health data in a decentralized manner. In their solution, the various nodes are connected to form the blockchain network such that the data is distributed among the nodes. This gives each node the responsibility and the opportunity to handle its data which makes the solution patient-centric and gives the control of the data in the user's hands. The blockchain is a distributed ledger that allows organizations to store medical information. As the data is stored in the blockchain in a decentralized manner, it allows for trust to be created among the participants of this network. By integrating hyperledger fabric, all medical information is recorded in an untamperable manner and any medical organization that has added the details can be tracked easily. Every transaction is stored in the blockchain and all the medical details of an individual that have ever been recorded can be accessed using query tools. They have discussed the issues associated with centralized systems and tried to overcome them through their decentralized solution. This allows integrity and security of data, but their solution suffers from the issues of storing large amounts of data on the blockchain itself.

Dubovitskaya et al. [4] created a patient-centric blockchain-based EHR (electronic health record) data management system. They propose a permissioned blockchain-based system for EHR data sharing and integration. Their solution makes use of the hyperledger fabric blockchain. Each hospital acts as a node in the blockchain network with its integrated EHR system. These hospital nodes together form up the blockchain network. Patients and doctors are connected using a web-based interface. This interface is used to initiate transactions to share EHR. In this solution, a hybrid data management system is used. In this hybrid approach, only the management metadata is stored on the chain. An off-chain cloud-based storage, Health Insurance Portability and Accountability Act-compliant is used to store the EHR data. This data is also encrypted before storing. The shared EHR data is secured using digital signatures and public key infrastructure–based asymmetric encryption. In this solution, a distributed ledger is used as a shared immutable and transparent history of all the actions performed by the users; these actions include defining access control policies and sharing, accessing, and modifying the data, and though the data is stored in a secured manner using various encryption schemes, however, the storage is still a centralized one.

Sun et al. [5] discuss a blockchain-based framework for electronic medical records sharing with fine-grained access control. They have proposed a distributed electronic medical records system with a search feature to easily find the records using blockchain and smart contract technology. In this solution, firstly, the electronic medical data that is obtained undergoes a hash calculation. The corresponding hash value obtained is then stored on the blockchain. This is done to ensure the authenticity and integrity of data. Then, the InterPlanetary File System (IPFS) stores encrypted electronic medical record data. Then the Ethereum blockchain is used to store the encrypted keyword index information of the medical records. Besides that, keyword search is implemented using a smart contract deployed in the Ethereum blockchain.

Furthermore, they use an attribute-based encryption scheme to decrypt encrypted electronic medical records and access only by the attributes meeting the access policy. This solution uses blockchain and smart contracts to maintain security and integrity and IPFS for secure decentralized storage of data. This solution makes the storage and access control as decentralized as possible using IPFS, blockchain, and smart contracts. Still, the access control system is doctor-centric, wherein the doctor or the medical practitioner provides access control to various entities, even the patients themselves.

Our solution makes use of Ethereum and smart contracts to allow secure access control and IPFS for secure data storage besides using various encryption schemes to encrypt the data before storage and sharing. But our solution is completely patient-centric. All the power of access control is in the patient's hands, and no one can access any kind of data without the patient's approval. All this access control is encoded into the smart contract and, once deployed, cannot be tampered with in any way.

3 Concepts

Blockchain

Blockchain burst onto the scene when it was introduced in 2008 in a whitepaper by Satoshi Nakamoto, "Bitcoin: A Peer-to-Peer Electronic Cash System." He introduced a revolutionary new cryptocurrency named Bitcoin, wherein blockchain formed the underlying decentralized technology of the newly introduced currency. Blockchain is essentially a data structure wherein blocks are added one after the other and every new block is linked to the previous block through its cryptographic hash, forming a blockchain chain.

The top-most block on the blockchain defines the state of the blockchain at any given instance. Blockchain acts as a distributed ledger system. It consists of a connected peer-to-peer network of users. This distributed ledger is replicated over the whole network on every peer's system. It consists of transactions that are stored together in a block. As more transactions are carried out, they are grouped in a new block which is then added to the blockchain. To add a block to the blockchain, it must be confirmed by the majority of the peers that form the network through what is called a consensus protocol. Every peer sees and confirms the current state of the blockchain, and the state that is confirmed by the majority is accepted. These peers act according to the consensus algorithm, which allows choosing a peer who then adds a block to the blockchain and thus modifies the state of the blockchain.

The cryptographic linking of blocks ensures that the blockchain system is tamper-proof. To create a malicious transaction, a user would need to control the systems of more than half of the total peers connected to the network. Since a blockchain network can have millions of peers, it is practically impossible for a malicious entity to achieve its intentions. This makes sure that the blockchain cannot be corrupted.

Transactions are created by accounts owned by anonymous users on the network. Their account addresses identify these accounts. Two keys – a private key and a public key – are associated with each account. These keys can be stored and managed with a wallet [6]. There are several types of hardware and software wallets available. Users can access their accounts using their private key. The private key and the public key help create asymmetrically encrypted messages that could be sent to the network and create verifiable transactions.

The data stored on a blockchain is visible to everyone but cannot be changed or tampered with in any way. The distributed nature of blockchain combined with cryptography's power creates a resilient system against attacks while maintaining data privacy and sharing. Thus, the transparency, accountability, and immutability associated with the data stored on the blockchain solve the problems of data privacy and security related to centralized systems, and the sensitivity of medical records necessitate these traits. These traits make blockchain a suitable system for storing and managing electronic medical records.

Early cryptocurrencies used blockchain only for recording transactions as a distributed ledger, but Ethereum [7] took it a step further by introducing an application layer over the simple blockchain, allowing users to write and execute code. Ethereum thus not only stores transactions but can also run scripts, making it a distributed computation technology instead of just a ledger. This gave rise to the concepts of smart contracts and distributed apps that are at the forefront of decentralized technology trying to solve centralization problems and are discussed in more detail ahead in this section.

Cryptography Techniques

This section discusses symmetric and asymmetric cryptography techniques, which form the basis of security in our system.

Symmetric Cryptography Symmetric cryptography or private key encryption makes use of a single secret key which is used to encrypt and decrypt data. It ensures that the data cannot be decrypted using some other key. Advanced Encryption Standard (AES) is a widely used symmetric encryption technique. Good key management is important when using this technique.

Asymmetric Cryptography In asymmetric cryptography or public key encryption, a pair of keys – public and private – are used. The public key is distributed, while the private key is only known to the encryption initiator and is kept secret. Data once encrypted with the public key can only be decrypted with the associated private key. Asymmetric cryptography is heavily used in blockchain technology. RSA is one such common cryptography technique. It is based on the computational complexity involved in factorizing a large integer obtained from two large prime integers. Elliptic curve cryptography (ECC) is also a public key encryption technique better than RSA as it offers faster computation times and more robust encryption.

IPFS (InterPlanetary File System)

IPFS [8] is a peer-to-peer, content-addressed, version-controlled file system. This system is decentralized by nature; thus, no single entity has control over it. The files in IPFS are replicated over the network of IPFS nodes. Their content identifies the files in an IPFS. Instead of identifying data by their storage location, called location addressing as done by URLs in today's internet, IPFS uses the content itself in the form of a hash to identify the files in the IPFS system; this is called content-based addressing. Thus, IPFS uses a content ID (CID) or hash of the content, which represents the content itself, to address that data or file over the network. This allows data and files to be stored and served from anywhere by anyone.

It works by taking a file and cryptographically hashing it to obtain a small and reproducible representation of that file which ensures that every file has its unique hash, which can then be used to address the file. This CID is generated using Merkel-Tree Directed Acyclic Graphs (DAGs). Merkle DAGs allow distinctively identifying all content on IPFS as every data block part of the DAG has its unique hash, making the data tamper-proof as altering anything in the data would change the hash. Our solution uses IPFS to store all the medical records data.

Ethereum

Ethereum is a blockchain technology that also provides smart contracts integration. Ethereum uses Ether (ETH) as its cryptocurrency for transactions. It is a capable and robust blockchain platform to implement a decentralized medical records system. In the provided solution, smart contracts are used for access control of data, and to store the encrypted content ID (CID) or hash of the user data stored on the IPFS network. This is a unique feature of Ethereum [9] to create and run decentralized applications (dApps) and smart contracts [10]. Smart contracts are programs that can be used to describe legal documents, assets, and agreements and other services between different parties without the involvement of a middleman. They are essentially the code running over a distributed network known as the Ethereum Virtual Machine. These contracts are tamper-proof, ensuring that all the data they hold cannot be manipulated by any parties involved. They are visible to anyone who is connected to the Ethereum network.

Patient's Lifecycle

In general, a visit to the hospital begins with the patient going to the hospital and the hospital requesting the patient to share their previous records, which the hospital then uses for diagnosis, treatment, etc.

After providing the treatment, the hospital generates new records for the patient, including test results, prescriptions, bills, etc. These records are shared with the patient, who is then required to store them safely along with the patient's previous records, if any.

Now, the patient is required to settle the insurance claim about the current visit. For this, the insurance company requests the patient to share their medical bills and settle claims after due processing. This completes the life cycle for a patient's visit to the hospital.

Every patient has numerous visits to the hospital, each involving all the functions described above, which make the process difficult and prone to issues. To solve these problems in a decentralized manner, we propose the solution described in the next section.

4 Proposed Solution

The solution architecture (Fig. 1) shows the various entities, namely, patients (P), hospitals (H), and insurance agencies (A) (on the left) and technologies (on the right) that come about to form the solution.

The technological tools used are blockchain [11], IPFS [12], and an application interface. This application interface connects directly to the Ethereum blockchain and the IPFS storage to implement a secure, private, and transparent system which is the main focal point of this solution.

The application interface allows the patients to control which entity can access their records. The patient is the most powerful entity in the entire solution architecture, which is one of the big unique selling points of this solution as it gives the power in the patient's hands. The hospitals have the ability to add new records for the patients and to access their previous records. The insurance agencies have the functionality of being able to see the records and settle insurance claims. The patient controls these functionalities of hospitals and insurance agencies through an authorization mechanism implemented through blockchain transactions.

When authorized by a patient $P_i \in P$, the hospital $H_j \in H$ stores the records generated for the patient's current hospital visit (v) on the decentralized storage solution IPFS. The records stored on the IPFS are encrypted symmetrically with an identity number k_{P_i} of the patient P_i. After storage, the location or the hash of the record q_v for the current hospital visit (v) is returned by the IPFS.

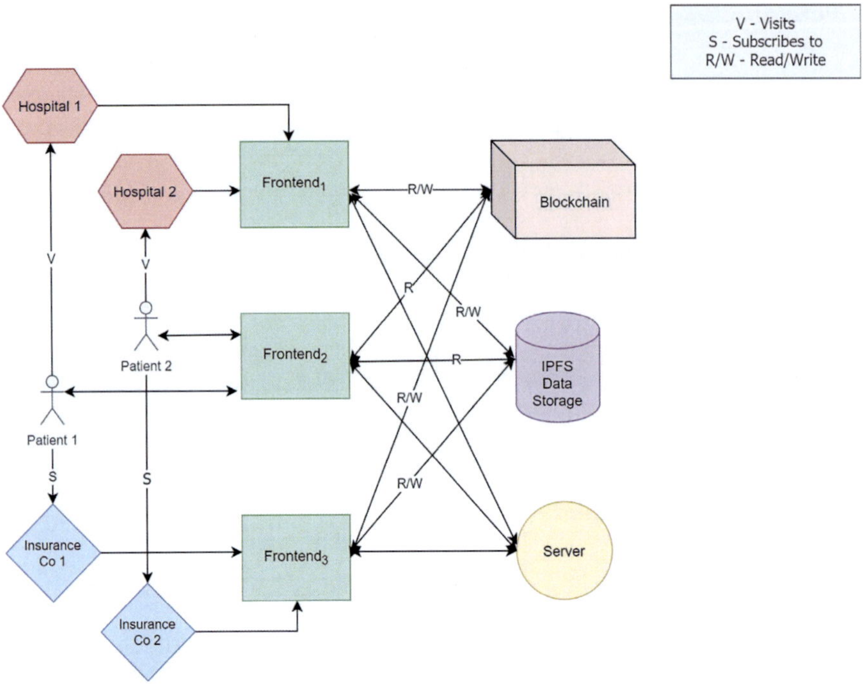

Fig. 1 Proposed solution architecture

Let D_v be the record generated due to the patient's current visit to the hospital, and AES be a symmetric cryptography function that takes data D_k and a key k_{Pi}, and returns the encrypted data. Thus, the encrypted records are

$$e_v = \text{AES}\,(D_v, k_{Pi})$$

If U is a functionality provided by IPFS, which takes a piece of data, uploads it onto IPFS, and returns a hash or CID, then

$$q_v = U\ \ (e_v)$$

This location or hash q_v of records is then added to a file that contains details of every hospital visit for the patient in an object in JSON (JavaScript Object Notation) format hereafter referred to as the master file M'_{Pi} of the patient P_i. An object J_v is created for every hospital visit which contains an identifier (name) of hospital H_j, the date of visit σ_v, the insurance settlement status δ_v, and the content hash q_v (also called location hash and content Id) of the stored record.

$$J_v = \left(H_j, \sigma_v, \delta_v, q_v\right)$$

Let the current master file be a set of all such J_k, where $k = 1, 2, 3, \ldots, n$ for n number of hospital visits,

$$M'_{Pi} = \{J_1, J_2, J_3, \ldots, J_n\}$$

After adding the data J_v to M_{Pi}, the new master file, $M_P{}^t$, will be

$$M'_{Pi} = \{J_1, J_2, J_3, \ldots, J_n, J_v\}$$

This newly generated master file M'_{Pi} is then stored on IPFS after symmetric encryption with the identity number K_P. This returns a new location for the master file itself.

$$E = \text{AES}\left(M'_{Pi}, K_{Pi}\right)$$

$$Q = U\ (E)$$

This location (Q) of the master file is then encrypted using asymmetric encryption (RSA) and the result (L) is stored onto the blockchain via a transaction. The various read/write access control functionalities are also managed through several blockchain transactions.

RSA takes a piece of data Q and a public key α_{Pi} of the patient P_i and returns an encrypted value.

$$L = \text{RSA}\,(Q,\ \ \alpha_{Pi})$$

This asymmetrically encrypted hash L can only be decoded by the secret private key of the patient, which is known only to the patient.

The writing functionality is controlled through the patient's authorization. It is limited to the hospital, wherein they can add new records for patients and the insurance company, which can settle insurance claims for the patient's bill incurred at the hospital visit. The read functionality is for all the entities, but the authorization is again in the patient's hands. Whenever an entity wishes to view the records, the application obtains the records through the blockchain and the IPFS, and then presents them on the user interface.

In each read operation, the encrypted location is first decoded by the patient using their private key and then shared with the reader entity by symmetrically encrypting the decrypted location again using the reader's public key and sharing via the blockchain.

5 Implementation Details

The following section contains the details of the two major components of the system: a patient smart contract and the structure of a patient master file.

Smart Contract Solidity Implementation Pseudo Code

```
contract Medical
   address private owner;
   string private medicalLocationHash;
   address public reader;
   address public writer;
   map(address => string) public readHash;
   map(address => bool) public canWrite;
   function RevokeRead(address memory readerAddress) public
      readHash[readerAddress] = "";
   function Read(address memory readerAddress) public view
   returns (string memory)
      return readHash[readerAddress];
   function Write(string memory locationHash,
              address memory
   writerAddress) public
      require(
        canWrite[writerAddress] == true,
        "You do not have write permission"
      );
      medicalLocationHash = locationHash;
      canWrite[writerAddress] = false;
   function AcquireReadPermission(address memory
              readerAddress) public
      reader = readerAddress;
   function AcquireWritePermission(address memory
              writerAddress) public
      writer = writerAddress;
      reader = writerAddress;
   function GrantWritePermission(string memory
           encryptedLocationHash) public
      require(owner == msg.sender, "You can not grant
           write permission");
      require(bytes(writer).length != 0, "No writer!");
      GrantReadPermission(encryptedLocationHash);
      canWrite[writer] = true;
      writer = "";
   function GrantReadPermission(string memory
           encryptedLocationHash) public
      require(owner == msg.sender, "You can not grant
           read permission");
      require(bytes(reader).length != 0, "No reader!");
      readHash[reader] = encryptedLocationHash;
```

```
   reader = "";
function CheckWritePermission(address memory
         writerAddress) public view
returns (bool)
   return canWrite[writerAddress];
function ViewLocationHash() public view returns (string
           memory)
   return medicalLocationHash;
function ViewReader() public view returns (string
           memory)
   return reader;
function ViewWriter() public view returns (string
           memory)
   return writer;
```

Structure of the Patient Master File

```
{
   patientName,
   patientDOB,
   patientSex,
   [
     visitOne : {
        hospitalName,
        dateOfVisit,
        claimSettlementStatus,
        dataOfVisit
     },                visitTwo : {
        hospitalName,
        dateOfVisit,
        claimSettlementStatus,
        dataOfVisit
     } ...
   ]
}
```

6 Workflows

There are two main operations that the entities in the proposed system perform. These are reading the records and writing the records.

Read Operation

The read operation is universal in the system which all the entities can perform. In it, various entities can search and access the records of a patient when authorized by the respective owner of the records. The read operation is a single-time process. When authorized, an entity can access the records only once and is required to obtain permission from the owner of the records every time they wish to access them again (Fig. 2).

The read operation begins with the hospital (or insurance company) searching for the patient in the application user interface and requesting their permission to view their records. This action stores the address of the hospital in a variable in the smart contract. Now, the patient grants the read permission. This first fetches the CID (Content Id), stored in a variable in the smart contract, of the master file stored on the IPFS. This CID, which was encrypted with the patient's public key can now be decrypted using their private key. This private key is never stored anywhere and is generated using a secret passphrase known only to the patient. Now, this decrypted CID of the master file is encrypted using the hospital's public key. This newly encrypted CID is now stored in the smart contract by the patient, through a transaction, in the form of a mapping that associates the hospital's public with this CID. This makes sure that now when the hospital goes on to access the records, only the hospital with the specific address which was granted permission by the patient can do so.

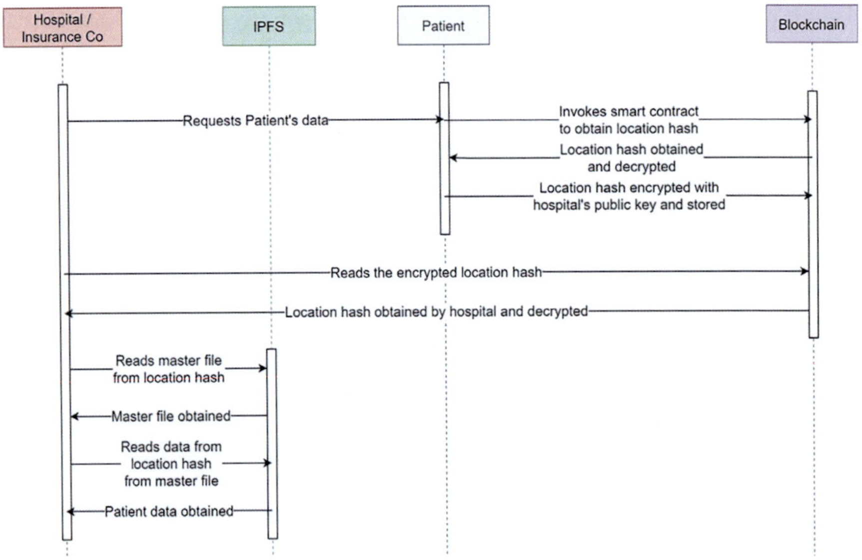

Fig. 2 Read operation

Now the hospital proceeds to view the patient's records. This is done through a smart contract interaction wherein the mapping variable is read. This matches the hospital's address and returns the CID stored in the mapping. This CID was encrypted using the hospital's public key and thus now can only be decrypted using their private key. The hospital decrypts the CID by generating their private key using their secret passphrase. This decrypted CID is now used to fetch the master file from the IPFS. Any of the previous records can now be brought using their CID stored in the master file. The obtained records are decrypted further as they were encrypted using symmetric encryption using the patient's unique ID number. So, the unique ID number is used to decrypt the records, and they are finally displayed on the application interface for the hospital/insurance company to view.

For the patients, read operation is quite simple as they have complete access control in their hands. A patient can see their records directly with a click in the application interface. The CID is obtained from the smart contract, which is owned and controlled by the patient and hence does not require any permission. The master file is then obtained from this CID from IPFS, and any record which the patient wishes to see is read from its CID stored in the master file.

Write Operation

The write operation is wherein the hospital writes the records generated and the insurance company settles the bill generated at the patient's visit to the hospital. The patient is not allowed to do any kind of write operation.

For the write operation, the hospital first searches for the patient in the application interface and requests the patient to be allowed to add new records. This authorization mechanism is written into the blockchain smart contract and thus cannot be tampered with. When the hospital asks for the write permission, it sends its address as a parameter to the smart contract where a mapping is then stored between the username and a Boolean value of true or false, which indicates if the username is allowed to write to the smart contract or not. After the hospital makes the write request, the patient logs into their account and grants the write permission wherein a call to the smart contract is created, which updates the mapping of the hospital's username to be true; this authorization can only be issued by the owner of the smart contract, that is, the patient. Now the hospital is ready to add new records for the patient (Fig. 3).

Once the permission is granted, the application interface on the hospital side redirects to a form wherein all details regarding the patient's visit can be filled and any photos or documents can be attached. When the hospital submits the record, it is first encrypted with symmetric encryption using the patient's unique ID number as the key and then stored on the IPFS, which returns the CID associated with this stored record. Now the master file is fetched from the IPFS. This CID of the stored record is added to the master file and other details like the date of visit and the hospital's name and the insurance settlement status in a JavaScript object and the

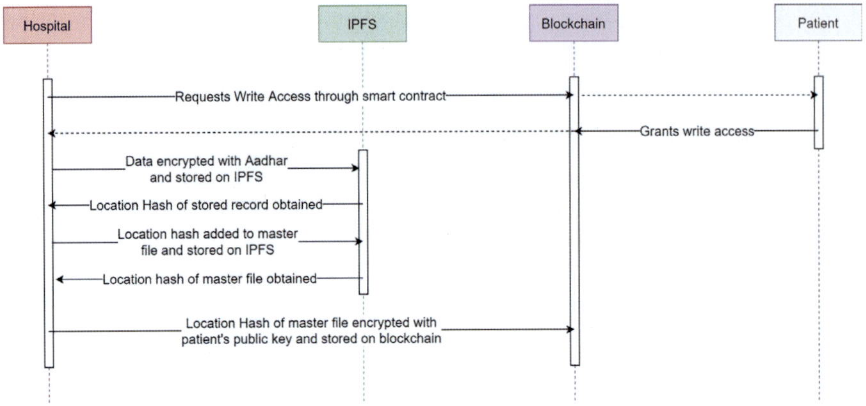

Fig. 3 Write operation for hospital

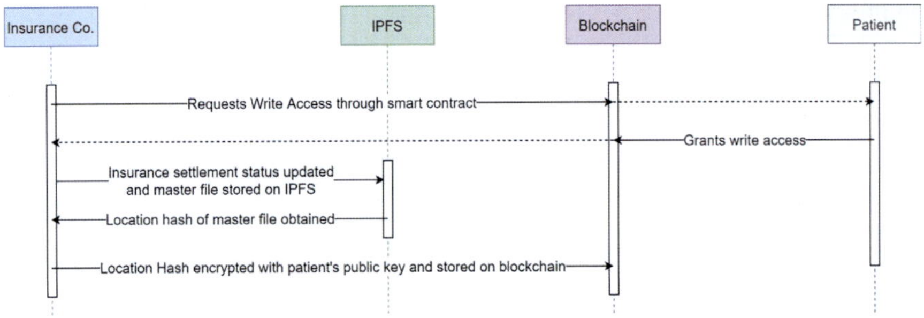

Fig. 4 Write operation for insurance company

master file is in JSON format. Now, this master file is stored on IPFS and a CID is obtained. This CID of the master file is then encrypted using the patient's public key and stored in the smart contract. This completes the write operation for the hospital. Note that a write operation permission grant comes with the permission to read records inherently.

For the insurance company, the permission requesting and allowing procedure is similar to that of the hospital. The difference is that when allowed by the patient, the hospital is redirected to a record page wherein they can add a new record. In contrast, the insurance company is redirected to a settle claim page wherein they can simply view and settle the insurance claim for the patient. The insurance claim settlement status is saved as a mapping of true or false in the master file in an object pertaining to that visit. When the insurance company settles the claim, the mapping is made to be true and the claim is expected to have been settled. This changing of mapping generates a new CID when the changed master file is stored on IPFS. This new CID is stored in the smart contract, encrypted with the patient's public key. This completes the write operation for the insurance company (Fig. 4).

7 Analysis

Security and Accessibility

In our solution, we have used blockchain and IPFS for the storage of data, combined with various encryption schemes at every step. The blockchain allows secure, untameable recording of transactions and integrated with smart contract functionality. It produces one of the most efficient and secure systems to date. The system is more transparent due to the public ledger, more secure as it is a distributed system without any central point of failure, and more accessible and interoperable as it is always available, for anyone, anywhere around the world.

IPFS as the storage solution, a distributed system like blockchain, provides similar accessibility and interoperability benefits. The content addressing feature of IPFS provides integrity and security to the system, ensuring that the data is not changed or tampered with.

Encryption schemes are used for secure storage and sharing of data. The patient data is symmetrically encrypted with a key known to the patient and the location of the data is again asymmetrically encrypted with a private key of the patient. And every time it is shared, it is encrypted such that only the person who has been allowed by the patient can decrypt and access the data.

The use of blockchain, IPFS, and encryption techniques at every step of data storage and sharing makes the data almost impossible to tamper with in any way and makes the system more secure, accessible, and interoperable.

Privacy

Our solution has privacy-aware access control with the patient at its center. From data generation, to storage and sharing, every step has access control in the hands of the patient. We have stored the data using a combination of symmetric and asymmetric encryption techniques with keys decided and owned by the patient, making the patient the real owner of their data. Giving the power of access control in the hands of patients, through untamperable smart contract code, makes our solution completely patient-centric. No one can access or use any kind of data belonging to a person without their explicit permission. To gain access to data, any entity needs the decrypted CID and the unique patient ID, which only resides with and can be managed by the patient. Thus, patient privacy and data ownership are the main USP of our solution and are ensured in the strictest ways possible.

8 Conclusion and Future Work

Conclusion

This chapter examined and developed a system for applying blockchain technology in maintaining medical health records more securely and privately than the current systems permit. Blockchain is the technology at the forefront of the current decentralization revolution, which aims to provide more security and privacy to every field that uses technological solutions. Medical records is one field that is in dire need of such revolutionary solutions, given the sensitivity and importance of such records. Our solution of electronic medical records based on blockchain improves security, accessibility, and interoperability, over traditional records-keeping systems, all at the same time.

Future Work

The current version of the application uses Ethereum for blockchain purposes. The time and cost associated with each Ethereum transaction is very high due to the large network of nodes that come to a consensus over each transaction, and at the current growth of the technology, these costs and time are bound to increase more so in the future. This high cost reduces the feasibility of the current solution. A different blockchain technology that uses a different consensus algorithm might help reduce these costs. The balance between costs and decentralization needs to be assessed to find the most optimal solution for these other blockchain technologies.

Apart from that, the current solution uses asymmetric encryption based on RSA keys derived from a passphrase that needs to be remembered by the entity. Remembering the passphrase along with having strong passphrases is essential in keeping data secure in the application. However, remembering such passphrases becomes challenging, so encryption techniques like biometric encryption could improve the solution and make it easier to use. Effectiveness with different encryption techniques without compromising security needs to be assessed to reach the most optimal solution.

Also, the current solution provides complete functional support for the previously listed entities, that is, patient, hospitals, and insurance agencies. However, this solution can be further extended to providing bulk data for research purposes and keeping patients in control of their data.

References

1. Z. Ying, L. Wei, Q. Li, X. Liu, J. Cui, A lightweight policy preserving EHR sharing scheme in the cloud. IEEE Access **6**, 53698–53708 (2018). https://doi.org/10.1109/ACCESS.2018.2871170
2. X. Liang, J. Zhao, S. Shetty, J. Liu, D. Li, Integrating blockchain for data sharing and collaboration in mobile healthcare applications, in *2017 IEEE 28th Annual International Symposium on Personal, Indoor, and Mobile Radio Communications (PIMRC)*, (IEEE, 2017). https://doi.org/10.1109/PIMRC.2017.8292361
3. V. Vijayakumar, K.M. Sabarivelan, J. Tamizhselvan, B. Ranjith, B. Varunkumar, Utlization of blockchain in medical healthcare record using hyperledger fabric. Int. J. Res. Advent Technol. **7**, 414–419 (2019). https://doi.org/10.32622/ijrat.74201922
4. A. Dubovitskaya, F. Baig, Z. Xu, R. Shukla, P.S. Zambani, A. Swaminathan, M.M. Jahangir, K. Chowdhry, R. Lachhani, N. Idnani, M. Schumacher, ACTION-EHR: Patient-Centric Blockchain-Based EHR Data Management for Cancer Care. J. Med. Internet Res. **22**, e13598 (2019). https://doi.org/10.2196/13598
5. J. Sun, L. Ren, S. Wang, X. Yao, A blockchain-based framework for electronic medical records sharing with fine-grained access control. PLoS One **15**(10), e0239946 (2020). https://doi.org/10.1371/journal.pone.0239946
6. L. Mearian, What's a crypto wallet (and how does it manage digital currency)? Computer World. Access Link: https://www.computerworld.com/article/3389678/whats-a-crypto-wallet-and-does-it-manage-digital-currency.html. Accessed date: 05 Feb 2021
7. V. Buterin, A next-generation smart contract and decentralized application platform [White Paper], 2014, Ethereum. Access Link: https://ethereum.org/en/whitepaper/. Accessed date: 24 Jan 2021
8. Y. Psaras, D. Dias, The interplanetary file system and the filecoin network, in *2020 50th Annual IEEE-IFIP International Conference on Dependable Systems and Networks-Supplemental Volume (DSN-S)*, (2020), pp. 80–80. https://doi.org/10.1109/DSN-S50200.2020.00043
9. A.M. Antonopoulos, G. Wood, Mastering ethereum. Access Link: https://github.com/ethereumbook/ethereumbook. Accessed date: 20 Jan 2021
10. N. Szabo, Formalizing and securing relationships on public networks. First Monday **2**(9) (1997)
11. S. Nakamoto, Bitcoin: a peer-to-peer electronic cash system [White Paper], cryptography mailing list at https://metzdowd.com, 2009. Access Link: https://bitcoin.org/en/bitcoin-paper. Accessed date: 24 Jan 2021
12. J. Benet, IPFS – content addressed, versioned, P2P File [White Paper], IPFS. Access link: https://github.com/ipfs/papers/raw/master/ipfs-cap2pfs/ipfs-p2p-file-system.pdf. Accessed date: 10 Feb 2021

Security of IoT-Based e-Healthcare Applications Using Blockchain

Sachin Gupta, Babita Yadav, and Bhoomi Gupta

1 Introduction

The Internet of Things (IoT) has become a leading agent of growth in e-healthcare and supply chain [1] applications being developed lately. During the nascent years of telemedicine [2], challenges based on technology lacunae kept its adaptation low. Still, during recent years, enhanced e-healthcare solutions are seeing light of the day with improved network communication over IoT devices like IoT-based wearables and sensor-based monitoring devices [3]. IoT has thus emerged as the key technology for smart healthcare applications [4], but the focus of challenges faced by e-healthcare has shifted its base from lack of technology to lack of user privacy and security. The challenges can be understood in the context that the nature of the data involved in any e-healthcare application is considered very personal, and maintenance of its privacy is a legal requirement in several countries. The underlying use of a public communication network adds new security and privacy challenges to healthcare applications using IoT [5]. This is compounded by the fact that IoT devices use small-sized low-performance components with limited resource capacity, making it extremely difficult to apply the conventional security algorithms and models designed for resource-rich systems.

A potential solution to the multiple security challenges explained above in the pursuit of IoT-based e-healthcare security is the use of blockchain technology [6]. Blockchain technology is gaining popularity in innumerable technological use cases owing mainly to its decentralized, distributed, and immutable transaction record

S. Gupta (✉) · B. Yadav
SoET, MVN University, Aurangabad, Haryana, India
e-mail: sachin.gupta@mvn.edu.in; babita.yadav@mvn.edu.in

B. Gupta
Maharaja Agrasen Institute of Technology, Delhi, India
e-mail: bhoomigupta@mait.ac.in

Y. Maleh et al. (eds.), *Advances in Blockchain Technology for Cyber Physical Systems*, Internet of Things, https://doi.org/10.1007/978-3-030-93646-4_4

keeping. In a blockchain-based application, a shared ledger residing at the heart of record keeping maintains a chronological record of transactions distributed as shared copies within stakeholders.

The choice of name behind blockchain technology makes its purpose self-evident. It uses a block of information for storage, chained with cryptographic signatures of the previous data blocks called hash, stored in shared ledgers supported by distributed network processes called nodes. Each node maintains a copy of the complete chain. We propose the concept of "one patient one record chain" to be maintained across all the potential organizations as nodes interacting with the proposed IoT-based e-healthcare application using blockchain [7].

Contributions

The main contributions of this study are as follows:

- Presentation of a comprehensive literature review survey for the applications of IoT and the sensors being used in healthcare services as proposed by contemporary researchers.
- Survey blockchain as an enabling technology for security of IoT-based healthcare services.
- A theoretical sample secure architecture for maintaining healthcare data in a blockchain approach.

The remainder of this chapter is divided into five sections. Section 1 describes the three underlying technologies being discussed in the chapter. Section 2 presents a thorough survey of the sensors being used in healthcare and associated security challenges faced by healthcare systems due to the multitude of sensors being deployed. Section 3 presents the conceptual knowledge of blockchain technology as a decentralized ledger and a comprehensive survey of its applications as suggested in the healthcare domain. Section 4 discusses the usefulness of blockchain for the healthcare domain along with open issues of interoperability, and this chapter concludes with Sect. 5 which analyzes the security provisions for healthcare data made using a blockchain approach using a proposed sample architecture of application data flow integrating blockchain and IoT in e-healthcare. Section 6 analyzes security provisions available in a blockchain-based system as compared to a centralized database management service based on the architecture proposed in Sect. 5, and this chapter concludes with potential future research directions in the Sect. 7.

2 Evolution of Telemedicine Through Text to IoT-Based e-Healthcare

Telemedicine is conventionally defined as remote monitoring of patients and giving them treatment according to diagnosis using information communication technology. The original idea included sharing diagnostic reports through emails and tele-consultation thereafter, which evolved into a complete audio-visual consultation through teleconferencing services with streaming video services. While the service proved useful for remote locations with a scarcity of medical experts, the idea did not gain much traction due to several technological and adaptation issues. Healthcare traditionally has been an in-person diagnostic and care giving service. The absence of an "in-person" factor made it difficult for patients and doctors to adapt to the new paradigm of remote healthcare. The streaming video call services on popular messengers and communication platforms like WhatsApp, Facetime, Facebook Live, and others made the masses comfortable with the idea of "assumed" closeness despite being physically distant. This social transformation helped the cause of telemedicine well. For the healthcare workers like doctors however, the confidence in technology was boosted by real-time health parameter monitoring through sensor-based devices which they find is close to their way of working at the hospitals. The e-healthcare based on IoT made this transition far easier and adaptable for the healthcare providers. The immense opportunities arising from the permeation of smartphones equipped with healthcare apps and wearable sensors providing real-time data for analytics are taking this evolution forward at a rapid pace today.

Benefits of Telemedicine

Through all the phases of its evolution, telemedicine has provided immense benefits to the stakeholders [8]. The most prominent benefits being offered by telemedicine technology include the following but are not limited to:

- *Reducing Cost*: Telemedicine ensures cost savings in terms of travel expenses and in terms of time. Telemedicine also lowers secondary expenses.
- *Improved Accessibility*: People like older unattended adults, geographically isolated patients who cannot access the medical services physically can consult through telemedicine [9].
- *Preventive Care*: In case of long-term health issues, preventive care can be provided through telemedicine easily.
- *Convenience*: Patients can access care at home in their comfort zone. Working people do not have to take time off from their work.
- *Avoid Spreading Infection*: When going to hospitals or clinics, there is a probability of meeting more sick people which translates to increased risk of infection. In

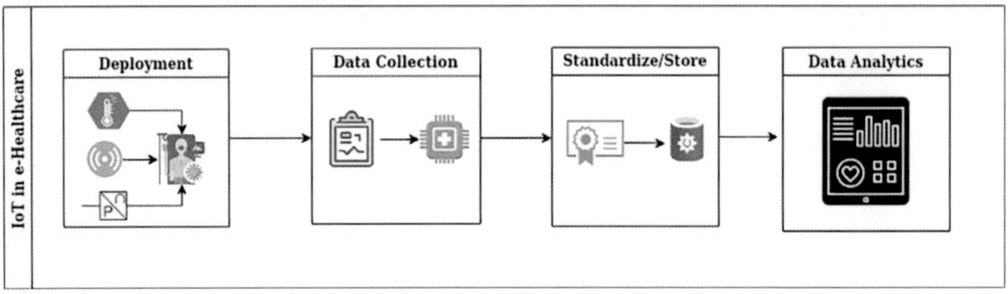

Fig. 1 IoT in e-healthcare

case of a less sick patient with low immunity, this becomes dangerous, instead of getting care there is a chance that patient might bring infection from the doctor's clinic [10].

Redefinition of e-Healthcare by IoT

The new paradigm introduced in e-healthcare could not have been envisioned without the technological advancements made through the Internet of Things. IoT-based devices and products have brought in immense opportunities in telemedicine which are instrumental in transforming healthcare [4]. We have discussed in the previous sections how before the existence of IoT devices only text and voice-based communication was possible between patient and doctor. With the advent of IoT solutions, however, technology related to remote patient monitoring systems witnessed propulsions to the future [11, 12]. The potential security problems however also increase and can be handled with security at the transport layer using datagrams [13]. The applications of IoT-based systems in healthcare can be understood in four successive stages as shown in Fig. 1.

Stage 1: Deployment of IoT

The first stage consists of the deployment of devices like sensors which includes temperature sensors, ECG sensors, blood pressure sensors, fluid sensors and other related sensors, and IoT-based 3D camera system in an internet-enabled environment which collects actual data related to the patient. A comprehensive list of sensors being used in IoT-based healthcare management systems and specific to the diseases that a patient is being treated for is listed in Table 1. The table also lists the purpose/usage of IoT devices.

Table 1 IoT sensors deployed in healthcare services and their use

Sensor type	Healthcare application	Purpose
Motion, accelerometer, and camera	Motion disability – Wheelchair [37–38]	The sensor is used to detect abnormal speed / orientation of the wheelchair and sends alerts to the gateway
Spo2 device	Lung infections [39–40]	Alerts generated on depleting blood oxygen or falling pulse rate to gateway
Robotic MEMS	Robotic surgery [41]	Microcontroller-based remote movements to control robotic arms
ECG, blood pressure, heart rate sensors	Heart monitoring [42–51]	Uses smartphones / device gateways to continuously transmit monitoring data
Wearable temperature sensors	Hypothermia [51–53]	Raw data transmission for monitoring using gateway
Pulse and temperature sensors	Asthmatic conditions [54–55]	Processed signals stored on cloud using http and analyzed
EMG sensors	Neuromuscular defects [56–57]	Raw data is collected for processing and analyzed for neuromuscular disorders
Glucose sensors	Blood sugar management[58–59]	Abnormality data transmitted through mobile apps

Stage 2: Healthcare Data Collection and Processing

The data collected by the abovementioned healthcare sensor-based IoT devices is continuous physical data and is usually available in analog form as per the source. Since the eventual requirement is transmission, this data must be aggregated and converted to digital form via sampling and digitization and then compressed using suitable lossless algorithms to enable transmission and processing by computer systems. Having so many input sources, transmission sources, and various formats of data brings several unforeseen challenges to security of healthcare data.

Stage 3: Data Standardization

Data originating through a variety of devices being used for sensing and monitoring varies in nature and is not necessarily ready for analytical and storage purposes. It is thus imperative to ensure that while the data is aggregated from various sources, it must be digitized and then preprocessed to finally be standardized [14]. The data may then be moved to persistent storage.

Stage 4: Data Analytics

The recent advancements in the fields of Data Analysis using Artificial Intelligence (AI) and related technologies have ushered data-driven intelligence and decision-making into a different realm. The insights being provided by these automated models were not observable through basic statistical analysis or human observations previously. This newfound elixir of analytics is the primary reason for the challenges surrounding healthcare data privacy and security. This is also precisely the reason why data is being touted as the "new oil."

Security Challenges with IoT-Based e-Healthcare

Medical practitioners' increased reliance upon the new age healthcare systems based on assistive technologies like IoT wearables, remote monitoring devices for telehealth has invariably proved successful in delivering healthcare to more people, and has increased the medics reach to remote areas and needy people at affordable cost. IoT sensor-based devices used for e-healthcare services collect and transmit individual patient's diagnostic data to the central monitoring systems associated with the patients registered through mobile apps or web-based applications. The data being transferred by these IoT devices in e-healthcare systems can be easily attacked, hacked by various intruders over the internet to obtain personal information. Most attacks on Wireless Sensor Networks (WSN) [15] can be easily modified for IoT-based services, and Table 2 lists a comprehensive survey of the types of attacks on healthcare applications using IoT.

At times, the remote monitoring and transferring devices may include information like location or background images of their personal space which a patient does not want to share with anyone else except the doctor. Whenever data will be transferred using the internet, the risk of data breach increases manifold. Security of a patient's medical and personal data, both at rest and in motion is always desired and recommended. The key areas where security challenges potentially affect e-healthcare raise some pertinent questions for its implementation strategists:

- How to overcome authentication problems related to IoT-based wearable devices and other connected devices.
- How to monitor tampering of data in transit.
- How to deal with privacy and integrity of data.
- How to deal with patient specific information storage privacy.

Various researchers are pursuing research on the security and privacy issues arising from the use of IoT-based e-healthcare.

One potential solution in the form of blockchain storage-based access control to medical data of a patient is presented in the remaining part of this chapter.

Table 2 Common attacks on IoT-based healthcare

References	Type of attacks	Description	Category
[60–61]	Blackhole attack	Invalid path advertised as a good path for data and travels through adversary	Interception
[62–63]	Grayhole attack	Partial interception using only a node or two unlike whole network in blackhole to avoid suspicion	Interception
[64–65]	DoS and DDoS	Damage nodes' battery lives by keeping them active and rendering them useless	Interruption
[66–67]	Radio jamming	Intentionally using the same radio frequency for other transmissions in the vicinity to jam original signal	Interruption and fabrication
[68–69]	Wormhole attacks	Attacks the routing protocols to capture and replay	Interruption, interception, and modification
[70–71]	Sybil attack	Fake identities to participate in routing	Interception and modification with fabrication
[72]	Message injection attacks	Deliberate misleading messages injected in network	Fabrication
[73–74]	Flooding	Fake identity devices flood hello packets to disrupt genuine network	Fabrication and interruption
[75]	Message replication attacks	Replication and forwarding of captured packets	Interception
[76]	Node stealing / damage	Destroy multiple sensors to disable network	Interruption
[77]	Sinkhole attacks	Provides fastest path to gateway for intercepting packets	Fabrication and interception

3 Blockchain Records: Decentralizing the Trust

Blockchain technology is based on the concept of a distributed ledger with an underlying premise that data security is ensured while bringing transparency to the system through the elimination of the requirement of third parties for trust, storage, and key certification [15]. We can understand the technology by observing that before the existence of blockchain, there was a complete reliance on a single centralized authority or server for storage and security. Any breach happening at the central server level could compromise data of each individual stored at the machine. We can also visualize it as a decentralized data structure where data is distributed across all nodes within the sanctioned distributed network. Each stakeholder can become a storage/processing node and may keep the complete or partial state of the blockchain data on itself [16].

This technology allows for a shared copy of the data, replicated at all participating nodes and updated only after establishing consensus among the majority of stakeholders, thereby providing security from a corrupted/compromised central authority. The consensus protocols are a standard part of blockchain architecture and are needed to add transactions on ledger. Various implementations of the consensus algorithms [17] have been proposed and standardized recently. Blockchain technology constructs the equivalent of an electronic ledger by adding blocks in chronological order on a chain of secure data blocks. In technical terms, the blockchain offers decentralized data [18] with immutability property and a cryptographic hash-based consensus.

A logical extension of blockchain technology is the concept of smart contracts. Smart contracts are self-executing assertions that are used to control the exchanges or redistributions of digital assets between involved stakeholders on the basis of certain preconditions to be met. Smart contracts are used as member functions applicable to stored data in the form of objects while defining possible operations on data. The technology explained in this section can be understood better with a complete explanation of the keywords as listed below.

Genesis Block

It is the first block of any blockchain and forms the foundation of the implementation.

Hash

We can understand hash as a representative of each bit present in the block, effectively identifying the complete block. The hash chaining procedure in blockchain ensures that any change in block information propagates through the chain, and can be easily detected since not only is the hash unique for each block, it is also dependent upon the hash of the previous block-making blockchain secure despite being decentralized. The process of hash chaining can be understood with the help of Fig. 2.

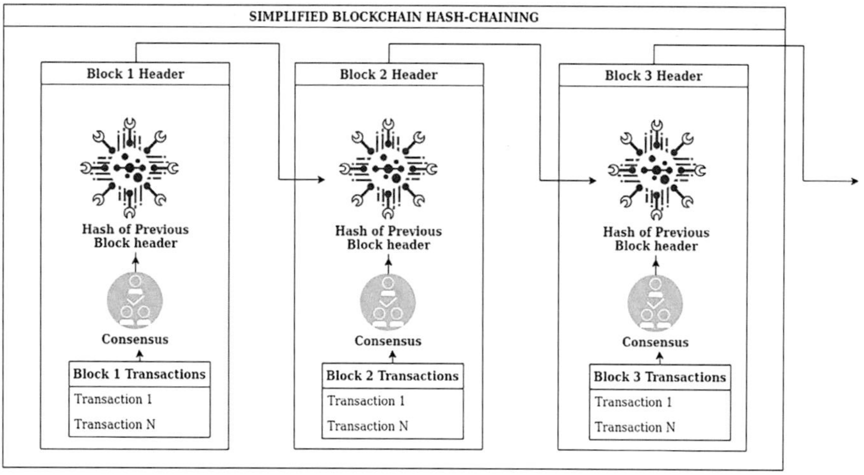

Fig. 2 Hash chaining the blocks

Consensus Mechanism

Consensus is the generic term for the protocol required by all the participating nodes to agree on the legitimacy of the transactions that are being added to the distributed ledger. The coordinated mechanism of group-based consensus is however dependent upon the assumption that there shall be malicious nodes and a possibility of faults in the process, despite of which the mechanism should work.

A miscreant node may deliberately create messages which can lead to chaos and conflict between the group members, making consensus harder to obtain thereby breaking down the protocol. Such chaos-led failures to reach consensus are called byzantine faults. To be acceptable as a generic consensus protocol, a mechanism must have provision for fault tolerance.

Proof of Work

Whenever a participating node wants to add a new block of transactions to an existing chain, then some work is to be done. It is easier to understand it by a challenge-based analogy. To enter a new block the participant has to solve a hard puzzle, only after that new content can be added. The process of solving this "puzzle" is called mining and the participants who are vying to add a valid block have to perform a complete mining process, which exists in varying degrees of hardness depending upon the application. The participants trying to add blocks are called miners, and the mining process includes a lot of hits and trials. All the nodes which are part of the network create a consensus, that is, a type of agreement to accept a block or not upon checking the transaction status.

Proof of work has been a popular consensus mechanism but due to the resource-intensive nature of the protocol, it is considered to be unsustainable, leading to several other consensus mechanisms being proposed.

Proof of Stake

Proof of Stake is a consensus mechanism based on the inherent assumption of more stakeholder interest based on more stake /ownership of coins or the value held. This mechanism is not only environment friendly but also maintains the network better. While using Proof of Chain, within each fixed slot of time, a validator is pseudo randomly selected with authority to create a block, and append it to the longest existing chain (CS) but with no penalty clauses, the validator may add blocks on multiple competing chains for multiple rewards, without extra cost, there it is very hard to reach consensus despite no of attackers.

Remember chain splits are penalized in the Proof of Work system where the longest chain is preferred, and suboptional addition causes implicit loss in rejection. To remedy the "no-stake" problem in Proof of Stake mechanism, *Casper Proof of Stake* in Ethereum ensures in case of malicious activity, the validator loses their ETHs put as stake for participation.

DPoS

Delegated Proof of Stake is a distributed voting system based on stakes owned by numbers. They vote for an organization or person to produce blocks for the network and in turn delegates get rewards. The mechanism runs much faster than Proof of Stake.

The above three are the most common consensus mechanisms but as blockchain and their functionality/use cases are evolving, the consensus mechanisms are also witnessing evolution.

BFT Compliant Consensus Mechanism

Both Proof of Stake and Proof of Work fail to perfectly address the BFT problem, making their application difficult to the noncurrent solution of blockchain. AlgoRAND is a Byzantine fault tolerance consensus mechanism with guaranteed security even in the presence of adversaries. It chooses officials from the complete user base randomly to maintain a two-thirds honest majority for a quick and low-computation cost consensus.

Reputation-Based Consensus

Some other consensus mechanisms worthy of mention include Proof of Authority which ensures very fast transactions based on the validator's accumulated reputation, and PoR (Proof of Reputation) which is a logical extension of Proof of Authority.

Distributed Ledger

In simple words, a ledger is a common record that has common information which will be used by all distributed nodes. Based on who can participate in the ledger transactions and who are responsible for its maintenance, we have three broad categories of blockchains being deployed.

- Public Blockchain: The public ledger allows anyone to join, add transactions based on challenges and the best example is bitcoin. There is a complete lack of privacy here clubbed with the very high computational costs of the challenges.
- Private Blockchain: This belongs to an organization that controls the participants, consensus protocols, and the maintenance of ledger storage. It is considered very trusted since traditional firewalling and on-premise secure installations are used.
- Permissioned Blockchains: The "permissioned" in the definition signifies an invitation-only blockchain network. This may be implemented across a public or a private blockchain as per the use case. Transaction-level access control can be implemented here.
- Consortium Blockchains: These are multiple organization permissioned blockchains with diverse stakeholders having a shared responsibility for the blockchain. This is one of the most ideal scenarios for a public-private partnership or governmental blockchains. Different types of blockchains based on ledger permissions have been illustrated in Fig. 3.

4 Blockchain and IoT-Based e-Healthcare

The most prominent among the major problem areas identified in the adaptation of telemedicine evolving toward IoT-based e-healthcare is trust deficit, that is, data security and individual privacy. Having an integration with blockchain technology, it can be envisioned that the e-healthcare process will gain mainstream importance. Blockchain can help in keeping all patient records like lab reports, patient treatment details including prescriptions, medical history of patient at one place and that too in a decentralized manner.

Electronic health records (EHR) are the cornerstone of the e-healthcare paradigm. Various stakeholders need on-demand access to patient EHR,

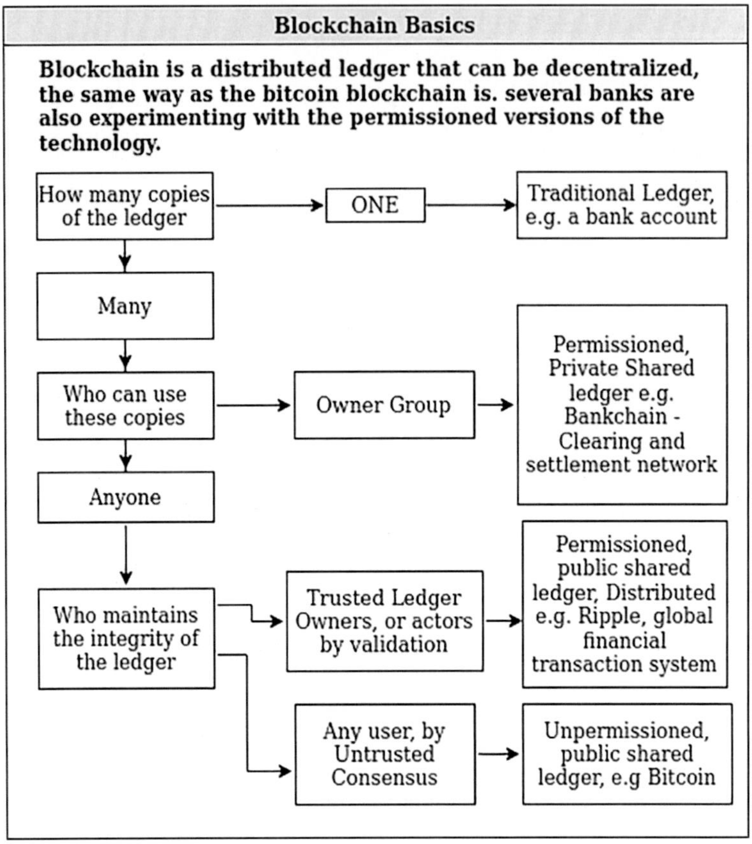

Fig. 3 Types of blockchains based on ledger permissions

necessitating that the provision of data security must be ensured in e-healthcare. Blockchains bring in the desirable technical aspects like immutability, transparency in transactions, and auditability of data access. Every stakeholder like the admitting hospital, assigned doctors, nurses, health insurance company, and lab staff can request granular control over access to the patient information, which is eventually the prerogative of the concerned patient or his representative who can be made the owner of his data blocks. At any stage, no stakeholders who intend to make any unauthorized changes to patient information are allowed by the underlying blockchain due to immutability features. Some common benefits of blockchain technology as envisioned by researchers are as follows [18–20].

Potential Blockchain Benefits in IoT-Based e-Healthcare

Peer-to-Peer Network

Blockchain's system is a decentralized hack proof system. Data does not exist as a single centralized copy but is replicated to ensure every participating and capable node has its own copy. This added advantage helps in eliminating any mishappening which can cause a complete system failure in the case of some disaster. The time and resource savings obtained here may be utilized as an investment to fund research and innovations in the treatment for diseases and other related work.

Single Block for Each Transaction

In blockchain technology, a single block is used to store data related to a group of transactions. Each block is added in the chain after the confirmation by all the stakeholders ensuring that the blockchain technology provides a 100% authentic as well as secure service. Nobody can alter or overwrite the data in the block once committed. This is the best way to preserve the security of patient personal data and other health records generated by e-healthcare. Privacy preservation can also be maintained by permissioning.

Maintenance of Chronologically Ordered Health Records

Blockchain for the healthcare industry can be extensively used for managing patient info, including data related to various parameters like blood pressure and pulse rate being generated by using IoT and other wearables. This may help doctors to remotely monitor their high risk or critical patients 24*7 and suggest appropriate actions. With the use of blockchain, the chronological order of records containing monitoring reports is automatically maintained [21].

Cost-Effective Deployment Solutions

Blockchain technology may prove to be one of the most cost-effective solutions in some large-scale implementations. There is a huge savings of costs due to no requirement of trust-establishing mediators and third parties. Removal of third parties also ensures that owing to this independence, data retrieval becomes faster and cost-effective.

Digital Rules Book Using Smart Contract

A smart contract is a digital protocol that verifies, implements, controls, and monitors pre-defined agreements to be implemented automatically. By using smart contracts, rule-based processes related to the patients, doctors, or organizations can be created. In simple words, it is an automatic contract enforcement code between various stakeholders, and if a violation is triggered by any party, then the system's smart contracts initiate necessary actions by itself [22].

Integrity of Data

Traditionally, data integrity has been defined about "verifiable" data security including timeliness, completeness, and accuracy to rule out any effect of outer forces which are not related to data. The definition highlights how important and challenging it is to maintain data integrity for electronic health records. Data integrity must be preserved by any technological means owing to its deep impact on medical as well as legal aspects.

Patient's Unique ID

Duplicates in health data are always possible especially with a country as populous as India, there can be several people with almost similar personal details, but in the healthcare industry, the hazards of duplication in patient records can result in life risk for the patient. In order to ensure risk avoidance, a single patient identification concept on the lines of Social Security numbers (SSN) or AADHAR is very beneficial, which is possible to maintain through blockchain. With such a system in place, data manipulation is impossible as already discussed.

Dynamic Business Models

Healthcare as a business can be impacted profoundly with the use of blockchain technology. With adaptation of this technology, the entire healthcare spectrum shall witness changes maximizing potential benefits for stakeholders. Integration with other business models including pharmacy and insurance shall also be very easy when using consortium blockchains.

Augmenting Storage Capacity Without Third Party

By using blockchain in healthcare, we can store individual patient's related records like X-ray reports, lab records, personal information, insurance details, prescriptions, consultant doctor's detail, appointment details all of which can be stored using

Table 3 Blockchain and IoT in healthcare applications

Reference	Healthcare challenges addressed	Security challenge addressed
[78]	Storing medical records data of big size	Cost-effective data security
[79]	Permanent storage of healthcare data	Immutability
[80, 90, 91]	BIoT for integrating sensor-based data on blockchains	Key management
[81–83]	Several healthcare use cases of blockchain have been proposed	Identity and access management
[84]	Blockchain uses in cardiac healthcare	Tamper-proof logs of events
[85, 86]	Smart contracts for telemedicine	Non-repudiation
[87]	Personal health record management	Integrity of EHRs
[88]	CryptoCurrencies in healthcare	Transaction management
[89]	Radiological data management on blockchain	Data management
[92–94]	BIoT (Blockchain and IoT) for clinical trials, trust management, privacy preservation	Privacy management
[95]	Peer-to-peer EMR storage	Distributed data security

a consortium permissioned blockchain with the patient having ownership rights. Each stakeholder in the consortium maintains its own copy based on access granted by the patient [23].

The comprehensive survey of the research work focusing on the use of blockchain in healthcare services shows useful insights into the domain applicability as summarized in Table 3. The table also shows the cross-domain research work across the IoT and blockchain in healthcare applications.

Paradigm Shift from Service-Based Care to Patient-Based Care

Traditionally, the healthcare industry commonly practiced to serve more patients and earn on volumes. This model is popularly close to the service-centric care in industry. The practitioners' remuneration model is calculated as per service rendered. Incentives for the employees of healthcare institutions were also proportional to the number of patients or the quantity of care sessions. In the present scenario of e-healthcare, which is service based, some issues like interoperability and patient's control over their own personal information [24] are still not considered. The value-based care model allows for these provisions and is discussed below.

The new e-healthcare model brings in a paradigm shift, it being a value-based care model rather than service based. The focus of e-healthcare is client-centered, that is, value-based care for patients. In a value-based care model, the patient is involved in decision-making, sometimes indirectly through IoT for example, where the patient may directly send the vital statistics using wearables, generating continuous medical history records, on the basis of which experts make decisions and provide requisite treatment accordingly.

It is very likely that in this scenario, the patient might visit different consultants or different healthcare centers for the health issues, thereby generating across different locations or databases. To avoid patient's related information being fragmented across multiple database locations, all related patient data should be centralized, and available with the patient on-demand in a single click. The patient should have online instant access to his medical records to ensure access without communication delays while avoiding communication errors. With this model, continued care even by any other medical practitioner besides the regular family doctor can be provided to a patient based on his collective medical history.

In a fully functional deployment of this model, a patient can get every information related to their treatment including the lab and diagnostic reports immediately at the same time while they are made available to the doctor. In addition, the patient should have full access control on when to share medical records and to whom, and even which part of their medical data should be given access. This complete model is envisioned to be scalable for more users, that is, if a new doctor is supposed to join the treatment panel, then he can be given access to the patient's data.

Interoperability as a Challenge in e-Healthcare

Interoperability in healthcare implies varied types of ICT-based systems and software-based applications, and various information formats that are being used by healthcare organizations within and outside their boundaries, to exchange data and to use exchanged data.

This kind of Interoperability allows information systems to work together, which not only makes better business sense but also improves the effectiveness of the e-healthcare systems for individuals and communities.

Data sharing and collaborative treatment are essential in many e-healthcare cases. Consider the case of a local caregiver working in association with an expert consultant available remotely toward the treatment of a patient. There has to be a lot of document and medical record sharing back and forth between the two. This sharing obviously should be secure as well as scalable. A case in point may be a team of experts from various locations participating collaboratively in a critical patient's treatment, say some oncologists hypothetically from different states or some foreign expert sharing data.

Another use case advocating interoperability and fast interchange of information can be put across with a patient admitted to the emergency room of the nearest hospital for urgent treatment with a severe to worsening condition. The hospital, in this case, is not the regular treatment center but an intermediate stopover for stabilizing the patient. In such cases, even primary care to a patient can be provided only if his/her treatment and prescription history can be urgently made available to the treating doctor [25]. The procurement and sharing of medical records become absolutely difficult and time critical here, especially if only a hard copy of the records is available at the patient's home. The challenge becomes compounded if medical history data of the patient is scattered across his home and at the treating hospital. It is a complete breakdown of the medical systems in the absence of medical history records despite the presence of doctors and possibility of treatment and the patient. Medical record location-based fragmentation and the lack of interoperability together form the biggest bottleneck in the adaptation of IoT-based e-healthcare universally.

Interoperability, as explained above, can be considered as a basic requirement for patient-centric architecture and, if implemented, can be a great boon for patients worldwide [26–28]. Hurdles in the technical infrastructure of existing healthcare systems either disallow, or slow down interoperability, thereby hampering the value-centered care models. Some technical challenges in the patient medical data interoperability are discussed below.

- Data breach problems related to security and privacy are presumed to be high in an interoperable environment.
- It becomes very difficult for a patient to trust other health providers besides the regular doctor, and it is even more challenging to maintain trust relationships among healthcare and allied organizations.
- Issues related to data scalability shall arise as hospitals as well as patient's diagnostic data is huge due to medical images.

The most interesting aspect of the above discussions on interoperability, security, privacy, and medical records' locational fragmentation is that in each case, the "simplest solution" as per Occam's Razor "is the most feasible and the best." The solution lies in placing the data online, decentralizing it with consensus-based updates, and giving access control of a person's medical data to the owner of this data. This leads us to use blockchain as a solution as it has shown potential to cater to all the requirements for maintenance of patient's electronic health records. The opportunities are exciting, but it would not be out of place to mention the potential challenges introduced by using blockchain for the specific purpose of storing the electronic health records data. The next section introduces the readers to some storage issues applicable to blockchain in general but specific to storing electronic health records on a blockchain.

5 Blockchain Storage Challenges for IoT-Based e-Healthcare Data

To get a deeper insight into the complete data lifecycle for the healthcare records, we have to take a holistic view of the system. A blockchain-based record-keeping system can be fairly represented by four phases based on data lifecycle.

- Generation of data.
- Enrichment of data.
- Data storage on blockchain.
- Data consumption by smart contracts.

The basic phases remain similar in all applications using blockchain as a storage technology, but the volume, velocity, and variety of the data being generated in an IoT-based e-healthcare application make it unique from a storage perspective. Each of the data lifecycle phases listed above can be understood from the perspective of a blockchain-based record-keeping system as per the discussion in the following subsections.

Generation of e-Healthcare Data

This is the first phase of planning for the implementation of blockchain in IoT-based e-healthcare. The developer should assess all sources of data, its properties, and other related data information including how data is generated (using IoT devices/wearable devices), where it is being generated, and the scope of data being generated. A quick domain survey of the healthcare processes indicates that at every step of medical treatment like consultation, diagnosis, and treatment including surgical processes, sensitive and crucial medical data is generated [29]. This medical data may include any or all of the following:

- The doctor's prescriptions.
- Blood test reports.
- X-rays.
- CT scans.
- Sonography reports.
- Radiography.
- Endoscopy.

The above is not an exhaustive list, it is only a small representative subset of the vast amount of data that is possibly being generated. Healthcare data, as explained above, can exist in any form as images, scanned images, text, videos, 3-D images as multimedia-based data, and different formats of each type may be further envisioned.

The Sensitive Nature of Medical Data

Generally, all patients' medical data is stored digitally on a centralized server. Medical records have traditionally been considered low risk, making it very easy for a hacker to attack the centralized storage. There can be instances of social engineering where administrators having control over centralized data can share personal data of a patient without seeking the consent of the data owner. Many studies have indicated that employees working in the healthcare sector sell sensitive data very easily to any unauthorized person or organization for a low incentive, without realizing its implications. A patient's health record contains sensitive and personal information which, if leaked or lost, can be used to exploit the patient. Electronic health records (EHRs) may contain some or all of exploitable information like:

- Patient identity.
- Address.
- Aadhar or social security identifiers.
- Office address.
- Prescribed medicines.
- Allergic medicines.
- Frequency of doctor visits.
- Payment modes and payment details.

Further, a medical record may contain some sensitive information which the patient does not want to reveal to anybody, for example, psychological conditions, STDs, or long-term but curable illness like tuberculosis treatment [30, 31].

Enrichment of Data

Enrichment of data refers to the process of converting raw data into some meaningful or structured form by adding value to it. Before we commit the patient's medical and diagnostic history to persistent storage, we may pre-process the same. Patient's health records need to necessarily have the following basic data attributes:

- Accurate.
- Secure.
- Understandable.
- Time-stamped.
- Structured.

If data in raw or unstructured formats is stored, it could lead to inconsistency which in turn can cause the treatment process to be delayed, and the storage itself may become inefficient.

To avoid the problems of inconsistent or insufficient storage, a three-step process is used to clean and secure data before putting on blockchain.

1. Hiding or obscuring identity: replacing private identity data with one-way secure hash.
2. Meta data computation: computing metadata for original data. This allows the choice of storing either metadata or actual data (accessible using metadata) on the blockchain as explained in the next subsection.

3. Information compliance or data compliance: means data conforms to the applicable rules in smart contracts.

Data Storage: Storing Electronic Health Records on Healthcare Blockchain

Storage in blockchain is different from traditional relational and NoSQL databases. While considering IoT-based e-healthcare, it becomes important to understand whether blockchain storage is appropriate for our use case or not. Blockchain technology can be visualized as a data structure with a chained group of blocks which solves the problem of transparency and immutability, but for data storage the main concern of a developer is how block data is stored and retrieved when searching information. Even though we have established the premise that blockchain has the potential to eliminate all risks occurring in a centralized data storage, it is important to understand the requirements that a blockchain-based storage should fulfil as elaborated further [32].

Data Ownership and Compatibility

Various stakeholders in the healthcare industry can store health records, necessitating that the ownership of personal data has to be established for the patient. It can be managed by using the patient's private key to encrypt personal records before storage on the blockchain. The information is stored with data compliance, that is, it is compatible with smart contracts, the structure of which may be provided by doctors, health insurance companies, or medical centers using the blockchain storage on the basis of permissions given.

Access Control with Complete and Unique Record of Transactions

Patients' health records are saved on the blockchain with a unique ID and their public key. The blockchain access process maintains completeness. If any stakeholder wants to see a patient's public data after seeking access permissions from the patient, then this data access is recorded as a new transaction with a unique ID and only after this, the information is displayed.

Querying a Blockchain System

In the traditional healthcare system doctors, TPA department and insurance companies need to approach patients to access prior treatment data records at the time of admission and multiple other times, but in blockchain, the patient just has to allow access permissions as necessary, and thereafter the medical record blockchain can be queried by the permission holder. In order to implement these permissions, smart contracts must be there to safeguard privacy and security concerns of all stakeholders.

While choosing the medium of storage for the blockchain, certain key considerations have to be made on the basis of use case. We have already discussed the types of blockchain available as per the participants in Sect. 3. Another classification for the implementation of storage is based on how much data is to be stored in a block. Depending upon the implementation, data on a blockchain may be stored "on-chain" or "off-chain." The distinguishing feature is that in on-chain storage, all data gets stored in blocks on chain one after the other. Every new transaction will add a new block with updated data on the chain. This type of storage is useful only when data is limited, but faces huge challenges with increase in record sizes. It becomes a very expensive option for storage. The costs rise exponentially even when the network gets overloaded.

In contrast, the off-chain storage solutions do not store complete data, provide a low-cost option, but only metadata is stored on the chain. The actual data may possibly be stored using traditional data storage techniques. It makes the application susceptible to data breach and once lost, the data cannot be recovered. In IoT-based e-healthcare applications, some data is kept on-chain and some off-chain varying according to requirement.

The key participants to the smart contract for any e-healthcare application based on blockchain technology are shown in Fig. 4.

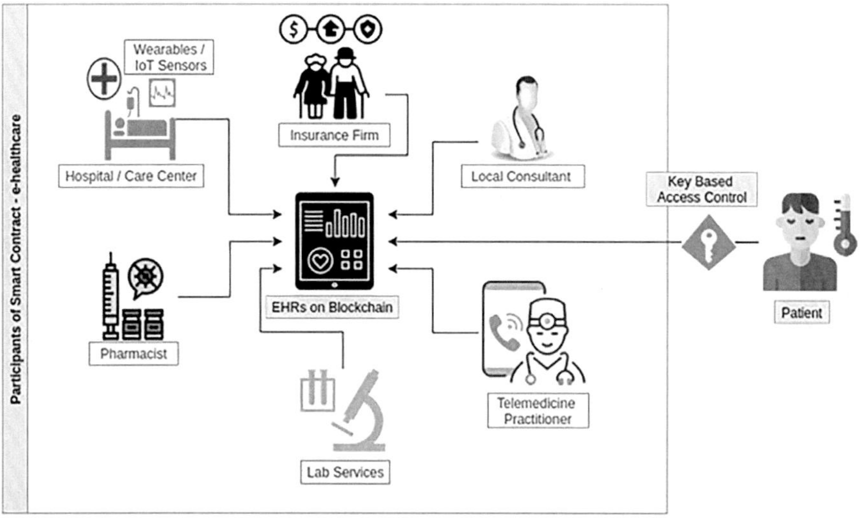

Fig. 4 Key participants of an IoT-based e-healthcare blockchain application

6 A Simple IoT-Based e-Healthcare Model Using Blockchain

We have come a long way from text-based telemedicine consultations to present-day advancements. The opportunities associated with IoT-based e-healthcare are enormous, but there exists a mismatch between the technical growth vis-a-vis the adoption of this technology.

The reason for blockchain technology's success is due to its distributed storage, common ledger, individual control, distributed computation, and the only append mode functionality leading to trust establishment can become a key driving force for e-healthcare [33]. We propose a basic framework for the IoT-based e-healthcare system using blockchain as per Fig. 5 to summarize the chapter.

To keep the explanation simple, we are considering only the six key stakeholders as per the following:

- Patients.
- Physicians or doctors.
- Medical center.
- TPA and insurance.
- Lab services.
- Pharmacist.

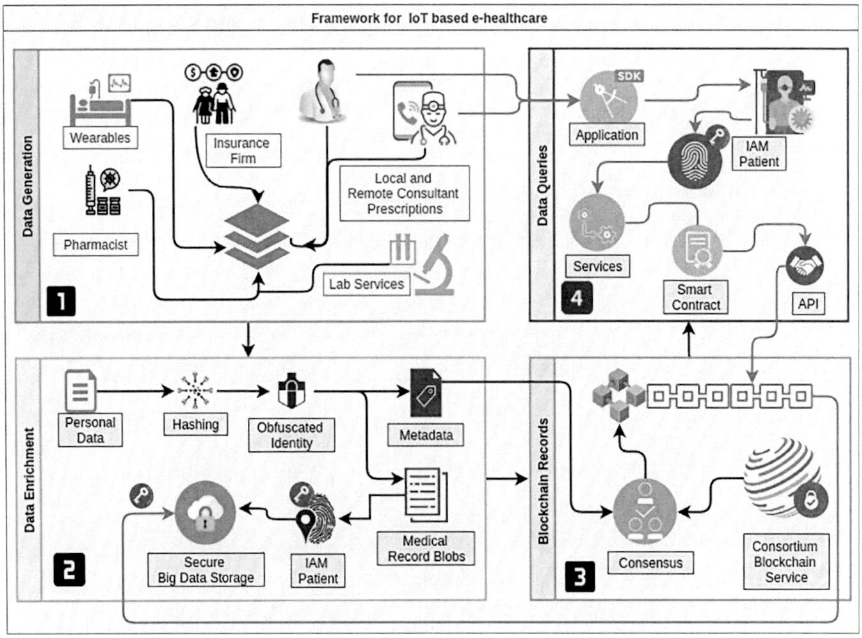

Fig. 5 A simple framework for IoT-based e-healthcare system using blockchain

The process flow starts with data collection from the stakeholders. In the IoT-based e-healthcare system, data from wearable sensors and IoT-based devices along with prescriptions, insurance policy data, and laboratory reports as shown in the data generation module is depicted in part 1 of Fig. 5.

The collected set of patient records are sent for data pre-processing as shown in part 2 of the image. This module is responsible for removing privacy information from the documents before registering the same for blockchain-based storage. A secure one-way hash is applied to obfuscate the privacy data, and then metadata is generated. The data-intensive medical records are sent to be stored as blobs on an appropriate secure medium under the patient's key using an identity and access management module (IAM) as shown.

The metadata information is used for maintaining the blockchain information base. All collected metadata is sent to a blockchain network as a single transaction on a single block after seeking due consensus from the stakeholders as per the consortium participants as shown in part 3 of Fig. 5. This process can be repeated after regular intervals of time in case of critical patients, where one new block is added for every transaction. It is noteworthy here that blockchain works only in append mode, and any stakeholder cannot modify the details contained in any block added post consensus. We have considered the ownership and access management of the block with the patient; however, the owner of the chain can be patients, doctors, any department, or even the medical organization as well, depending on the use case and implementation.

This chain remains distributed among related stakeholders at all times and there is a completely controlled access of the records, based on IAM as shown in part 4 of Fig. 5. Medical centers are usually permitted the highest access clearance because they need the complete data of the patient, doctor, emergency services, and data for TPA and insurance. Every e-healthcare system has some regulatory body that can also have full access rights, maintainable across blockchain but without modification rights. Access to medical record blobs is permitted only through API access to the blockchain, and even that happens after the execution of designated smart contracts requested by application services.

The use case discussed above has been oversimplified for the sake of conceptual clarity, but a practical implementation shall have far more complexity [34–36]. The comparative analysis from the perspective of security for a blockchain-based healthcare application vis-a-vis that running on a centralized database is presented in Table 4.

7 Conclusions and Future Directions

The establishment of trust using blockchain in the industrial use case of IoT-based e-healthcare shows very promising opportunities and is sure to go a long way in the adaptation of this essential technology. There are some pertinent issues however, which relate to the establishment and implementation of a blockchain consortium

Table 4 Comparative security provisions of blockchain-based versus centralized database healthcare applications

Parameter	Blockchain	Centralized database
Decentralization of trust	Default	Not available
Identity and access management	By patient	Hospital management
Ownership of data	High granularity, up to individual patient record	Organization based, low granularity
Regulatory audit	Automated – Consortium blockchain	Not automated
Immutability	Available	Not by default
Privacy preservation	Easy to manage	Difficult by default

and maintenance of the infrastructure. It would be interesting to see how the cost-effectiveness of these solutions is brought to the tipping point, as this might make for an easier and early adoption of the technology. The role of some governmental agencies as the certifying authority for e-healthcare documents shall also be an interesting development to track in the future. The debate for on-chain versus off-chain storage solutions in the case of medical records also needs to be settled, and the ownership of data and adherence of record maintenance to statutory data privacy compliance laws like HIPAA is another potential area for researchers.

Another potential area for research is the development of application-based consensus mechanisms. For example, new consensus mechanisms are needed for healthcare-based blockchain applications. The traditional consensus mechanisms based on anonymity and high cost of computing resources are contrary to the requirements of quick response and user verification.

References

1. V.K. Saini, S. Gupta, Blockchain in supply chain: journey from disruptive to sustainable. J. Mech. Continua Math. Sci. **14**(2) (2019). https://doi.org/10.26782/jmcms.2019.04.00036
2. X. Wang, Z. Zhang, J. Zhao, Y. Shi, Impact of telemedicine on healthcare service system considering patients' choice. Discret. Dyn. Nat. Soc. (2019). https://doi.org/10.1155/2019/7642176
3. O. Sonnis, A. Sunka, R. Singh, T. Agarkar, IoT based telemedicine system, in *IEEE International Conference on Power, Control, Signals and Instrumentation Engineering, ICPCSI 2017, d*, (2018), pp. 2840–2842. https://doi.org/10.1109/ICPCSI.2017.8392239
4. S. Chen, A. Cheng, K. Mehta, A review of telemedicine business models. Telemed. E-Health **19**(4), 287–297 (2013). https://doi.org/10.1089/tmj.2012.0172
5. T. Tekeste Habte, H. Saleh, B. Mohammad, M. Ismail, IoT for Healthcare, in *Ultra Low Power ECG Processing System for IoT Devices. Analog Circuits and Signal Processing*, (Springer, Cham, 2019). https://doi.org/10.1007/978-3-319-97016-5_2
6. A. Hameed, A. Alomary, Security issues in IoT: a survey, in *2019 International Conference on Innovation and Intelligence for Informatics, Computing, and Technologies, 3ICT 2019*, (2019), pp. 1–5. https://doi.org/10.1109/3ICT.2019.8910320

7. J.L. Hall, D. Mcgraw, For telehealth to succeed, privacy and security risks must be identified and addressed. Health Aff. **33**(2), 216–221 (2014). https://doi.org/10.1377/hlthaff.2013.0997
8. S. Pinto, J. Cabral, T. Gomes, We-care: an IoT-based health care system for elderly people, in *Proceedings of the IEEE International Conference on Industrial Technology*, (2017), pp. 1378–1383. https://doi.org/10.1109/ICIT.2017.7915565
9. U. Albalawi, S. Joshi, Secure and trusted telemedicine in Internet of Things IoT, in *IEEE World Forum on Internet of Things, WF-IoT 2018 – Proceedings, 2018-Janua*, (2018), pp. 30–34. https://doi.org/10.1109/WF-IoT.2018.8355206
10. A.A. Siyal, A.Z. Junejo, M. Zawish, K. Ahmed, A. Khalil, G. Soursou, Applications of Blockchain technology in medicine and healthcare: challenges and future perspectives. Cryptography **3**(1), 3 (2019). https://doi.org/10.3390/cryptography3010003
11. B. M. In, *Business Models In Telemedicine Improving Last Mile Health Delivery*. December (2017)
12. IoT to redefine healthcare ecosystem – eHealth Magazine [Online], Available: https://ehealth.eletsonline.com/2019/03/iot-to-redefine-healthcare-ecosystem/
13. Y. Maleh, A. Ezzati, M. Belaissaoui, An enhanced DTLS protocol for Internet of Things applications, in *2016 International Conference on Wireless Networks and Mobile Communications (WINCOM)*, (2016), pp. 168–173. https://doi.org/10.1109/WINCOM.2016.7777209
14. G. Ateniese, M.T. Goodrich, V. Lekakis, C. Papamanthou, E. Paraskevas, R. Tamassia, Accountable storage. IACR Cryptology ePrint Archive **2014**, 886 (2014)
15. Y.A. Bangash, Y.E.A. Al-Salhi, Y. Maleh, Security issues and challenges in wireless sensor networks: a survey. IAENG Int. J. Comput. Sci. **44** (2017)
16. R. Somasundaram, M. Thirugnanam, Review of security challenges in healthcare internet of things. Wirel. Netw (2020). https://doi.org/10.1007/s11276-020-02340-0
17. S. Tamang, *Decentralized Reputation Model and Trust Framework*, (2018) December., https://www.diva-portal.org/smash/record.jsf?pid=diva2:1352089
18. L. van Velsen, M. Tabak, H. Hermens, Measuring patient trust in telemedicine services: development of a survey instrument and its validation for an anticoagulation web-service. Int. J. Med. Inform. **97**, 52–58 (2017). https://doi.org/10.1016/j.ijmedinf.2016.09.009
19. X. Zhu, Research on blockchain consensus mechanism and implementation. IOP Conf. Ser. Mater. Sci. Eng. **569**(4) (2019). https://doi.org/10.1088/1757-899X/569/4/042058
20. A. Hasselgren, K. Kralevska, D. Gligoroski, S.A. Pedersen, A. Faxvaag, Blockchain in healthcare and health sciences—a scoping review. Int. J. Med. Inform. **134**, 104040 (2020). https://doi.org/10.1016/j.ijmedinf.2019.104040
21. U. Opportunities, *The Internet of Things and Blockchain: Unique Opportunities for Healthcare*. March (2018)
22. I. Abu-elezz, A. Hassan, A. Nazeemudeen, M. Househ, A. Abd-alrazaq, The benefits and threats of blockchain technology in healthcare: a scoping review. Int. J. Med. Inform. **142**(February), 104246 (2020). https://doi.org/10.1016/j.ijmedinf.2020.104246
23. Blockchain in Healthcare | Healthcare Blockchain [Online]. Available: https://www.leewayhertz.com/healthcare-blockchain-how-medical-records-secured-blockchain
24. A. Khatoon, A blockchain-based smart contract system for healthcare management. Electronics (Switzerland) **9**(1) (2020). https://doi.org/10.3390/electronics9010094
25. S. Ali, G. Wang, B. White, R.L. Cottrell, A Blockchain-based decentralized data storage and access framework for PingER, in *Proceedings – 17th IEEE International Conference on Trust, Security and Privacy in Computing and Communications and 12th IEEE International Conference on Big Data Science and Engineering, Trustcom/BigDataSE 2018*, (2018), pp. 1303–1308. https://doi.org/10.1109/TrustCom/BigDataSE.2018.00179
26. eHealth Technologies | The Benefits and Challenges of e-Health Applications [Online]. Available: https://www.scientificworldinfo.com/2019/09/the-benefits-and-challenges-of-e-health-technologies.html
27. J. Wang, K. Han, A. Alexandridis, Z. Chen, Z. Zilic, Y. Pang, G. Jeon, F. Piccialli, A blockchain-based eHealthcare system interoperating with WBANs. Futur. Gener. Comput. Syst. **110**, 675–685 (2020). https://doi.org/10.1016/j.future.2019.09.049

28. What Is Value-Based Healthcare? [Online]. Available: https://catalyst.nejm.org/doi/full/10.1056/CAT.17.0558
29. H.Y. Paik, X. Xu, H.M.N.D. Bandara, S.U. Lee, S.K. Lo, Analysis of data management in blockchain-based systems: From architecture to governance. IEEE Access **7**, 186091–186107 (2019). https://doi.org/10.1109/ACCESS.2019.2961404
30. Y. Zhu, C. Lv, Z. Zeng, J. Wang, B. Pei, Blockchain-based decentralized storage scheme. J. Phys. Conf. Ser. **1237**(4) (2019). https://doi.org/10.1088/1742-6596/1237/4/042008
31. P. Zhang, D.C. Schmidt, J. White, G. Lenz, Blockchain technology use cases in healthcare, in *Advances in Computers*, vol. 111, 1st edn., (Elsevier Inc., 2018). https://doi.org/10.1016/bs.adcom.2018.03.006
32. V.K. Saini, S. Gupta, B. Gupta, Traceability and optimization over Merkle Tree. J. Xi'an Univ. Archit. Technol. **XII**(III) 2020 [Scopus] https://www.xajzkjdx.cn/gallery/389-mar2020.pdf. (n.d.)
33. Telemedicine and Cybersecurity: Keeping Health Data Safe. Available: https://prognocis.com/telemedicine-and-cybersecurity-keeping-health-data-safe
34. H. Kaur, M.A. Alam, R. Jameel, A.K. Mourya, V. Chang, A proposed solution and future direction for Blockchain-based heterogeneous Medicare data in cloud environment. J. Med. Syst. **42**(8) (2018). https://doi.org/10.1007/s10916-018-1007-5
35. Blockchains From a Distributed Computing Perspective. Available: https://cacm.acm.org/magazines/2019/2/234355-blockchains-from-a-distributed-computing-perspective/fulltext
36. V.K. Saini, S. Gupta, B. Gupta, Blockchain framework for cyber-physical systems: reengineering approach. Int. J. Future Generat. Commun. Netw. **13**(4), 1461–1467 (2020)
37. A. Ghorbel, S. Bouguerra, N.B. Amor, M. Jallouli, Cloud based mobile application for remote control of intelligent wheelchair, in *Proceedings of the 2018 14th International Wireless Communications & Mobile Computing Conference (IWCMC)*, Limassol, Cyprus, 25–29 June 2018, pp. 1249–1254
38. Y.K. Lee, J.M. Lim, K.S. Eu, Y.H. Goh, Y. Tew, Real time image processing based obstacle avoidance and navigation system for autonomous wheelchair application, in *Proceedings of the Asia-Pacific Signal and Information Processing Association Annual Summit and Conference (APSIPA ASC)*, Kuala Lumpur, Malaysia, 12–15 December 2017, pp. 380–385
39. Y. Fu, J. Liu, System design for wearable blood oxygen saturation and pulse measurement device. Proc. Manuf. **3**, 1187–1194 (2015)
40. Y. Xie, Y. Gao, Y. Li, Y. Lu, W. Li, Development of wearable pulse oximeter based on Internet of Things and signal processing techniques, in *Proceedings of the European Modelling Symposium (EMS)*, Manchester, UK, 20–21 November 2017, pp. 249–254
41. N. Shabana, G. Velmathi, Advanced tele-surgery with IoT approach, in *Intelligent Embedded Systems*, (Springer, Berlin/Heidelberg, 2018), pp. 17–24
42. Z. Yang, Q. Zhou, L. Lei, K. Zheng, W. Xiang, An IoT-cloud based wearable ECG monitoring system for smart healthcare. J. Med. Syst. **40**, 286 (2016)
43. U. Satija, B. Ramkumar, M.S. Manikandan, Real-time signal quality-aware ECG telemetry system for IoT-based health care monitoring. IEEE Internet Things J. **4**, 815–823 (2017)
44. C. Beach, S. Krachunov, J. Pope, X. Fafoutis, R.J. Piechocki, I. Craddock, A.J. Casson, An ultra low power personalizable wrist worn ECG monitor integrated with IoT infrastructure. IEEE Access **6**, 44010–44021 (2018)
45. D. Sobya, S. Muruganandham, S. Nallusamy, P. Chakraborty, Wireless ECG monitoring system using IoT based signal conditioning module for real time signal acquisition. Indian J. Public Health Res. Dev. **9**, 294–299 (2018)
46. J. He, J. Rong, L. Sun, H. Wang, Y. Zhang, J. Ma, D-ECG: a dynamic framework for cardiac arrhythmia detection from IoT-Based ECGs, in *Proceedings of the International Conference on Web Information Systems Engineering*, Dubai, UAE, 12–15 November 2018, pp. 85–99
47. M. Bansal, B. Gandhi, IoT based smart health care system using CNT electrodes (for continuous ECG monitoring), in *Proceedings of the 2017 International Conference on Computing, Communication and Automation (ICCCA)*, Greater Noida, India, 5–6 May 2017, pp. 1324–1329

48. Q. Xin, J. Wu, A novel wearable device for continuous, non-invasion blood pressure measurement. Comput. Biol. Chem. **69**, 134–137 (2017)
49. P.C.P. Chao, T.Y. Tu, Using the time-domain characterization for estimation continuous blood pressure via neural network method, in *ASME 2017 Conference on Information Storage and Processing Systems collocated with the ASME 2017 Conference on Information Storage and Processing Systems*, San Francisco, CA, USA, 29–30 August 2017, p. V001T02A003
50. A. Dinh, L. Luu, T. Cao, Blood pressure measurement using finger ECG and Photoplethysmogram for IoT, in *Proceedings of the International Conference on the Development of Biomedical Engineering in Vietnam*, Ho Chi Minh, Vietnam, 27–29 June 2017, pp. 83–89
51. M. Huang, T. Tamura, Z. Tang, W. Chen, S. Kanaya, A wearable thermometry for core body temperature measurement and its experimental verification. IEEE J. Biomed. Health Inform. **21**, 708–714 (2017)
52. Q. Li, L.N. Zhang, X.M. Tao, X. Ding, Review of flexible temperature sensing networks for wearable physiological monitoring. Adv. Healthc. Mater. **6**, 1601371 (2017)
53. H. Ota, M. Chao, Y. Gao, E. Wu, L.C. Tai, K. Chen, Y. Matsuoka, K. Iwai, H.M. Fahad, W. Gao, et al., 3d printed "earable" smart devices for real-time detection of core body temperature. ACS Sens. **2**, 990–997 (2017)
54. T.G. AL-Jaf, E.H. Al-Hemiary, Internet of Things based cloud smart monitoring for asthma patients, in *Proceedings of the 1st International Conference on Information Technology (ICoIT'17)*, Erbil, Iraq, 10 April 2017, p. 380
55. A. Raji, P.K. Devi, P.G. Jeyaseeli, N. Balaganesh, Respiratory monitoring system for asthma patients based on IoT, in *Proceedings of the 2016 Online International Conference on Green Engineering and Technologies (IC-GET)*, Coimbatore, India, 19 November 2016, pp. 1–6
56. K. Subhash, P. Pournami, P.K. Joseph, Census transform based feature extraction of EMG signals for neuromuscular disease classification, in *Proceedings of the 2017 IEEE 15th Student Conference on Research and Development (SCOReD)*, Putrajaya, Malaysia, 13–14 December 2017, pp. 499–503
57. A. Subasi, E. Yaman, Y. Somaily, H.A. Alynabawi, F. Alobaidi, S. Altheibani, Automated EMG signal classification for diagnosis of neuromuscular disorders using DWT and bagging. Proc. Comput. Sci. **140**, 230–237 (2018)
58. T.N. Gia, M. Ali, I.B. Dhaou, A.M. Rahmani, T. Westerlund, P. Liljeberg, H. Tenhunen, IoT-based continuous glucose monitoring system: a feasibility study. Proc. Comput. Sci. **109**, 327–334 (2017)
59. S. Sunny, S.S. Kumar, Optical based non invasive glucometer with IoT, in *Proceedings of the 2018 International Conference on Power, Signals, Control and Computation (EPSCICON)*, Thrissur, India, 6–10 January 2018, pp. 1–3
60. S.N. Mohammad, R. Singh, A. Dey, S.J. Ahmad, ESMBCRT: enhance security to MANETs against black hole attack using MCR technique, in *Innovations in Electronics and Communication Engineering*, (Springer, Berlin/Heidelberg, 2019), pp. 319–326
61. V. Kumar, R. Kumar, An adaptive approach for detection of blackhole attack in mobile ad hoc network. Proc. Comput. Sci. **48**, 472–479 (2015)
62. S. Gurung, S. Chauhan, A novel approach for mitigating gray hole attack in MANET. Wirel. Netw **24**, 565–579 (2018)
63. N. Schweitzer, A. Stulman, R.D. Margalit, A. Shabtai, Contradiction based gray-hole attack minimization for ad hoc networks. IEEE Trans. Mob. Comput. **16**, 2174–2183 (2017) [CrossRef]
64. V. Adat, A. Dahiya, B. Gupta, Economic incentive based solution against distributed denial of service attacks for IoT customers, in *Proceedings of the 2018 IEEE International Conference on Consumer Electronics (ICCE)*, Las Vegas, NV, USA, 12–14 January 2018, pp. 1–5
65. Q. Chen, H. Chen, Y. Cai, Y. Zhang, X. Huang, Denial of service attack on IoT system, in *Proceedings of the 2018 9th International Conference on Information Technology in Medicine and Education (ITME)*, Hangzhou, China, 19–21 October 2018, pp. 755–758

66. H.B. Salameh, S. Almajali, M. Ayyash, H. Elgala, Securing delay-sensitive cognitive radio IoT communications under reactive jamming attacks: spectrum assignment perspective, in *Proceedings of the 2018 Fifth International Conference on Software Defined Systems (SDS)*, Barcelona, Spain, 23–26 April 2018, pp. 20–24
67. N. Namvar, W. Saad, N. Bahadori, B. Kelley, Jamming in the internet of things: a game-theoretic perspective, in *Proceedings of the Global Communications Conference (GLOBECOM), 2016 IEEE*, Washington, DC, USA, 4–8 December 2016, pp. 1–6
68. P. Pongle, G. Chavan, Real time intrusion and wormhole attack detection in internet of things. Int. J. Comput. Appl. **121**, 1–9 (2015)
69. D. Giri, S. Borah, R. Pradhan, Approaches and measures to detect wormhole attack in wireless sensor networks: a survey, in *Advances in Communication, Devices and Networking*, (Springer, Berlin/Heidelberg, 2018), pp. 855–864
70. A.K. Mishra, A.K. Tripathy, D. Puthal, L.T. Yang, Analytical model for Sybil attack phases in the internet of things. IEEE Internet Things J. **6**, 379–387 (2019)
71. M. Jamshidi, E. Zangeneh, M. Esnaashari, A.M. Darwesh, M.R. Meybodi, A novel model of Sybil attack in cluster-based wireless sensor networks and propose a distributed algorithm to defend it. Wirel. Pers. Commun. **105**, 145–173 (2019)
72. A. Abdallah, X.S. Shen, Efficient prevention technique for false data injection attack in smart grid, in *Proceedings of the 2016 IEEE International Conference on Communications (ICC)*, Kuala Lumpur, Malaysia, 22–27 May 2016, pp. 1–6
73. R.K. Gill, M. Sachdeva, Detection of hello flood attack on LEACH in wireless sensor networks, in *Next-Generation Networks*, (Springer, Berlin/Heidelberg, 2018), pp. 377–387
74. T. Bhatia, A. Verma, G. Sharma, S. Bala, A novel defense scheme against flooding attack in mobile adhoc networks. Recent Patents Eng. **12**, 15–22 (2018)
75. T. Amah, M. Kamat, K. Bakar, S. Rahman, M. Mohammed, A. Abali, W. Moreira, A. Oliveira-Jr, The impact of message replication on the performance of opportunistic networks for sensed data collection. Information **8**, 143 (2017)
76. S.P. Singh, S. Sharma, Secure clustering protocols in wireless sensor networks. J. Wirel. Sens. Netw. **3**, 1–10 (2016)
77. C. Cervantes, D. Poplade, M. Nogueira, A. Santos, Detection of sinkhole attacks for supporting secure routing on 6LoWPAN for Internet of Things, in *Proceedings of the 2015 IFIP/IEEE International Symposium on Integrated Network Management (IM)*, Ottawa, ON, Canada, 11–15 May 2015, pp. 606–611
78. E. Karafiloski, A. Mishev Blockchain solutions for big data challenges: a literature review, in *Proceedings of the IEEE EUROCON 2017—17th International Conference on Smart Technologies*, Ohrid, Macedonia, 6–8 July 2017, pp. 763–768
79. T. Ahram, A. Sargolzaei, S. Sargolzaei, J. Daniels, B. Amaba, Blockchain technology innovations, in *Proceedings of the 2017 IEEE Technology Engineering Management Conference (TEMSCON)*, Santa Clara, CA, USA, 8 June 2017, pp. 137–141
80. M. Conoscenti, A. Vetro, J.C.D. Martin, Blockchain for the Internet of Things: a systematic literature review, in *Proceedings of the 2016 IEEE/ACS 13th International Conference of Computer Systems and Applications (AICCSA)*, Agadir, Morocco, 29 November–2 December 2016, pp. 1–6
81. M. Mettler, Blockchain technology in healthcare: the revolution starts here, in *Proceedings of the 2016 IEEE 18th International Conference on e-Health Networking, Applications and Services (Healthcom)*, Munich, Germany, 14–17 September 2016, pp. 1–3
82. P. Zhang, D.C. Schmidt, J. White, G. Lenz, Blockchain technology use cases in healthcare, in *Advances in Computers*, (Elsevier, Amsterdam, 2018)
83. T.T. Kuo, H.E. Kim, L. Ohno-Machado, Blockchain distributed ledger technologies for biomedical and health care applications. J. Am. Med. Inform. Assoc. **24**, 1211–1220 (2017)
84. S. Angraal, H.M. Krumholz, W.L. Schulz, Blockchain technology: applications in health care. Circ. Cardiovasc. Qual. Outcomes **10**, e003800 (2017)
85. M. Engelhardt, Hitching healthcare to the chain: an introduction to Blockchain technology in the healthcare sector. Technol. Innov. Manag. Rev. **7**, 22–34 (2017)

86. Z. Alhadhrami, S. Alghfeli, M. Alghfeli, J.A. Abedlla, K. Shuaib, Introducing blockchains for healthcare, in *Proceedings of the 2017 International Conference on Electrical and Computing Technologies and Applications (ICECTA)*, Ras Al Khaimah, UAE, 21–23 November 2017, pp. 1–4
87. A. Roehrs, C.A. da Costa, R. da Rosa Righi, R. Alex, C.A. Costa, R.R. Righi, OmniPHR: a distributed architecture model to integrate personal health records. J. Biomed. Inform. **71**, 70–81 (2017)
88. M. Borge, E. Kokoris-Kogias, P. Jovanovic, L. Gasser, N. Gailly, B. Ford. Proof-of-personhood: redemocratizing permissionless cryptocurrencies, in *Proceedings of the 2017 IEEE European Symposium on Security and Privacy Workshops (EuroS&PW)*, Paris, France, 26–28 April 2017, pp. 23–26
89. European Coordination Committee of the Radiological. Blockchain in Healthcare; Technical report; European Coordination Committee of the Radiological: Brussels, Belgium, 2017
90. R. Guo, H. Shi, Q. Zhao, D. Zheng, Secure attribute-based signature scheme with multiple authorities for Blockchain in electronic health records systems. IEEE Access **6**, 11676–11686 (2018)
91. H. Zhao, Y. Zhang, Y. Peng, R. Xu, Lightweight backup and efficient recovery scheme for health Blockchain keys, in *Proceedings of the 2017 IEEE 13th International Symposium on Autonomous Decentralized System (ISADS)*, Bangkok, Thailand, 22–24 March 2017, pp. 229–234
92. F. Angeletti, I. Chatzigiannakis, A. Vitaletti, The role of blockchain and IoT in recruiting participants for digital clinical trials, in *Proceedings of the 2017 25th International Conference on Software, Telecommunications and Computer Networks (SoftCOM)*, Split, Croatia, 21–23 September 2017, pp. 1–5
93. A. Dubovitskaya, Z. Xu, S. Ryu, M. Schumacher, F. Wang, Secure and trustable electronic medical records sharing using Blockchain. AMIA Annu. Symp. Proc. **2017**, 650–659 (2017)
94. G.G. Dagher, J. Mohler, M. Milojkovic, P.B. Marella, Ancile: privacy-preserving framework for access control and interoperability of electronic health records using blockchain technology. Sustain. Cities Soc. **39**, 283–297 (2018)
95. C. Mcfarlane, M. Beer, J. Brown, N. Prendergast, *Patientory: A Healthcare Peer-to-Peer EMR Storage Network v1.1* (Entrust Inc, Addison, 2017)

Privacy-Preserving k-Means Clustering over Blockchain-Based Encrypted IoMT Data

Rakib Ul Haque, A. S. M. Touhidul Hasan, Tasnia Nishat, and Md Akhtaruzzaman Adnan

1 Introduction

In the modern healthcare sector, a rapidly emerging technology is the Internet of Medical Things (IoMT) [1], which is based on enormous volumes of data being continuously collected from health monitoring devices. Software applications and medical devices are combined for providing health services and building health care systems [2]. For which it is possible to create a wave of stand-alone devices for remote patient monitoring [3]. The union of the Internet-connected health devices with patient information and sensor-based tools such as wearables have eventually established the ecosystem of IoMT [4]. IoMT can generate, store, investigate, or transfer medical data or images to healthcare service provider's networks and retain data to either an internal database or cloud repository [5]. This connection within

R. U. Haque
School of Computer Science & Technology, University of Chinese Academy of Sciences, Beijing, China

Institute of Automation Research and Engineering, Dhaka, Bangladesh
e-mail: rakibulhaqueraj@mails.ucas.ac.cn

A. S. M. T. Hasan (✉)
Department of Computer Science and Engineering, University of Asia Pacific, Dhaka, Bangladesh

Institute of Automation Research and Engineering, Dhaka, Bangladesh
e-mail: touhid@uap-bd.edu

T. Nishat
Daraz Bangladesh Ltd, Dhaka, Bangladesh
e-mail: nishat.tasnia@daraz.com.bd

Md. A. Adnan
Department of Computer Science and Engineering, University of Asia Pacific, Dhaka, Bangladesh
e-mail: adnan.cse@uap-bd.edu

Y. Maleh et al. (eds.), *Advances in Blockchain Technology for Cyber Physical Systems*, Internet of Things, https://doi.org/10.1007/978-3-030-93646-4_5

healthcare devices and sensors is the streamline of medical workflow administration and directing to the overall improvement in medical care. IoMT devices are demonstrating higher accuracy in diagnosis, ensuring fewer errors but are cheaper in terms of costs. Nowadays, diversified Machine Learning (ML) [6, 7] algorithms are used to train these vast amounts of data to build the prediction model.

The dataset required for training various ML models generally comes from entities like health care service providers or hospitals. For ML model classifiers to produce predictive results with higher accuracy, it is essential to have data sample distinctiveness along with data amounts [8]. This is effectively achieved by using a process of unifying different sample sets gathered from various entities. But many entities often disapprove to provide their datasets for training as there are many privacy concerns. The most common issues are regarding ownership, data integrity, and data privacy [9]. During training phases, medical data from IoMT devices are handled by other associates or can be manipulated causing loss of privacy of sensitive and private data. Unauthorized data modification by altering or tampering done by dormant invaders during data sharing can lower the data integrity, resulting in a faulty classification of the ML model. The ownership authority of data providers may be lost during replicating or reusing the shared datasets by many associates.

Data privacy issues are handled in the past by using cryptographic and differential privacy (DP) [10]. While cryptographic methodologies are heavy and time-consuming, DP does not ensure data utility. In order to make sure data utility this work focuses on cryptographic methods [11]. Recent works on cryptography ML methods are expensive in terms of space and time-consuming. Among ML methods k-mean is one of the most simple and lightweight unsupervised algorithms, but past work on k-mean does make sure all privacy concerns. Those methods lack data authenticity.

In order to solve these issues, this study proposes secure k-means. A partially Homomorphic known as Paillier is applied with Blockchain technology. IoMT data of each data owner is encrypted using Paillier and then recorded on a distributed ledger. The secure building blocks are developed in order to handle the classification tasks with encrypted data, i.e., Secure Polynomial Operation (SPO) (addition/subtraction), and Secure Comparison (SC). No trusted third party is needed. The secure k-means can employ k-means classifiers with the loss of insignificant accuracy and faster than other cryptographic methods.

The rest of the paper is articulated as follows. Sections 2 and 3 describe related work and preliminaries, respectively. System overview and Model constructions are outlined in Sects. 4 and 5. Section 6 represents experiment and result evaluation. Finally, this paper is concluded in Sect. 7.

2 Related Work

Numerous research focused on privacy issues with various methods, i.e., cryptographic [12], differential privacy [13], and data publishing in privacy-preserving manner [14–17], where cryptographic and differential privacy is time-consuming and provides less data utility. On the contrary, ML training is not considered in privacy-preserving data publishing. In addition, these methods cannot keep data owner and data analyst information at the time of data sharing. Recent solutions to protect the data owner's privacy when training ML techniques are secure support vector machine (SVM) [18], secure k-nearest neighbor (k-nn) [19, 20], and secure linear regression (LR) [21]. All these methods consolidate Blockchain to keep the information related to any transaction of the data owner and the data analyst into ML training with encrypted IoMT data. These methods achieve the most proximal correctness to standard SVM, k-nn, and LR. However, secure SVM, secure k-nn, and secure LR need several comparisons and calculations that result in higher space and time complexity for analyzing the health data.

Previous research on secure k-means mainly focused on any specific domain [22], some do not consider Blockchain [23] and only a few sets of research utilizing Blockchain [24–28]. None of them are versatile as they are based on a specific setup and address all the privacy concerns related to data integrity, authenticity, and privacy. In this study, a Cryptosystem, which is partially Homomorphic (Paillier) is applied along with Blockchain technology to address the above concerns when employing k-means classifiers with IoMT data of the owners. Paillier is employed in order to encrypt IoMT data of various data owners. An immutable distributed ledger is used to record all transactions. Secure k-means can be employed by the data analysts after associating with the individual data owner in order to obtain encrypted data. No participant can infer the original data of other participants from the Blockchain as only the hash of the transaction is saved in the ledger. SPO and SC are employed as the secure protocol for polynomial calculation of encrypted data in k-means. These are also noted as a secure building block, where a trusted third party is not necessary. The proposed method achieves higher efficiency with minimal loss of accuracy.

3 Preliminaries

In this section, all background technologies and mathematical equations are presented. Dataset D has m records, where x_i and y_i are the ith attributes and after classification they get label l_i. The distance d represents the interval between two points and the k-means' model parameters are $\sum_{i=1}^{k} (c_{x_i}, c_{y_i})$. The symbols P stands for data owner, and A stands for data analyst, respectively. The encrypted messages under Paillier are represented as $[[m]]$.

Homomorphic Cryptosystem

Three methods combinedly develop Cryptosystems: key generation $(KeyGen)$, encryption of data (Enc), decryption of data (Dec). $(PK; SK)$ are pair of keys known as (public key; private key) used in public-key cryptosystems. These key pairs are used for encryption and decryption. A cryptosystem can be Homomorphic if and only if its feature can map the calculation over ciphertext to the respective plaintext without knowing the decryption key. In the proposed model, polynomial operations (secure mathematical addition and subtraction) are operated based on Paillier. p, q, and N are n-bit prime numbers, where, $N = p\ q$. The public and private keys are denoted by N and $(N,\ \phi(N))$.[1] The encryption function of Paillier is $c := [[(1+N)^m\ r^N\ mod\ N^2]]$, and decryption function is $m \in Z_N$ and $m := [[\frac{[c^{\phi(N)}\ mod\ N^2-1]}{N} \times \phi(N)^{-1}\ mod\ N]]$. Paillier is elaborately discussed in [11].

Blockchain

Blockchain is a continuously expanding list of transactions, known as blocks connected and protected utilizing cryptography [29]. In order to avoid the single point of failure, a Peer-to-Peer (P2P) architecture is adopted in Blockchain. The consensus mechanism ensures robust unambiguous control of blocks and transactions. It also assures the consistency and integrity across distributed nodes of the Blockchain, i.e., Auditability, Integrity, and Decentralization.

- Public Blockchain (Bitcoin and Ethereum).
- Consortium Blockchain (Hyperledger, Ripple).
- Private Blockchain. Blockchain labors as the stage to be hosted and executed on for smart contracts.

k-Means Algorithm

k-Means [30] is an unsupervised ML algorithm mainly utilized for the classification task. In order to identify centroids (c_{x_j}, c_{y_j}) calculation of distance d is necessary. Popular methods to identify distances are $Euclidean\ d_e$ (Eq. (1)), $Manhattan\ d_m$ (Eq. (2)), etc. In this study, we will use $Manhattan\ d_m$. Let, $(x_1,\ y_1), (x_2,\ y_2), \ldots, (x_n,\ y_n) \in D$.

---

[1] Let, an integer be $N > 1$. Then, multiplication modulo N over abelian group Z_N^*. Define $\phi(N)$ def $|Z_N^*|$ as the order of the group Z_N^*.

$$d_e = \sqrt{(c_{x_j} - x_i)^2 + (c_{y_j} - y_i)^2} \tag{1}$$

$$d_m = |(c_{x_j} - x_i)| + |(c_{y_j} - y_i)| \tag{2}$$

4 System Overview

This section discusses models related to the system, threat, and security definitions.

System Model

The objective of the proposed model is to make sure privacy and guarantee protected data sharing between A and P. A gets encrypted IoMT data from each P. All the shared data are recorded in a distributed ledger of Blockchain by forming transactions in order to keep authenticity. A assembles its Ml model (k-means) by acquiring recorded data from the public ledger of Blockchain. A erects a protected method utilizing protected building blocks (SC and SPO). At the time of employing the secure ML model, it is important to have moderate interaction between A and P in order to share intermediate results. When sharing the intermediate data for comparison, a tiny amount of bias δ is added by P. It reduces the possibility of model inversion attacks. It also diminishes the algorithm's space and time complexity, and the performance of the model is not affected. Its goal is to make sure the privacy of the data owner at the time of SC. The entire process is illustrated in Fig. 1.

- **IoMT Devices:** These are accountable for collecting and transferring important IoMT data by the wireless medium.
- **Data Owners P:** It gathers every part of data from the IoMT devices.
- **Data Analyst A:** It wants its ML model to be trained on the dataset of various P.

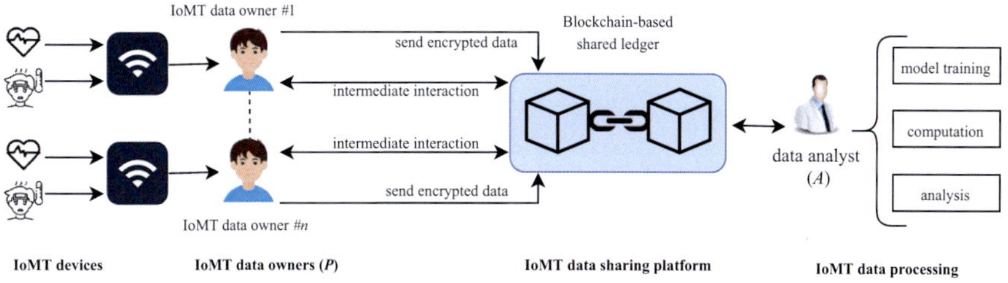

Fig. 1 Data driven IoMT ecosystem

The considered scenario has n number of $P := P_i$ $(i \in (1, \ldots, n))$ with dataset D_i consisting of sensitive information and an untrusted data analyst A. Horizontal data sharing [31] method is utilized for n datasets $[D_i]_{i=1}^{n}$ with alike feature space but distinct in sample sets. A collects n encrypted data sequentially and k-means model is applied on the dataset $D := (D_1\ U \ldots .U\ D_n)$, where, $|D| = \sum_{i=1}^{n} |D_i|$. A can obtain its model after the execution of the secure protocol π.

Security Goals The protocols π meets the points discussed subsequently.

- A will not be able to infer any sensitive information of P from D.
- P failed to learn A's private information.
- P failed in acquiring the private information of another P.

Threat Model

All the participants are acknowledged as adversaries, who are honest but curious (semi-honest) and they do not trust each other. A is fair in obeying protocol π and also interested in the raw data of other participants. A also tries to infer further information of P from the shared intermediate data. On the other hand, P might infer in A's private information. Following threats are considered:

- Encrypted information registered in the blockchain is hardly accessible to A but can record the intermediate results (iteration steps) at the time of data sharing.
- A is considered to be aware of the details, which can be extracted from the public encryption model. A is expected to plot with diverse P to acquire another P's private information.

Encrypted Data Sharing via Blockchain

All alike instances of data are allocated to the corresponding feature vectors, and all of them are pre-processed locally. Input and output are the two domains, which are developed for transaction formation. The input field consists of:

- Sender's address
- Encrypted data
- Name of IoMT device (Source)

The output field (corresponding) consists of:

- Receiver's address
- Encrypted data
- Name of IoMT device (Source)

This study employs the proposed model in the Hyperledger Fabric platform. It is known as a permissioned blockchain platform. Sender and receiver's addresses are hash values. Paillier determines the encrypted data. Only the hash of the transactions is recorded in the Blockchain. The length of them and private key are 128 bytes. The segment length is 4 bytes for the type of IoMT device. The sender node assembles the transaction and broadcasts it in the Peer to Peer (P2P) [32] system of the Blockchain network. The operation's correctness is validated by miner nodes. The transaction is packaged into a block by a specific miner node. Each block may record various transactions. Common protocols for consensus mechanisms are used such as Proof of Work (PoW) or Byzantine Fault Tolerance [33, 34].

Security Definitions

Secure Two-party Computation [12, 35, 36] and Modular Sequential Composition [37] are employed in order to assemble the secure building blocks for deploying the protocols of privacy-preserving ML in a modular design.

Secure Two-Party Computation To guarantee the security for two-party protocols, it is important to confirm that X (Y) can be calculated from its interactions with Y (X), which is also calculated from the input and output. Ultimately, it points toward secure two-party computation [12, 35, 36]. Let a probabilistic polynomial function be $G = (f_X, f_Y)$ and G be computed by protocol π. X's and Y's inputs are x and y and X and Y compute $G(x, y)$. X's view at the time of executing π is the tuple $view_X^{\pi}(x, y) = (x, c, a_1, b_2, \ldots, a_b)$ where $a_1, a_2, \ldots, a_b$ are the messages received from Y. X's random tape is c. The view of Y is defined similarly.

Modular Sequential Composition Modular Sequential Composition [37] is employed for confirming the protection proofs of protocols.

Definition 1 (Modular Sequential Composition [37]) Let $g_1, \ldots, g_b$ be two-party probabilistic polynomial-time functionalities , which is securely calculated by $\rho_1, \ldots, \rho_b$ in semi-honest adversaries' presence. Let G be a probabilistic polynomial-time functionality and π a protocol that securely computes G with $g_1, \ldots, g_b$ in the presence of semi-honest adversaries. Then G is securely computed by $\pi^{\rho_1,\ldots,\rho_b}$ in semi-honest adversaries' presence.

5 Model Construction

The construction details of the proposed system are presented in this section. The aim is to secure the privacy of distinct P and A at the time of k-means classification.

Secure Polynomial Operations (SPO)

Secure addition and subtraction are developed to develop the secure k-means based on Paillier. It can ensure reliability at the time of addition and subtraction on encrypted data. Additional homomorphic property of Paillier is: $[[ma_1 + ma_2]] = [[ma_1]] \times [[ma_2]]\ mod\ n^2$ and subtraction is : $[[ma_1 - ma_2]] = [[ma_1]] \times [[ma_2]]^{-1}\ mod\ n^2$.[2] Secure addition and subtraction are statistically indistinguishable as Paillier is alike [11].

Secure Comparison (SC)

It aims at a privacy-preserving comparison of encrypted numbers. Protocol π based upon which A and B participant in SC to compare $[[m_1]]$ and $[[m_2]]$. None of the participants can know original m_1 and m_2. SC is illustrated in Algorithm 1.

Proposition 1 (Sanctuary of SC) *Algorithm 1 is protected in a semi-honest scenario.*

Proof of Proposition 1 Two entities (P and A) are involved in Algorithm 1.

The view of P is:

$$view_P^{\pi} = ([[m_1^{'}]]_A,\ [[m_2^{'}]]_A,\ PK_A)$$

Algorithm 1: Secure comparison

P' **Input**: $D = [\ m_1,\ m_2\ ]$
A' **Input**: $PK_A,\ SK_A$
P' **Output**: $flag$
P compute $[[m_1^{'}]],\ [[m_2^{'}]]$ by SPO as $[[m_1 + \delta]],\ [[m_2 + \delta]]$;
P send $[[m_1^{'}]],\ [[m_2^{'}]]$ to A;
A decrypt and compare $[[m_1^{'}]]$ and $[[m_2^{'}]]$;
if $[[m_2^{'}]] \leq [[m_1^{'}]]$ **then**
  | P gets $flag = 0$ from A;
end
else
  | P gets $flag = 1$ from A;
end

[2] The modular multiplicative inverse is represented as $[[ma]]^{-1}$. Based on Paillier, it can calculate $[[ma]] \times [[ma]]^{-1}\ mod\ N^2 = 1$. $\phi(N)$ function can calculate $[[m]]^{-1}$, where, $[[m]]^{-1} = [[m]]^{\phi(N)-1}$.

Consequently, the simulator S:

$$S_P^{\pi}((m_1, m_2);\ F(m_1, m_2)) = view_P^{\pi}([[m_1^{'}]]_A,\ [[m_2^{'}]]_A,\ [[\delta]]_A,\ PK_A)$$

Security of $[[m_1]]_A$ and $[[m_2]]_A$ is same as Paillier as $[[m_1^{'}]]_A$ and $[[m_2^{'}]]_A$ are encrypted by PK_A based on Paillier. Therefore, P will never be able to deduce the original $(m_1, m_2)_A$. A's view:

$$view_A^{\pi} = ([[m_1^{'}]]_A,\ [[m_2^{'}]]_A,\ PK_A, SK_A)$$

Then, S_A^{π} runs as follows:

$$F(m_1^{'}, m_2^{'}) = view_A^{\pi}\ (m_1^{'},\ m_2^{'},\ PK_A,\ SK_A)$$

Any attempt of A to infer original m_1 and m_2 from $m_1^{'}$ and $m_2^{'}$ will fail as A is not knowledgeable of δ. A returns 0 or 1 depending on the case $m_1^{'} \geq m_2^{'}$ or $m_1^{'} < m_2^{'}$ as A is trustworthy in following protocols. □

Training Algorithm of Secure k-Means

Protocols are employed for secure k-means classification, where all participants' parameters are protected. This study assumes one A and n amount of P. Secure k-means' protocol $protocol_{\pi}$ is specified in Algorithm 2. In Algorithm 2, all participants (A and P) parameters are secret. At the time of facing any semi-honest collusions, no participant will lose data privacy from intermediate results of the algorithm.

Algorithm 2: Proposed $protocol_{\pi}$

1 P' **Input**: $D = [(x_1,\ y_1),\ \ldots\ (x_m,\ y_m)]$
2 A' **Input**: $PK_A,\ SK_A,\ D_{cen} = [(c_{x_1},\ c_{y_1}),\ \ldots\ (c_{x_n},\ c_{y_n})]$
3 A' **Output**: $D_l = [(l_1,\ x_1,\ y_1),\ \ldots\ (l_m,\ x_m,\ y_m)]$
4 A send $[[D_{cen}]]_{PK_A}$ to P;
5 P compute $[[d_m]]_{PK_A}$ by SPO;
6 P identify minimum $[[d_m]]_{PK_A}$ by SC;
7 P compute $[[data_{i=1}^{n}]]_{PK_A}$ by SPO;
8 $[[data_{i=1}^{n}]]_{PK_A} = [[\ [\sum(c_{x_1}, c_{y_1}\, withnearest(x, y))]_{i=1}^{n}\ ]]_{PK_A}$
9 P send $[[data_{i=1}^{n}]]_{PK_A}$ and $[[length_{i=1}^{n}]]_{PK_A}$ to A;
10 A decrypt $[[data_{i=1}^{n}]]_{PK_A}$ and $[[length_{i=1}^{n}]]_{PK_A}$ using SK_A;
11 A compute centroids $D_{cen}^{'}$ by $[\frac{data}{length}]_{i=1}^{n}$;
12 A update centroids as $D_{cen}^{'}$;
13 Repeat from 1 to 7 until centroids remain same.

Proposition 2 (*$Protocol_\pi$*'s **Security**) *$protocol_\pi$ in Algorithm 2 is secured in the semi-honest scenario.*

Proof of Proposition 2 In $protocol_\pi$ two entities P and A are involved. Each P follows the same protocol, so security satisfaction for one P is enough to cover all P. Individual P's view is:

$$view_P^{protocol_\pi} = ([[D_{cen}]]_{PK_A},\ PK_A,\ D_P)$$

where, $[[D_{cen}]]_{PK_A}$ are encrypted by PK_A, the confidentiality of $[[D_{cen}]]_{PK_A}$ is alike to Paillier. Therefore P cannot learn $(D_{cen})_{PK_A}$.

A's view:

$$view_A^{protocol_\pi} = ([[data_{i=1}^n]]_{PK_A},\ [[length_{i=1}^n]]_{PK_A},\ d_m^{'},\ D_{cen},\ PK_A,\ SK_A)$$

Now, the confidentiality of $[[data_{i=1}^n]]_{PK_A}$ and $[[length_{i=1}^n]]_{PK_A}$ needs to be analyzed. Whether P's private D_P can be inferred from shared data by A. Clearly, $[[data_{i=1}^n]]_{PK_A}$ and $[[length_{i=1}^n]]_{PK_A}$ are not the resulted values from secret D_P. A might aim to compute D_P utilizing the distance $d_m^{'}$ and centroids D_{cen} at the time SC. D_P is added with bias δ by P and δ's exact value is unknown to A. Therefore, A will not be able to infer D_P. At the time of division, A has the summation of each centroid from $[[data_{i=1}^n]]_{PK_A}$. A also gets the exact number of points $[[length_{i=1}^n]]_{PK_A}$, which are added together. Still, A will fail to guess the exact D_P of P. Without brute force cracking, a genuine value of D_P cannot be perceived by anyone, which is not the realistic possibility to achieve[11]. So, in a scenario of semi-honest adversaries $protocol_\pi$ is secured. □

6 Experiment and Result Evaluation

Dataset and performance analysis are showed in this section.

Dataset

Three medical datasets are used namely Diabetes dataset (DD), Breast cancer Wisconsin data (BCWD), and Heart disease data (HDD) [18, 19]. BCWD and DD have 9 numeric attributes. On the other hand, HDD has 13 Discrete attributes. 80% of the dataset is used for training and 20% of the dataset is used for testing. Table 1 summarizes the utilized datasets.

Table 1 Statistics of datasets

	Datasets		
Measures	BCWD	HDD	DD
Instances	699	303	768
Attributes	9	13	9

Float Format Conversion

Cryptosystems can only operate on whole numbers. Therefore, the conversion of format is a must, and all numbers are converted into an integer. Based on the IEEE 754 global standard format (floating-point binary number) D is $D = (-1)^s \times M \times 2^E$, where s, M, and E are sign bit, significant number, and exponential bit, respectively. This study considers a key size of 1024-bit for the Paillier cryptosystem.

Evaluation Parameters

There are three most popular metrics, i.e., accuracy (3), precision (4), and recall (5).

$$accuracy = \frac{t_p + t_n}{t_p + t_n + f_p + f_n} \tag{3}$$

$$precision = \frac{t_p}{t_p + f_p} \tag{4}$$

$$recall = \frac{t_p}{t_p + f_n} \tag{5}$$

Here, the positive or relevant classes are represented as t_p. These classes are precisely labeled. The negative or irrelevant classes that are labeled correctly are represented as f_p. f_n and t_n represent the number of relevant but mislabeled and the number mislabeled but irrelevant, respectively, in the test result. Table 2 shows the outcomes.

Table 2 shows that secure k-nn achieve highest performance and followed by secure SVM and secure k-means, where k-nn with a threshold value $t = 8$ and $k = 2$ cluster for k-means. Most importantly the difference of correctness among these ML models is in the range between 1% to 4%. The proposed secure k-means achieved 78.10%, 81.88%, and 94.95% of accuracy on DD, HDD, and BCWD datasets, where the state-of-the-art technique provides 77.00%, 81.00%, and 96.60%, respectively. Therefore, secure k-means performers are approximately alike compared to conventional k-means and slightly differ from state-of-the-art secure k-nn [18].

Table 2 Summary of performance

		Datasets		
Measure	Model	BCWD	HDD	DD
Accuracy	SVM	96.60%	81.00%	77.00%
	Secure SVM	95.25%	80.89%	76.67%
	k-nn ($t = 8$)	96.96%	83.50%	79.00%
	Secure k-nn ($t = 8$)	97.80%	82.33%	78.00%
	k-Means ($k = 2$)	95.23%	82.54%	78.55%
	Secure k-Means ($k = 2$)	94.95%	81.88%	78.10%
Precision	SVM	96.16%	81.79%	75.00%
	Secure SVM	96.02%	81.25%	74.80%
	k-nn ($t = 8$)	96.54%	83.85%	77.00%
	Secure k-nn ($t = 8$)	96.26%	82.30%	76.00%
	k-Means ($k = 2$)	95.95%	82.75%	76.23%
	Secure k-Means ($k = 2$)	95.01%	81.58%	75.85%
Recall	SVM	96.48%	80.38%	71.00%
	Secure SVM	95.65%	79.65%	70.91%
	k-nn ($t = 8$)	96.85%	83.85%	75.90%
	Secure k-nn ($t = 8$)	96.67%	82.66%	75.10%
	k-Means ($k = 2$)	96.01%	82.91%	74.69%
	Secure k-means ($k = 2$)	95.87%	81.76%	74.25%

Efficiency

The scalability analysis of SPO is showed in Table 3. The proposed method consumes minimal time compared to other methods based on Table 3. Several Ps' are linearly simulated. SPO in k-means takes the 2500 s, 1000 s, 1790 s on BCWD, HDD, DD datasets, respectively, which is better than other methods. Facing diverse datasets with numerical attributes, and discrete attributes, the proposed method shows enough efficiency scalability. The scalability performance comparison is summarized in Fig. 2 and the x-axis holds the datasets (BCWD, HDD, and DD), and the y-axis holds time (seconds). It is also clear that secure k-means is more realistic.

The proposed secure k-means is efficient and practical than the state-of-the-art techniques. In our designed protocol, secure k-means need a single iteration consisting of two interactions to calculate the new clustering points, and no trusted third party is required since it employs Blockchain for secure data sharing. Therefore, the proposed method is more atomic in scalability than other methods and covers all the security and privacy features. Table 3 and Fig. 2 illustrate that secure k-means achieve the best possible computation time for all datasets. Some insignificant fluctuations exist between the proposed methods and other techniques in the case of correctness. However, Table 2 proves that secure k-means achieve almost similar performance like secure SVM and secure k-nn.

Table 3 Summary of time consumption

Dataset	Time	Methods: Secure SVM	Secure k-nn	Secure k-means
BCWD	Total	3674 s	3357 s	2200 s
	P	2789 s	2534 s	1500 s
	A	1066 s	860 s	500 s
	SPO	3462 s	3113 s	2500 s
HDD	Total	2735 s	2534 s	1500 s
	P	1761 s	1520 s	700 s
	A	924 s	765 s	300 s
	SPO	2333 s	1922 s	1000 s
DD	Total	3959 s	3709 s	2605 s
	P	3199 s	2920 s	1580 s
	A	1045 s	995 s	507 s
	SPO	3773 s	3527 s	1790 s

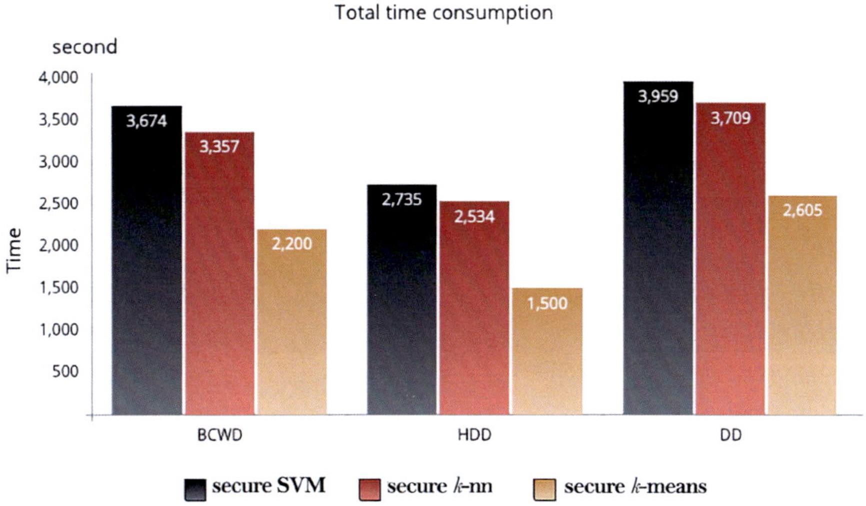

Fig. 2 Comparison of time consumption

7 Conclusion

This study introduces a secure protocol for training the k-means algorithm. It mainly focuses on data authenticity, data integrity, and data privacy issues of P. It employs Blockchain technology to record all intermediate data. A multi-party scheme is considered for training the algorithm, where involved entities are n number of P and a A. A reliable method is achieved by employing Paillier for cryptographic polynomial operations. The introduced approach encompasses approximately comparable correctness compared to the state of the arts but outperforms them in terms of time consumption. Generally, cryptographic methods are secured but take exponentially higher time than straightforward ML techniques. These methods only allow operations on integers, and there also exist some limitations in the case of division operations. Future work includes developing lightweight Cryptographic privacy-preserving ML algorithms.

Acknowledgments The authors thank the school of computer science and technology of the University of Chinese Academy of Science, Beijing, China; Institute of Automation Research and Engineering, Dhaka 1205, Bangladesh; and the Department of Computer Science and Engineering of University of Asia Pacific, Dhaka, Bangladesh for their support toward this study.

References

1. G.J. Joyia, R.M. Liaqat, A. Farooq, S. Rehman, Internet of Medical Things (IoMT): applications, benefits and future challenges in healthcare domain. J. Commun. **12**(4), 240–247 (2017)
2. C.F. Breidbach, D. Antons, T.O. Salge, Seamless service? On the role and impact of service orchestrators in human-centered service systems. J. Serv. Res. **19**(4), 458–476 (2016)
3. L.P. Malasinghe, N. Ramzan, K. Dahal, Remote patient monitoring: a comprehensive study. J. Ambient. Intell. Humaniz. Comput. **10**(1), 57–76 (2019)
4. T. Zhang, J. Lu, F. Hu, Q. Hao, Bluetooth low energy for wearable sensor-based healthcare systems, in *2014 IEEE Healthcare Innovation Conference (HIC)* (IEEE, Piscataway, 2014), pp. 251–254
5. S. Ahmad, L. Hang, D.H. Kim, Design and implementation of cloud-centric configuration repository for DIY IoT applications. Sensors **18**(2), 474 (2018)
6. A. Rasool, R. Tao, M. Kamyab, S. Hayat, GAWA–a feature selection method for hybrid sentiment classification. IEEE Access **8**, 191850–191861 (2020)
7. A. Likas, N. Vlassis, J.J. Verbeek, The global *k*-means clustering algorithm. Pattern Recogn. **36**(2), 451–461 (2003)
8. C. Sun, A. Shrivastava, S. Singh, A. Gupta, Revisiting unreasonable effectiveness of data in deep learning era, in *Proceedings of the IEEE International Conference on Computer Vision* (2017), pp. 843–852
9. P. Voigt, A. Von dem Bussche, *The EU General Data Protection Regulation (GDPR). A Practical Guide*, vol. 10, 1st edn. (Springer, Cham, 2017), p. 3152676
10. P. Mohassel, Y. Zhang, SecureML: a system for scalable privacy-preserving machine learning, in *2017 IEEE Symposium on Security and Privacy (SP)* (IEEE, Piscataway, 2017), pp. 19–38
11. J. Katz, Y. Lindell, *Introduction to Modern Cryptography* (CRC Press, Boca Raton, 2020)
12. R. Bost, R.A. Popa, S. Tu, S. Goldwasser, Machine learning classification over encrypted data, in *Proceedings of NDSS*, vol. 4324 (2015), p. 4325
13. M. Abadi, A. Chu, I. Goodfellow, H.B. McMahan, I. Mironov, K. Talwar, L. Zhang, Deep learning with differential privacy, in *Proceedings of the 2016 ACM SIGSAC Conference on Computer and Communications Security* (2016), pp. 308–318
14. A.S.M. Hasan, Q. Qu, C. Li, L. Chen, Q. Jiang, An effective privacy architecture to preserve user trajectories in reward-based LBS applications. ISPRS Int. J. Geo-Inf. **7**(2), 53 (2018)
15. A.S.M. Hasan, Q. Jiang, H. Chen, S. Wang, A new approach to privacy-preserving multiple independent data publishing. Appl. Sci. **8**(5), 783 (2018)
16. A.S.M. Hasan, Q. Jiang, C. Li, An effective grouping method for privacy-preserving bike sharing data publishing. Future Internet **9**(4), 65 (2017)
17. A.T. Hasan, Q. Jiang, J. Luo, C. Li, L. Chen, An effective value swapping method for privacy preserving data publishing. Secur. Commun. Netw. **9**(16), 3219–3228 (2016)

18. M. Shen, X. Tang, L. Zhu, X. Du, M. Guizani, Privacy-preserving support vector machine training over Blockchain-based encrypted IoT data in smart cities. IEEE Internet Things J. **6**(5), 7702–7712 (2019)
19. R.U. Haque, A.S.M. Hasan, Q. Jiang, Q. Qu, Privacy-preserving K-nearest neighbors training over blockchain-based encrypted health data. Electronics **9**(12), 2096 (2020)
20. R.U. Haque, A.S.M. Hasan, Overview of Blockchain-based privacy preserving machine learning for IoMT, in *Big Data Intelligence for Smart Applications*. Studies in Computational Intelligence (2022). eBook ISBN 978-3-030-87954-9. https://doi.org/10.1007/978-3-030-87954-9
21. R.U. Haque, A.S.M.T. Hasan, Privacy-preserving multivariant regression analysis over blockchain-based encrypted IoMT data, in *Artificial Intelligence and Blockchain for Future Cybersecurity Applications, Studies in Big Data*, vol. 90, Chap. 3, 1st edn. (Springer, Cham, 2021). https://doi.org/10.1007/978-3-030-74575-2
22. J. Sakuma, S. Kobayashi, Large-scale k-means clustering with user-centric privacy-preservation. Knowl. Inf. Syst. **25**(2), 253–279 (2010)
23. C. Saranya, G. Manikandan, A study on normalization techniques for privacy preserving data mining. Int. J. Eng. Technol. **5**(3), 2701–2704 (2013)
24. Y. Zhu, X. Li, Privacy-preserving k-means clustering with local synchronization in peer-to-peer networks. Peer-to-Peer Netw. Appl. **13**(6), 2272–2284 (2020)
25. E.M. Abou-Nassar, A.M. Iliyasu, P.M. El-Kafrawy, O.-Y. Song, A.K. Bashir, A.A. Abd El-Latif, DITrust chain: towards blockchain-based trust models for sustainable healthcare IoT systems. IEEE Access **8**, 111223–111238 (2020)
26. K. Abbas, L.A.A. Tawalbeh, A. Rafiq, A. Muthanna, I.A. Elgendy, A. El-Latif, A. Ahmed, Convergence of blockchain and IoT for secure transportation systems in smart cities. Secur. Commun. Netw. (2021). https://doi.org/10.1155/2021/5597679
27. G.N. Nguyen, N.H. Le Viet, M. Elhoseny, K. Shankar, B.B. Gupta, A.A.A. El-Latif, Secure blockchain enabled Cyber–physical systems in healthcare using deep belief network with ResNet model. J. Parallel Distrib. Comput. **153**, 150–160 (2021)
28. A.A.A. El-Latif, B. Abd-El-Atty, I. Mehmood, K. Muhammad, S.E. Venegas-Andraca, J. Peng, Quantum-inspired blockchain-based cybersecurity: securing smart edge utilities in IoT-based smart cities. Inf. Process. Manag. **58**(4), 102549 (2021)
29. S. Nakamoto, Bitcoin—a peer-to-peer electronic cash system (2008). https://bitcoin.org/bitcoin.pdf. Accessed 19 Dec 2020
30. M. Ghadiri, S. Samadi, S. Vempala, Socially fair k-means clustering, in *Proceedings of the 2021 ACM Conference on Fairness, Accountability, and Transparency* (2021), pp. 438–448
31. R. Shokri, V. Shmatikov, Privacy-preserving deep learning, in *Proceedings of the 22nd ACM SIGSAC Conference on Computer and Communications Security* (2015), pp. 1310–1321
32. A. Angrish, B. Craver, M. Hasan, B. Starly, A case study for Blockchain in manufacturing: "FabRec": a prototype for peer-to-peer network of manufacturing nodes. Procedia Manuf. **26**, 1180–1192 (2018)
33. M. Vukolić, The quest for scalable Blockchain fabric: proof-of-work vs. BFT replication, in *International Workshop on Open Problems in Network Security* (Springer, Cham, 2015), pp. 112–125
34. N. Stifter, A. Judmayer, E. Weippl, Revisiting practical byzantine fault tolerance through Blockchain technologies, in *Security and Quality in Cyber-Physical Systems Engineering* (Springer, Cham 2019), pp. 471–495
35. O. Goldreich, *Foundations of Cryptography: Volume 2, Basic Applications* (Cambridge University Press, Cambridge, 2009)
36. M.D. Cock, R. Dowsley, A.C. Nascimento, S.C. Newman, Fast, privacy preserving linear regression over distributed datasets based on pre-distributed data, in *Proceedings of the 8th ACM Workshop on Artificial Intelligence and Security* (2015), pp. 3–14
37. R. Canetti, Security and composition of multiparty cryptographic protocols. J. Cryptol. **13**(1), 143–202 (2000)

Blockchain for Smart Transport Applications

Palak Bagga and Ashok Kumar Das

1 Introduction

The world has witnessed immense evolution in transportation, in the last couple of decades. Not only has there been a devastating increase in several vehicles in the last few years, but also technology has turned traditional vehicles into smart vehicles, capable of sharing and collecting data. The smart vehicles are assigned with IP addresses to connect to Internet and the fellow vehicles on the fly. Also, many IoT sensors are installed within vehicles to gather information. Moreover, the rapid growth in the usage of vehicles, IoT devices, has increased the need for vehicle-to-vehicle communication and upgraded vehicle to infrastructure, vehicle to Internet, vehicle to pedestrian, vehicle to personal devices, and intra vehicular communications.

As a result, normal transportation has undergone an intense innovation and got transformed into smart transportation framework called Internet of Vehicles (IoV). IoV is a network of different entities, such as vehicles, pedestrians, roads, parking lots, and city infrastructure, allowing the entities to communicate by sending messages over open channels. The message contains information flown regarding the road conditions or the drivers' travel information. It might include some private details of the driver or the passenger, such as identity or biometrics, which might question privacy. Each node uses the information from neighboring nodes to provide services related to traffic management, road conditions, lost and found vehicle locating, speed control, etc.

A smart transportation system comprises intelligent vehicles, equipped with smart devices, sharpened processing capabilities, heightened communication tech-

P. Bagga · A. K. Das (✉)
Center for Security, Theory and Algorithmic Research, International Institute of Information Technology, Hyderabad, India
e-mail: palak.bagga@research.iiit.ac.in; ashok.das@iiit.ac.in

Y. Maleh et al. (eds.), *Advances in Blockchain Technology for Cyber Physical Systems*, Internet of Things, https://doi.org/10.1007/978-3-030-93646-4_6

nologies, creating an intelligent scenario to support extended services for large applications. IoV-enabled smart transportation is combined with HWSN (Heterogeneous Wireless Sensor Network) technology to assist vehicles in performing data compilation and transferring information easily and seamlessly. IoV uses 802.11p protocol and DSRC (Dedicated Short-Range Communications) [1] to support enhanced communication and handoff schemes, making the network highly secured and providing end-to-end authentication simultaneously.

IoV-enabled smart transportation [2] manages traffic; improves transportation, energy consumption, and efficiency; saves cost and time of customers; and reduces fatal occurrences saving lives. Vehicles upload traffic-related information in the data center for traffic analysis. Vehicles can even fetch information from the data center via *RSU* to take traffic-related decisions. Wrong information might lead to loss of life, time, and economy. Therefore, the authenticity of vehicles and *RSU*s is important to ensure the correctness of the information stored and preserve the customers' privacy.

Many authentication Schemes [3–9] have been suggested to ensure privacy, integrity, unlinkability, along with secure communication in an open channel. But the schemes face many issues, like central registration authority, privacy issues, long certificate mechanism, fabricated hardware problems, excess storage overheads, and large computation and communication costs. These issues make the applications of such schemes in real-time smart transportation difficult and impossible as they may lead to errors, consequences, and threats. For this reason, various blockchain-based solutions are provided recently which mitigate all the issues faced by traditional authentication mechanism and makes it efficient for smart transportation. This chapter focuses on efficient blockchain-based authentication protocols and has less computation and communication costs with extra security and functionality features.

The Architecture of IoV-Enabled Smart Transport System

Figure 1 represents the architecture of the IoV-enabled smart transportation system. Smart or Intelligent Transportation System (ITS) consists of many vehicles (V_i, $i = 1, 2, 3, \ldots, n$) and roadside units (RSU_j, $j = 1, 2, 3, \ldots, n_r$). Vehicles collect sensitive information about their surroundings via installed multiple sensors. IoT devices like mass airflow sensor, engine speed sensor, oxygen sensor, spark knock sensor, coolant sensor, Global Positioning System (GPS), forward and rear sensors, speed sensor, smart card device, and fingerprint device collect information via *onboard unit (OBU)* which is placed in the vehicle. The collected information is stored in the vehicle's *tamper-proof device (TPD)*.

The *RSU*s are installed at fixed places and are responsible for managing instantaneous vehicles in their scope. They also regulate the flow of information among the entities within their zones. A smart transportation system also has a trusted authority (*TA*), which is fully trusted and is used to register vehicles and *RSUs*. Some architectures might also deploy many *TAs* to decrease the latency and

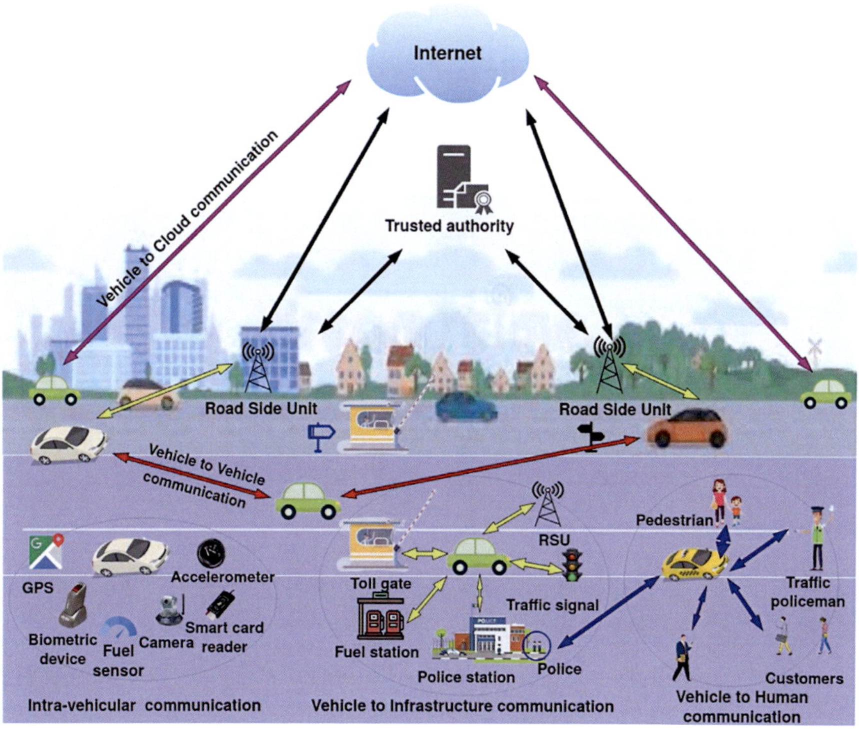

Fig. 1 IoV-enabled smart transport system

load of one TA. With multi *TA* architecture, a cross *TA* authentication problem occurs, when a vehicle goes from one *TA* zone to another [10]. *RSU* and *TA* generally have dedicated connections that are secured. Still, the communication between the vehicles and *RSU* happens over vulnerable, open channels using the IEEE 802.11p protocol and Dedicated Short-Range Communications (DSRC) [1].

The new advancements allowed intelligent vehicles to communicate with other vehicles and collaborated vehicles with infrastructure and the Internet by exchanging messages. With the increased population and boost in number of vehicles, IoV has become one of the most stretched incentives in today's world [11].

A smart transportation system allows vehicles to communicate to other entities via many communication systems, as shown in Fig. 1 [12].

- *Intra-Vehicle system*: Intra-vehicle communication allows a vehicle to communicate with pre-installed IoT devices and sensors like camera, smart card reader, and fuel sensor. A vehicle collects the information from within and uses that to broadcast it to other vehicles or *RSU*.
- *Vehicle-to-Vehicle (V2V) system*: Under V2V system, every vehicle is allowed to communicate with other fellow vehicles in its communication range. This data is used to form clusters or give traffic updates to *RSU*.

- *Vehicle-to-Infrastructure (V2I) system*: Smart transportation comprises infrastructure capable of communicating with vehicles. The communication between a vehicle and infrastructure (e.g., Parking station, fuel station, police station roadside units (RSUs)) is performed under V2I communication system.
- *Vehicle-to-Cloud (V2C) system*: The V2C system allows vehicles to interact with the cloud directly. This might help vehicles to store and retrieve sensitive data securely.
- *Vehicle-to-Sensor (V2S) system*: The V2S system provides access to the data gathered from the sensors installed in the vehicles.
- *Vehicle-to-Human (V2H) system*: The V2H system allows the users like drivers, pedestrians, traffic policeman, cab customers, and cyclists to communicate with vehicles. This increases the awareness and spontaneity of the application.

Applications of IoV-Enabled Smart Transport System

With the rapid increase in the number of vehicles, urban cities and roads are scummed under the pressure of the growing population. As a result, managing and controlling traffic is one of the biggest and vital applications of ITS. Smart transportation has improved the quality of transportation and has added luxury to the on-road experience. It provides various services to the users on the road and meanwhile, it also promotes safety applications to take care of users' (drivers and customers) safety and privacy [13–16].

- *Safety applications:* Accident-prone areas notification, alert turn notification, pedestrian crossing notification, overtaking vehicle notification, collision avoidance application.
- *Real-time applications of smart transportation:* Some real-time applications are automatic road administration, emergency vehicle notification, notification when the wrong way is taken, condition of the traffic on the road, notification on breaking a signal, alerts on the speed limit, collection of tolls digitally.
- *Driver assistance applications:* Fuel limit warnings, mileage notification, door and window open signaling, seat belt sign, general Internet access, parking management, GPS navigation signaling, nearby fuel station information, pre-crash systems to avoid hazardous crashes on the road [17], brakes monitoring, proximity with objects warnings, merging road warnings.
- *Passenger services:* Video streaming, sharing multimedia, general Internet access, locating Automated Teller Machine (ATM), hotels, restaurants on the way, availability of cabs, etc.
- *Cost-effective services:* Better use of resources, accident prevention mechanism, minimum energy consumption, inexpensive public transit, saving fuel cost, road maintenance [18], etc.

Chapter Motivation

Smart transportation has lots of benefits and various real-time applications. However, various vulnerabilities need strong consideration while designing the security schemes. The communications among these entities (e.g., vehicles, pedestrians, fleet management systems, and roadside infrastructure) occur via open channels. The adversary can target an insecure communication to eavesdrop, modify, insert fabricated (or malicious) messages, or delete data in transit, resulting in replay, impersonation, man-in-the-middle, or privileged-insider attacks, among others. Therefore, strong security solutions or authentication protocols are required that fulfill the security and functionality aspects. It is also important for the security solution to be cost-effective and time-efficient. Recently, blockchain technology has come out as a flying color to provide security solutions in managing smart transportation. Blockchain technology is decentralized, immutable, and transparent, making it appropriate to use as a security solution. In this chapter, we mainly focus on blockchain-based security protocols for smart transportation. The main objective of this chapter is to provide a brief introduction to the latest blockchain technology and its implementation to provide security solutions for smart transportation.

Chapter Contributions

The main contributions of this chapter are as follows:

- We provide a brief explanation of blockchain and its types. Further, we elaborate on various consensus algorithms that can be implemented in smart transportation. We also state the advantages of blockchain implementation in smart transportations.
- We emphasize various security and privacy issues in the smart transport system. The section describes security aspects and functionality requirements on Internet of Vehicles-enabled smart transportation. We also list several attacks that can be performed and threats on smart transportation, due to the vulnerability of open communications between the entities.
- A detailed description of recent schemes that have provided security solutions for blockchain-based smart transportation systems is provided. To ease the understanding, we summarize their advantages, limitations, and other characteristics in a table.
- We also provide a detailed performance comparison of the schemes. We compare and analyze the schemes based on their computation cost, communication cost, and other security functionality features.

Chapter Organization

The organization of this chapter is as follows:

- In Sect. 2, we briefly explain blockchain and its types and elaborate on various consensus algorithms implemented in smart transportation. We also state the advantages of blockchain implementation in smart transportations.
- In the next section (Sect. 3), we provide security aspects, functionality requirements, attacks, and threats in Internet of Vehicles-enabled smart transportation.
- Further in Sect. 4 we describe recent schemes that have provided security solutions for the blockchain-based smart transportation system.
- Later in Sect. 5, we provide a detailed performance comparison of the schemes described in Sect. 4.
- Section 6 concludes this chapter.

2 Blockchain Technology and Its Evolution in IoV-Enabled Smart Transport System

Intelligent Transportation System (ITS)-based IoV consists of entities that do not trust each other and communicate over an open channel. The lack of trust and open communication forms an insecure channel, making it easy and approachable for an adversary to launch various security attacks. Not only this, but in a real-time application like IoV, an attacker can even trace the users' messages and identities, and might hamper the privacy of the customers and drivers, which can be life-threatening. Therefore, blockchain-based solutions are one of the most optimal approaches to providing security in an IoV-enabled smart transport system. It also maintains functionality features like traceability and anonymity.

Blockchain is distributed database whose copy exists in parallel on different nodes in the network. The blocks are added one after the other in a chain such that each block is linked to the previous block's hash value. The root block in the blockchain is known as the genesis block. Every blockchain block consists of a version of the block, the previous block's hash value, timestamp value, a random nonce value, and number of transactions within the block. After the block is formed, every node validates the block and the validated block is added to the blockchain and is linked to the previous block by the parent hash value. Therefore, any block added in the chain is impossible to tamper with, and no block can be added between two already added blocks. This way, the records stored in the block are simultaneously open and secure.

Types of Blockchain

Blockchain technology can be categorized into three types: public blockchain, private blockchain, and consortium blockchain.

- *Public blockchain*: Public blockchain, also known as a permissionless blockchain, works in an open environment like Ethereum and Bitcoin where anyone can join and write the shared blocks. Every participant in the public blockchain is given equal privilege in drawing a consensus in the consensus mechanism. Public blockchains abide entirely by the properties like non-repudiation, transparency, and traceability. Scalability is an issue in such blockchains as the rate of validation of blocks increases with an increase in the number of nodes.
- *Private blockchain*: Private blockchain [19] like Hyperledger, multichain fabric works in a closed environment where all the participants allowed in the process are well known. A private blockchain is also known as a business blockchain [20]. Public and private blockchains differ in enabling users to access, store, modify, send, and receive transactions. Public blockchains are open to all; anybody can access the blockchain, whereas only trusted entities are allowed to access the blockchain in private blockchain, thus forming a trusted network. In a private blockchain, only the authoritative entity assigns specific tasks to the trusted entities to perform. Private blockchains are more scalable than public as a centralized group monitors the users.
- *Consortium blockchain*: A hybrid approach combining public and private blockchain to reach consensus in a peer-to-peer network is called consortium blockchain. The access in consortium blockchain is given to a predefined set of nodes. Any new node that wishes to join the network should be authenticated and authorized. Private and consortium blockchain are known as permissioned blockchain.

Consensus Mechanism

Blockchain technology does not rely on a third party for validation and verification. Therefore, a mechanism is followed to validate the information and add the transactions to the block and the blockchain. The mechanism is called a *consensus mechanism* [21]. Consensus means a process to agree with a decentralized or distributed network platform where the nodes cannot trust each other. A consensus mechanism is a procedure like a state machine running on every node in the network so that every individual concludes on the same output. A consensus mechanism is an algorithm that helps the miners validate a transaction and decide to add or drop a block in the blockchain. It ensures a tamper-free environment where one version of the truth should be agreed upon. It solves the problem of trust in blockchain, as all the non-trusted miners participating in the process undergo a similar algorithm to agree on the block's validity. The consensus algorithm also mitigates the effect of the presence of faulty nodes in the network. All the nodes must reach an agreement about the state of the blockchain.

A consensus mechanism should have the following properties:

- *Consistency*: The result of a consensus algorithm is that all nodes should agree on the same block.
- *Validity*: The agreed block should be the block that receives the majority consensus.
- *Liveliness*: Eventually, the algorithm should terminate; the nodes should decide on some block.

Choosing an appropriate consensus algorithm is the most important part of the implementation of an effective blockchain solution. The choice of the consensus algorithm is based on various factors like type of blockchain: public, private, or consortium, scalability of the network, tolerance to withstand attack or failures like node failures, partition failure, or byzantine failure. The consensus mechanism should produce high throughput and incur low latency. IoV is a real-time application with many vulnerabilities; therefore, a consensus mechanism should be less complex and should consume low bandwidth with minimum energy consumption [22].

Consensus algorithm basically can be classified into two types: (1) Proof based.

(2) Voting based. In *a proof-based consensus algorithm*, the nodes with the highest computational power are given the right to append the block to the blockchain. Proof-based consensus is used in public blockchains. *Voting-based algorithms* are preferred in private blockchains, where a block can be added to the blockchain only after a threshold number of nodes have agreed on it. Any node that wishes to append a block needs vote of its peer nodes to get the consensus to add it [23]. Some effective consensus algorithms are briefed as follows:

- *Byzantine Fault Tolerance (BFT)* [24, 25] algorithm helps a group of nodes within a closed network to reach a consensus even in the presence of faulty nodes. The algorithm runs in pre-prepare, prepare, and commit phases. Once the message sent in the pre-prepare phase is accepted by $(2f + 1)$ nodes where f: number of faulty nodes, the message is accepted.
- *Practical Byzantine Fault Tolerance (PBFT)* [26] algorithm is a variant of BFT and it reaches to consensus with $(3f + 1)$ accepting nodes. The consensus is reached in pre-prepare, prepare, and commit phases. PBFT has low scalability. Other variants of Byzantine Fault Tolerance algorithms are "Delegated Byzantine Fault Tolerance (DBFT)" and "Federated Byzantine Agreement (FBA)" [27, 28].
- *Ripple protocol* consensus algorithm is another voting-based consensus algorithm. All participating nodes maintain a list of trusted nodes called as "Unique node list." Participant nodes receive the transaction constantly throughout the process. If the transaction is valid, it is added to the candidate set. All the participating nodes exchange their candidate sets with each other as proposals. The transaction is checked for validity on receiving the proposal, only if it comes

from the trusted neighboring node. A transaction that gets more than 80 percent of votes is added to the block.

- In *Proof of Work* [29], the miner has to do some heavy computational work to calculate a nonce value based on the previous block's hash value to add the block to the blockchain. The work to add/tamper the block is based on all the blocks in the blockchain, and should be heavy and not be possible to be performed in generic environment to discourage an attacker. PoW requires heavy energy consumption which makes it infeasible to apply in IoT environment.
- *Proof of Stake (PoS)* [30, 31] mechanism chooses the miner based on economic stake or bitcoins that it holds. Adversaries can increase the number of transactions to increase stake, which might also lead to unfair method of choosing a leader.
- *Proof of Vote (PoV)* [32] consensus mechanism is proposed for consortium blockchains. The network nodes are categorized into four categories: (1) Butler, (2) Butler candidate, (3) Commissioner, and (4) Ordinary user. Several enterprises form a consortium network and commissioners are the members of the league. A butler is a node that can create a block like miners in PoW. A butler is chosen out of the butler candidates by a commissioner unlike PoW where they have to prove their power. A node can willingly become a candidate by registration and recommendations. A block is added to the blockchain based on the votes of commissioners. An ordinary user can only distribute the message but cannot take part in the block formation.

Advantages of Blockchain Implementation in Smart Transportation

Recent schemes use blockchain as the security solution in smart transportation because of the following advantages:

- *Transparency*: Any user can participate in adding or validating the block in the blockchain for a public blockchain. Similar to this, any transaction or block added to the blockchain is accessible to all the users. In a private blockchain, the data is only open to the private authorized users. Also, it is easy to track the transactions made by an entity even when its real identity is secured.
- *Immutability*: It means that once a block is inserted into public or private blockchain, it is impossible to modify or tamper it later. As the blocks consist of the previous block's hash value, any change of the value in a block would affect the validity of all the consecutive blocks. Moreover, the copy of blockchain is present with every network user, so copies could easily be identified.
- *Traceability*: The data is stored in blocks that are added to the blockchain. Verifying/tracing the data stored in blockchain is possible due to the presence of nonce and also the fact that the data is mapped to the timestamped value.
- *Interoperability*: Applications like IoV, Internet of Things (IoT), Internet of Drones (IoD), smart grids, and smart framings consist of heterogeneous devices.

These applications face one major challenge to interoperate with each other. Blockchain allows various IoT systems and devices to communicate among themselves by exchanging data.
- *Reliability*: The data stored within the blockchain blocks is valid and can be trusted. Various cryptographic techniques like hashing and encryptions form the underlying basis for storing data in the blockchains.
- *Decentralization*: Traditional database systems were dependent on any third party or agency for validation. Blockchain technology is unique and works independently using a distributed ledger that validates the nodes' transactions without consulting or requiring a third party. Using decentralized blockchain reduces the overall communication overheads and properly uses the shared resources within the network.
- *Non-repudiation*: When a transaction is added to the block, it is digitally signed using the private key of the miner, which the public key can only verify. So, no node can deny the digitally signed transaction added by it into the block.
- Blockchain technology reduces time, cost, dependency on the third party, and security of the data.

Due to the above-stated advantages, blockchain is implemented in applications like supply chains, financial administrations, medicinal services, governments and numerous different ventures, trailblazers, energy, health and medical care, Internet of Things, Internet of Vehicles, smart cities, digital asset trading, property right protection, and education. The recent schemes like Internet of Vehicles (IoV) [33], Intelligent Transportation Systems (ITS) [34], Internet of Intelligent Things (IoIT) [35], Software Defined Networks (SDN) [36], supply chains [37–39], smart grids [40], healthcare applications [41–44], Internet of Everything (IoE) [45], Internet of Drones (IoD) [46], smart farming [47, 48], IoT and industrial IoT [49, 50], and military applications [51] have implemented blockchain in order to increase their security features.

Figure 2 summarizes various applications that have implemented blockchain to enhance their security and functionality features.

3 Security and Privacy Issues in Smart Transport System

Security Requirements

The security requirements in smart transport applications are as follows:

- *Integrity:* One of the basic requirements of a smart transportation network is integrity. The integrity of the network is maintained, by ensuring that the data flown in the network is not manipulated or deformed. To be precise, the data received by the receiver should exactly be similar to what was sent by the sender. Possible attacks that might question the integrity of the network are malware

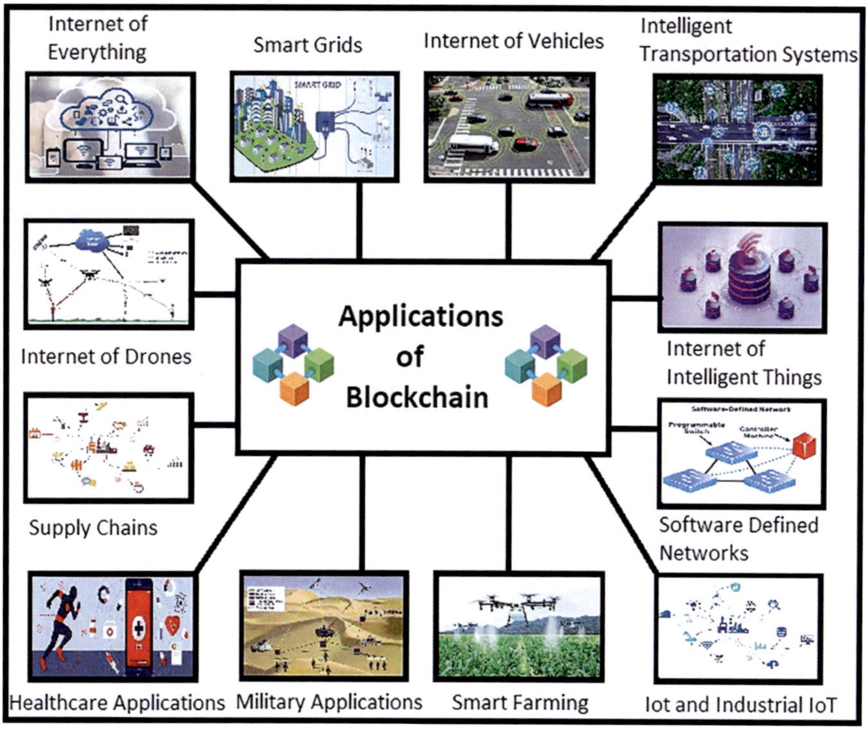

Fig. 2 Applications of blockchain technology

attacks, gray and black hole attacks, message tampering, or fabrication attacks, etc.

- *Authentication:* Authentication is one of the most important aspects of a successful smart transport network. The authentication of nodes like vehicles and *RSUs* in the IoV network ensures that no malicious vehicle can claim to be another true authenticated existing vehicle. A user is validated of who he/she claims to be before allowing him/her to send any message in the network. The authentication procedure also prohibits spoofing the receiver by the false sender of the data. It easily recognizes the fake crooked vehicles from the valid authorized ones. Sybil attack; Global Positioning System (GPS) spoofing; black, gray, and wormhole attack; fabrication attack; and replay attack are some of the attacks hampering the authenticity of messages or nodes in the network.
- *Confidentiality:* In an IoV-enabled smart transportation, certain information like the driver's identity and location or speed of the vehicle needs to be public. Therefore, it becomes important to safeguard the privacy of the customers or the business involved in IoV. So, a successful transportation network should hide delicate data so that it is not exposed to the adversary by using various cryptographic techniques like encryption or hashing. Having stated that, it is

also important to trace the identity of the vehicles that act fictitiously. So, conditional privacy or confidentiality should be maintained to avoid attacks like eavesdropping, ID disclosure, and traffic analysis.

- *Non-repudiation:* In real-time and vulnerable applications like smart transportation, fatal conditions like accidents, forgery, and thefts are prone to occur. Therefore, if any vehicle that falls in the range of accident or fatal condition sends any message in the network should not deny being the sender of the message. This would help the TA to identify the real offender of the situation easily.
- *Availability:* In the past few decades, the production of vehicles, and their users like drivers and customers are expanding. So, the smart transportation network needs to be available all time to all authenticated and authorized users. An attacker can attack a network via a denial of service (DoS) attack by sending numerous false illegal requests to congest or break down the network and make it unavailable for authenticated users [52]. Other possible attacks on the availability of the network are spamming, blackhole, grayhole, jamming, and malware attacks [53].
- *Scalability:* Scalability is an important security requirement of a smart transport network. Scalability as a security feature ensures the expansion of the vehicular network. An increase in the number of nodes or users in the network should not create or amplify other security issues [54].
- *Time constraint or freshness:* IoV-enabled smart transportation network is a real-time application where the messages like traffic updates, road conditions, accident warnings, and emergency warnings and signals should not reach with any delay to the intended user. Also, the attacker should not be able to use the rotten message to misuse the services provided by the network. Other requirements like authentication and confidentiality should be performed instantly without any delay to maintain the freshness of the network.
- *Forward secrecy:* Vehicles in smart transportation network constantly change their position within the network. The membership of a vehicle under a fixed RSU or TA changes continuously. Thus, to maintain the privacy of the messages, the network needs to be refreshed on every entry or exit of a vehicle. Forward secrecy ensures that the messages flowing in the network are not exposed to the vehicle once it has left the network.
- *Backward secrecy:* By backward secrecy, we ensure that the old messages flown before the entry of the new vehicle in the network should not be disclosed to the vehicle.

Threats and Attacks on Smart Transportation Network

IoV-enabled smart transportation network is a vulnerable network that is prone to numerous active or passive attacks. Few potential attacks and threats are represented in Fig. 3 and are also listed below [52, 55, 56].

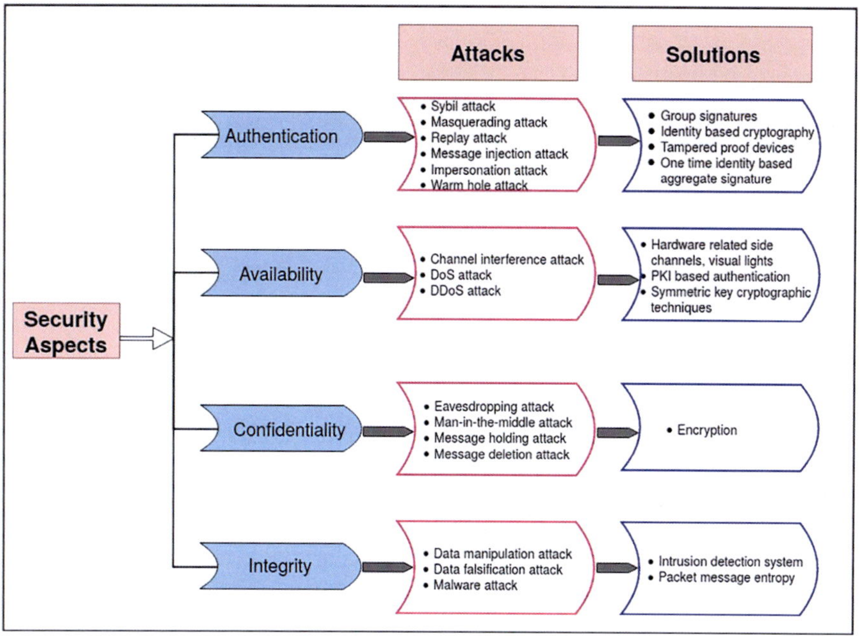

Fig. 3 Various possible potential attacks and their solutions

- *Flow of false message:* An attack on the message integrity, where an adversary creates a fake and false environment by sending bogus, rotten message to mislead the authenticated user.
- *Message injection attack:* A message injection attack is an attack on authentication, where an adversary tries to get access over the network by inducing an authorized message into the network. Further, the attacker uses the gained access to send tricky dangerous messages.
- *Replay attack:* An attack on authentication, where an attacker recapitulates an old rotten message already flown in the network to approach and access a network's services and resources.
- *Cookie theft attack:* Cookie theft attack is similar to a replay attack. A malicious user saves the earlier used authorized cookies (credentials like user name and password) to illegally access and consume the network resources in the future [57].
- *Sybil attack:* Sybil attack is one of the famous attacks on authentication. An adversary intentionally creates a vague environment by creating fake vehicles in the range of the targeted vehicle. The illusion of fake vehicles is created by using multiple false identities for a single vehicle. This creates an essence of a jam and triggers a fake jam signal even when the road is empty and compels the driver to change his/her route.

- *Impersonation attack:* An adversary removes an authorized user from the network and uses its credentials to illegally connect with innocent vehicles. After successful connection, an attacker can send hazardous misleading messages to existing authenticated vehicles.
- *Masquerading attack:* Masquerading attack is like an impersonation attack, where an attacker uses the real identity of an existing authenticated user within the network. This attack confuses the users of the network by creating two different senders with same identity.
- *Eavesdropping attack:* Eavesdropping attack is a passive attack on the confidentiality of the network. An adversary tries to fetch some confidential data illegally, by listening to the conversation of entities. An attacker uses this data in future, against their privacy without even letting them know.
- *Man-in-middle attack:* Man-in-middle attack is a combination of eavesdropping and impersonation attack where an adversary locates himself between the sender and the receiver (without letting them know) to either eavesdrop or impersonate one of them. By this, an attacker receives and can even fabricate all the messages from the sender before it reaches the receiver.
- *Denial-of-Service (DoS) attack:* The DoS attack attacks the network's availability work where an attacker intentionally throws multiple request messages to create a heavy legal message load on a particular communication channel more than its handling capacity to create congestion. This does not let the authorized user leverage the resources and services of the network. DoS attack does not require an attacker to know the network. In a smart transportation network like IoV, DoS attack can be performed on *RSUs* to refrain from the legal working of the network [58].
- *Wormhole attack:* A wormhole attack is also a tunneling attack, where an adversary advertises its wrong location to the targeted node to attract all the messages. The targeted node exposes all the messages to the attacker by sending them to the attacker node, assuming it to be a nearer node.
- *Message holding/manipulation/deletion attack:* Under this attack, an active insider node starts acting maliciously by intentionally holding back the received message or changing the message instead of sending it identically to the dedicated receiver. In other scenarios, the malicious node can even delete the message to halt the successful execution of services in the network. This can severely affect the network's security if the message contains some emergency warnings or signals [52].
- *Malware attack:* A malware attack is an attack on integrity. An adversary injects files in the network system containing worms or viruses that affect the functioning of the network [59].
- *Guessing attacks:* IoV is an open communication application. So, during the communication, an adversary might guess private credentials of the user like password or biometrics by intercepting or eavesdropping on messages. Lost/stolen OBU or smart card stolen attacks might form a basis for guessing attacks.

- *Data manipulation/falsification attack:* The message flown in the IoV network contains the data or information regarding traffic/road/network conditions. So, data manipulation/falsification attacks integrity that changes the content of the message or data to create fake signals like congestion or jams.

Functionality Requirements

The following are the functional requirements that are needed in smart transport network:

- A scheme should be able to deploy new nodes in the network. As an IoV is a network of heterogeneous entities, it should be able to deploy, add new nodes (vehicles or *RSU*) within the network whenever required.
- The entities or the device connected in an IoV network should mutually authenticate each other.
- There should be high connectivity within the nodes of network such that it should be easy for the nodes to derive a secret pair-wise session key to have secure communication.
- The scheme should inculcate low storage overhead on the entities.
- To implement a scheme in a practical environment, the number of messages flown in the network to mutually authenticate and establish a pair-wise session key should be minimal. Hence, an efficient scheme should have low communication overheads.
- Mutual authentication between the nodes followed by pair-wise key establishment should involve low computational overhead.
- IoV or smart transport network is growing as new nodes or devices are added to the network. So, no matter how many nodes are added and the network grows large, the communication and computation cost should remain low.
- To resist the attack by an adversary, a scheme should abide by anonymity and untraceability, which means that if an adversary gets exposed to the messages flown in the network, he should not be able to know about the real sender of the message by the content of the message data.
- It might be an instant, that the driver/passenger of the network/nodes/end devices lose their password or their password is known to an adversary. So due to security reasons, the entities should be able to change their password independently at any time.

4 Security Solutions for Blockchain-Based Smart Transportation System

Many researchers have implemented blockchain technology to provide security solutions for IoV-enabled smart transportation systems. The basic blockchain-based IoV-enabled smart transportation system model is represented in Fig. 4. It consists of the following entities:

- *Trusted authority:* A smart transportation system has a trusted authority (*TA*), which is a fully trusted entity and is responsible for managing the vehicles and *RSUs* in its area. *TAs* are assumed to have sufficient resources to perform intense computations. It also has large storage space. It registers all vehicles, RSUs before their deployment and provides them with certificates (in certificate-based schemes), and other credentials like public key and pseudo identities. TAs are the only entities that save the real identities of the entities, so they are also responsible for tracing the identity of the malicious vehicle on receiving any suspicious message.

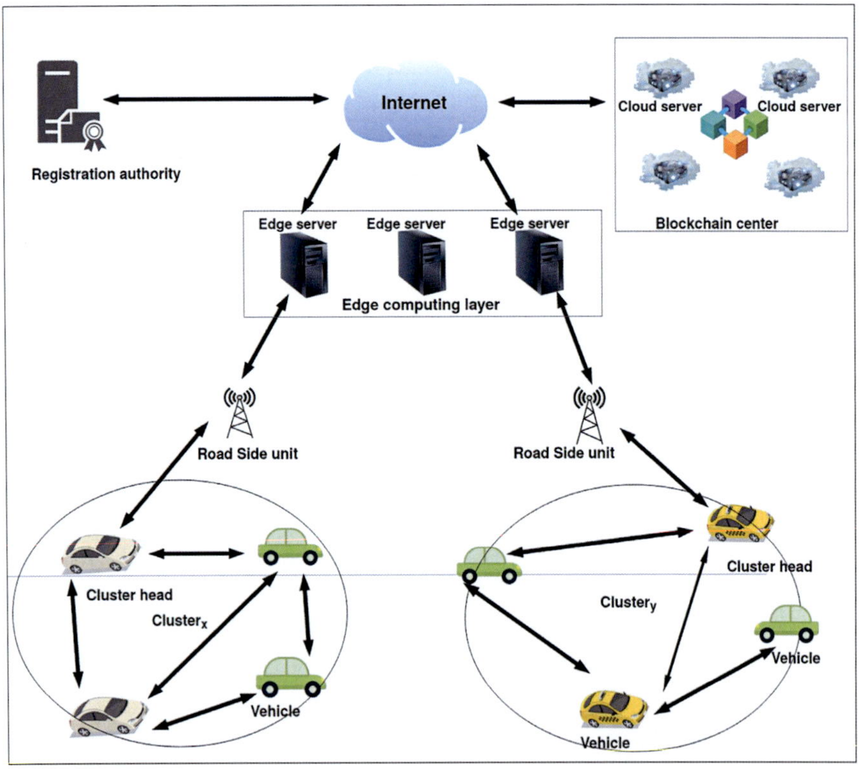

Fig. 4 Blockchain-based smart transportation system model

- *Vehicles:* A smart city has many vehicles (V_i, $i = 1,2,3 \ldots n$), forming clusters while on road. The clusters formed on the fly are dynamic as the vehicles are mobile nodes. A cluster consists of vehicles moving with similar speed, in the same direction and following the same path for quite a long time. Vehicles receive the certificates from the *TAs* during registration and use them in future communications. Vehicles collect sensitive information about the surroundings via installing multiple sensors. All smart vehicles are equipped with *on board unit (OBU)* that performs various computations throughout the authentication procedure. The collected sensitive information is stored in the vehicle's *tamper-proof device (TPD)*, which cannot be fabricated.
- *Roadside unit: RSU*s are installed at fixed places and are responsible for managing instantaneous clusters in their scope. They also regulate the flow of information among the entities within their zones. *RSUs* interact with vehicles over IEEE 802.11p protocol and Dedicated Short-Range Communications (DSRC) [1] and with *TAs* over secured wired channel. They are used to verify the authenticity of traffic update messages flown in the network. In blockchain-based schemes, *RSUs* form transactions containing the traffic events and other details and forward them to edge servers.
- *Edge server:* A smart transportation is a heavy data-driven application. So, an architecture also embeds an edge computing layer consisting of edge servers whose functionality is abstracted from Internet of Things (IoT). An additional edge computing layer in the architecture increases the efficiency as the data collected from the vehicles is processed here before reaching the blockchain center. Edge servers form partial blocks packaging the transactions received from *RSUs* and sending them to cloud servers. Anonymous privacy protection is added to the edge computing layers. The decentralization due to the edge computing layer, lowers the computation cost, increases scalability, and boosts the application's performance.
- *Cloud servers:* The cloud servers in the blockchain center layers receive the partially verified mined blocks from the edge computing layers or *RSU* (depending on the architecture), and are responsible for executing a consensus algorithm to mine and add a verified block the blockchain. Verified blocks, when added to blockchains, become accessible and immutable.

Access control and authentication are two important security services to secure different networking environments, like IoT, IoD, Internet of Vehicles (IoV), Wireless Sensor Networks (WSNs), cyber-physical systems, smart grids, and healthcare services [60–76]. The following describes the blockchain-based solutions for authentication and access control schemes based on the above-described model.

In smart cities, multiple TAs that manage the vehicles in their domain, avoiding bottleneck problems that could occur in one *TA* architecture. With multi *TA* architecture, cross *TA* authentication problem occurs when a vehicle goes from one *TA* zone to another. To support the decentralized nature of IoV and solve the above problems, Xu et al. proposed a blockchain-based roadside unit-assisted authentication and key agreement protocol for Internet of Vehicles [10]. Vehicle nodes (VNs) are equipped with onboard units (OBUs) which communicate with *RSU*s over open channel. System administrator initializes the system parameters used to register VNs before their deployment. VNs register themselves to their nearest TA. The registration information about VNs is stored in data center (DC). DC broadcasts the pointer/ block identifier and other information to TAs. TAs are the miners to construct private blockchain containing information about registration and encrypted traffic-related data. All TAs store the pointers to vehicle information in a block linked to the previous block and add a blockchain or distributed ledger based on the Proof of Stake consensus mechanism [30, 31]. During the authentication phase, VN sends an authentication request to *RSU* in its communication range. *RSU* forwards the request message to TA. *TA* checks for the presence of pointer to the vehicle in its blockchain and sends the authentication parameters of VN retrieved from DC to *RSU*. *RSU* authenticates *TA* and VN and sends the updated parameters to *TA* and VN. Next, *TA* authenticates *RSU* and vehicles authenticate *TA* simultaneously. *TA* updates the data center and sends an acknowledgment signal to *RSU*. Finally, both *TA* and VN agree on a session key for future communication. The scheme preserves anonymity and untraceability and is resistant to eavesdropping, impersonation, and replay attacks. The proposed scheme is efficient, as *RSU*s assist *TAs* during authentication to avoid bottleneck problems, also TAs maintain a common distributed ledger to store data to diminish cross-domain authentication problems.

Tan and Chung [77] proposed a mutual authentication and group key management scheme under cloud computing and edge computing layers for *RSU*s. The scheme does not use certificates; instead, it implements cloud-assisted infrastructure that improves the overall efficiency. The architecture of this certificate-less scheme is designed in three layers. *TA* a fully trusted entity and resides in the cloud layer of the architecture. It registers *RSU* and vehicles via offline registration and assigns them a unique identity and partial secret. Following this is the edge layer that contains *RSUs* dedicated to forming edge clusters for decentralization. Finally, the lowest layer is the user layer that contains the vehicles. Following this, *RSU* and vehicle compute a temporary session ID using the assigned unique ID and timestamp values. RSU and vehicles use the temporary session ID to communicate in the future. To initiate communication, *RSU* issues a public certificate. The cloud server acts as a database that saves all important information related to vehicles and *RSU*. Vehicles verify the *RSU*'s certificate and compute its signature to form a request. A batch verification is performed on the received signatures by *RSU*. Cloud servers perform the second step of verification. To upgrade the efficient communication, a group of vehicles (V2V) independently establish a group key via consortium blockchain and Chinese remainder theorem (CRT) between them and the nearest *RSU* an allocated channel.

Another blockchain-based solution for smart transportation system scheme proposed by [33] is a blockchain-based batch authentication protocol for IoV deployment (BBAS-IoV). The system model consists of TA, vehicles, *RSU*, fog servers, and cloud servers. Initially, the *TA* sets up the system parameters and registers vehicles and *RSU*s before their deployment. Before initiating the communication, the scheme performs a signing and two independent authentication phases, one between vehicles (V2V), and other between vehicles and *RSU*. In the signing phase, a vehicle broadcasts a hello message and signature to fellow vehicles and the nearest *RSU*. In V2V authentication, the hello message and the signature of the sender vehicle are verified by other vehicles of the same cluster. And finally, *RSU* authenticates the signatures of all the vehicles in the cluster in a batch to increase efficiency. At the end, a group key is established among the vehicles and *RSU* in their cluster for future communications. After successful authentication, *RSU* starts receiving and collecting traffic-related information from its vehicles securely and forms several transactions. The fog server partially forms the transactions and then forwarded to the blockchain center containing cloud servers. A Practical Byzantine Fault Tolerance (PBFT) [26] consensus algorithm is employed for block verification and addition in the blockchain. IoV produces huge data every day, so big data analytics and Artificial Intelligence (AI)/Machine Learning (ML) are also implemented to add security.

The development and increase in the production of vehicles in recent years have increased the probability of accidents on roads leading to fatal injuries to drivers and passengers and even deaths. In 2020, Vangala et al. designed an authentication scheme called blockchain-enabled certificate-based authentication for vehicle accident detection and notification in ITS environment (BCAS-VADN) [34]. The system model consists of a trusted registration authority (RA), vehicles flying in clusters, *RSUs*, edge servers, and a blockchain center containing cloud servers. RA sets up the system parameters and enrolls all the vehicles, *RSUs*, edge servers, and cloud servers and loads the credentials in their storage. Next, mutual authentication and key establishment phase occur between vehicle and their corresponding cluster head (CH) and between CH and *RSU*. The key established during the authentication phase ensures secure future communication. Vehicles are pre-installed with sensors that are capable of detecting accidents. On the occurrence of an accident, a message containing notification including the details of the accident like time, place, location, the ID of the vehicle in accident, ID of sender vehicle, level of accident, severity of the passengers, and its cause are signed and sent to the *RSU* via cluster head using the established session key. *RSUs* forward the verified details and transactions to the edge servers. Edge servers in the edge computing layer form the partial block and forward to the cloud servers present in the blockchain center. Cloud servers complete the block and form a peer-to-peer network to run the consensus algorithm to verify and add the block to the blockchain.

Zheng et al. [78] proposed a decentralized blockchain-based access authentication system with privacy preservation in VANETs. The scheme provides a trusted communication environment for Internet of Vehicles framework with a distributed cloud ledger. The system model consists of a certificate authority (CA), vehicles, *RSU*, and cloud server. CA is assumed to have enough space and issues the certificate containing a pseudo-identity, public/private key for the authenticated vehicle. It also maintains the data set to store the pseudo-identity with its corresponding real identity to track malicious vehicle computing hash value. A hash of pseudo-identity and the public key is also sent to the cloud server, which is used to verify the legality of vehicles in the future. *RSUs* form a peer-to-peer network to implement a consensus algorithm to form a blockchain. To begin with, the communication, when a vehicle comes in a range of *RSU* it sends an authentication request using its pseudo-identity. *RSU* checks the authenticity of the vehicle by verifying the request from the cloud server. The cloud server manages the pseudonyms issued by CA and saves traffic updates and messages flown by the vehicles in its database. To accomplish secure communication in the future, an integer negotiation process is implemented between *RSU* and a vehicle. A random number chosen by *RSU* is encrypted using vehicle's public key is sent to the vehicle. To confirm if the correct random number is received by the vehicle, a vehicle chooses another number, calculates a hash, and sends it to *RSU* along with its signature. The encrypted random number is decrypted by vehicle and used to send traffic updates and during future communication. Finally, when *RSU* hears any message update from vehicle, it forms a block containing the transaction, including pseudo-identity, the public key of the vehicle, traffic announcements by vehicles, hash value of the event transaction, and timestamp. The announced traffic update is notified to other vehicles by *RSU*. The vehicles make timely decisions after verifying the transaction details from the cloud server.

Lin et al. [79] proposed a blockchain-based conditional privacy-preserving authentication (BCPPA) protocol. During the system initialization phase, a key derivation algorithm is invoked to mitigate the risk and enhance security. According to the algorithm, a vehicle chooses a private root key and a chain code that derives a fresh private key for every communication. The corresponding root public key and chain code is sent to certificate authorities (CA), which can be used to generate certificates for every communication (corresponding to the private key derived). Although the scheme is based on a public key infrastructure model, it still incurs less storage cost as the public key certificates are not transmitted but are pre-recorded in the blockchain by CA. CA embeds certificates into a transaction using Ethereum (a public blockchain) and maps them to transactions using smart contracts. To send a traffic update to other vehicles, a vehicle first retrieves the transaction id of its certificate. Finally, a signing algorithm is triggered through which the sender vehicle signs the message, timestamp, and transaction id using the derived private key. Next, the receiving vehicle verifies the message by running a message verification algorithm. It fetches the certificate from the blockchain corresponding to the transaction id received in the message. The justified verification of the message signature via public key ensures the message's validity and authenticity of the sender vehicle (Table 1).

Table 1 Summary of characteristics of blockchain-based security protocols

Scheme	Techniques	Network model entities	Phases or steps	Benefits and limitations
Xu et al. [10]	ECC One-way hash function	Datacenter Vehicle nodes Trusted authority Roadside units	Initializations phase Registration phase Authentication phase	The proposed scheme is efficient, as *RSU*s assist TAs during authentication to avoid bottleneck problem, also TAs maintain a common distributed ledger to store data to diminish cross-domain authentication problems The authentication phase exchanges four communication messages. Thus, the scheme incurs some communication overheads
Tan and Chung [77]	ECC Modular exponentiation Bilinear pairing Hash functions Chinese remainder theorem	Access points TA RSU Vehicles	Offline registration phase Authentication phase V2V group key management Dynamic key updating	It is a lightweight certificate-less scheme that is unforged against chosen message attack The proposed scheme exhibits decentralization but does not support dynamic node addition Expensive in communication and computation The scheme suffers insider attack, and the session key is not secure under CK-adversary model

(continued)

Table 1 (continued)

Scheme	Techniques	Network model entities	Phases or steps	Benefits and limitations
Bagga et al. [33]	ECC Bilinear pairing Hash functions	TA RSU Vehicle Fog server Cloud server	Initial setup phase Vehicle and *RSU* registration Message signing and batch authentication phase Group key management phase Blockchain formation phase AI-based secure big data analytics phase Dynamic nodes addition phase	The scheme implements both V2V and batch authentications The use of big data analytics, AI/ML algorithms, and blockchain made the proposed BBAS-IoV efficient and smarter to work effectively in smart cities The scheme also supports the dynamic node addition phase
Vangala et al. [34]	ECC Hash functions	RA *RSU* Vehicle Edge server Cloud server	System initialization phase Enrollment phase Authentication phase Blockchain verification and addition phase Dynamic node addition phase	The scheme based on certificates Accomplishes mutual authentication between vehicle and vehicle and vehicle and *RSU* to establish secure communication The emergency information detailing on accident is shared using An established key

(continued)

Table 1 (continued)

Scheme	Techniques	Network model entities	Phases or steps	Benefits and limitations
Lin et al. [79]	Digital signatures (ECDSA)	CA Vehicle *RSU* Blockchain network	System initialization phase Message signing phase Message verification phase	A PKI-based authentication scheme that implements public blockchain to store certificates The scheme has low storage, computation, and communication cost The scheme proposes a dynamic key derivation algorithm that maintains the freshness of the keys
Zheng et al. [78]	ECC Public key infrastructure Hash functions	TA *RSU* Vehicle Fog server Cloud server	System initialization Vehicle registration Vehicle authentication Vehicle announcement Forwarding of message	A secure and anonymous authentication scheme that also provides privacy preservation Blockchain decentralizes the scenario and forbids the distribution of the malicious message The scheme does not add dynamic nodes during the process

5 Performance Comparison

The efficiency of any authentication scheme is measured in terms of communication and computational costs. The computational cost of a system is calculated as the total execution time of various cryptographic operations such as "elliptic curve point multiplication," "elliptic curve point addition," "map-to-point function," "bilinear pairing operation," "modular exponentiation," "one-way hash function," "symmetric key encryption/decryption," and "signature generation using the elliptic curve signature generation algorithm (ECDSA)".

The communication cost is defined as the total amount of data in bits transmitted via number of messages exchanged throughout the scheme.

In this section, we analyze the performance of the schemes Xu et al. [10], Tan and Chung [77], Bagga et al. [33], Vangala et al. [34], Lin et al. [79], Zheng et al. [78] (discussed in Sect. 4) in terms of communication, computational, and security features.

Communication and Computational Costs Comparison

The actual execution time of the scheme or the actual computational cost is computed by considering individual execution time taken by operations. For that, we have assumed that elliptic curve point multiplication, elliptic curve point addition denoted by T_{ecm} and T_{epa} takes 17.10 ms [80], 4.4 ms [81] to execute, respectively. A map-to-point function denoted by T_{mtp} takes 44.06 ms [82]. T_{bp}, T_{exp} represent bilinear pairing operation and modular exponentiation operation and they take 42.11 ms and 19.2 ms [80], respectively. A one-way hash function denoted by T_h and symmetric key encryption/decryption abbreviated as $T_{\text{enc dec}}$ takes 0.32 ms [81] each. Lin et al. [79] implement elliptic curve cryptography signature and verification denoted by $T_{\text{ecc sig}}$ and $T_{\text{ecc ver}}$ where, $T_{\text{ecc sig}} = T_{\text{ec}} + T_{\text{h}}$, which comes as 17.42 ms, and $T_{\text{ecc ver}} = 2T_{\text{ecm}} + T_{\text{eca}} + T_{\text{h,}}$ which is 38.92 ms approximately. Table 2 summarizes the detailed computation cost calculation of each scheme.

For calculating the communication cost, we calculate the lengths of the messages flown in bits by considering some assumed values such as the output of hash function such as SHA-1 is taken as 160 bits. ECC-based messages are assumed to be 160 bits. An elliptic curve point$P = (P_x, P_y)$ is $(160 + 160) = 320$ *bits*.

Where P_x and P_y are the x and y coordinates of the point P. The identities of the entities are assumed to be 160 bits. The random nonces and timestamp values used in all schemes are 160 and 32 bits, respectively. Also, for symmetric encryption or decryption, we assume the size of plain text/ciphertext to be 128 bits. Table 3 summarizes the detailed communication cost calculation of each scheme.

To ease the understanding of our comparative analysis, we have used the notations like *High* for the schemes with high/very high computational and communication costs, *Medium* for the schemes with average or medium computational and communication costs, *Low* for the schemes with low computational and communication costs. If the communication cost of more than 4000 bits is considered

Table 2 Comparative computational costs analysis

Scheme	Total cost	Estimated time (in milliseconds)
Xu et al. [10]	$19Th$	≈6.08 ms
Tan and Chung [77]	$12Th + 12T_{\text{ecm}} + 2T_{\text{eca}} + 2T_{\text{b p}} + 2T_{\text{ex p}}$	≈340.46 ms
Bagga et al. [33]	$6Th + 7T_{\text{ecm}} + 7T_{\text{eca}} + 3T_{\text{b p}}$	≈278.75 ms
Vangala et al. [34]	$10Th + 12T_{\text{ecm}} + 4T_{\text{eca}}$	≈226 ms
Lin et al. [79]	$T_{\text{ecm}} + T_{\text{ecc}} - sig + T_{\text{ecc}} - ver$	≈73.44 ms
Zheng et al. [78]	$^{4}Th^{+\,2T}enc/dec + T_{\text{ecc}} - sig + T_{\text{ecc}} - ver$	≈58.26 ms

Table 3 Comparative communication costs analysis

Scheme	Number of messages	Number of bits
Xu et al. [10]	4	4448
Tan and Chung [77]	$2n + 1$	$992 + 1344n$
Bagga et al. [33]	1	2912
Vangala et al. [34]	2	1856
Lin et al. [79]	4	640
Zheng et al. [78]	3	928

Table 4 Communication and computational costs comparison

Scheme	Communication cost	Computational cost
Xu et al. [10]	High	Low
Tan and Chung [77]	Medium	High
Bagga et al. [33]	Medium	High
Vangala et al. [34]	Low	Medium
Lin et al. [79]	Low	Low
Zheng et al. [78]	Low	Low

high, and less than 2000 bits is considered low. For the computational cost, the schemes that are based on heavy cryptographic operations like bilinear pairings and elliptic curves are stated High. The others with less time-consuming cryptographic operations are marked Low.

The computational and communication costs of the scheme are calculated and represented in Table 4.

Security and Functionality Features Comparison

In this section, we have compared security and functionality features of the schemes presented by Xu et al. [10], Tan and Chung [77], Bagga et al. [33], Vangala et al. [34], Lin et al. [79], and Zheng et al. [78]. Security and functionality features include various security aspects and other attacks discussed in Sect. 3. Xu et al. [10] resist various known attacks like eavesdropping, impersonation, and replay attacks. It also preserves anonymity, forward and backward secrecy, and untraceability. On the other hand, Tan and Chung's Scheme [77] and Vangala et al.'s [34] scheme resist various attacks and preserve anonymity and untraceability but do not support cross TA authentication. The scheme proposed by Bagga et al. in [33] is quite efficient as it also supports batch authentication, where multiple vehicles are authenticated simultaneously via *RSU*. It ensures conditional privacy preservation also resists man-in-the-middle attacks. Lin et al. [79] and Zheng et al. [78] provide unlinkability, authenticity, and integrity but do not agree on the session key for secure communication. Zheng et al.'s scheme does not even resist impersonation attacks.

The security features among the existing schemes are compared in Table 5.

Table 5 Comparative study on security features

Features	[10]	[77]	[33]	[34]	[79]	[78]
Privacy preservation	✓	✓	✓	✓	✓	✓
Integrity	✓	✓	✓	✓	✓	✓
Authenticity	✓	✓	✓	✓	✓	✓
Non-repudiation	✓	✓	✓	✓	✓	✓
Traceability or unlinkability	✓	✓	✓	✓	✓	✓
Cross TA authentication	✓	×	×	×	×	×
Key agreement	✓	✓	✓	✓	×	×
Replay attack	✓	✓	✓	✓	✓	✓
Man-in-the-middle attack	✓	✓	✓	✓	✓	✓
Impersonation attack	✓	✓	✓	✓	✓	×

Note: ✓: a scheme resists an attack or supports a feature; ×: a scheme does not resist an attack or does not support a feature

6 Conclusion

In this chapter, we focused on studying smart transportation that manages huge traffic in smart cities, making it luxurious and safe. We introduced blockchain and its types. Further, we described various consensus algorithms that can be implemented in smart transportation. We also listed some advantages of blockchain implementation in smart transportation. Next, we outlined security aspects, privacy issues along with several attacks and threats on smart transportation. Later in this chapter, a detailed description of recent schemes that have provided security solutions for blockchain-based smart transportation systems is provided. We compared the performance of the schemes based on their computational cost, communication cost, and other security functionality features. Few schemes have high computation and communication costs because of heavy cryptographic operations and huge message content, while others have average or low computation and communication costs. The schemes are also analyzed on various features, like privacy preservation, integrity, authenticity, non-repudiation, traceability or unlinkability, cross *TA* authentication and key agreement, replay, man-in-the-middle, and impersonation attacks.

References

1. J.B. Kenney, Dedicated Short-Range Communications (DSRC) standards in the United States. Proc. IEEE **99**(7), 1162–1182 (2011)
2. P. Bagga, A.K. Das, M. Wazid, J.J.P.C. Rodrigues, Y. Park, Authentication protocols in internet of vehicles: taxonomy, analysis, and challenges. IEEE Access **8**, 54314–54344 (2020)
3. J. Shao, X. Lin, R. Lu, C. Zuo, A threshold anonymous authentication protocol for VANETs. IEEE Trans. Veh. Technol. **65**(3), 1711–1720 (2016)

4. Y. Liu, Y. Wang, G. Chang, Efficient privacy-preserving dual authentication and key agreement scheme for secure V2V communications in an IoV paradigm. IEEE Trans. Intell. Transp. Syst. **18**(10), 2740–2749 (2017)
5. H.J. Jo, I.S. Kim, D.H. Lee, Reliable cooperative authentication for vehicular networks. IEEE Trans. Intell. Transp. Syst. **19**(4), 1065–1079 (2018)
6. J. Liu, Q. Li, R. Sun, X. Du, M. Guizani, An efficient anonymous authentication scheme for internet of vehicles, in *IEEE International Conference on Communications (ICC)*, (Kansas City, 2018), pp. 1–6
7. J. Cui, D. Wu, J. Zhang, Y. Xu, H. Zhong, An efficient authentication scheme based on semi-trusted authority in VANETs. IEEE Trans. Veh. Technol. **68**(3), 2972–2986 (2019)
8. L. Wu, Q. Sun, X. Wang, J. Wang, S. Yu, Y. Zou, B. Liu, Z. Zhu, An efficient privacy-preserving mutual authentication scheme for secure V2V communication in vehicular ad hoc network. IEEE Access **7**, 55050–55063 (2019)
9. H. Vasudev, V. Deshpande, D. Das, S.K. Das, A lightweight mutual authentication protocol for V2V communication in internet of vehicles. IEEE Trans. Veh. Technol. **69**(6), 6709–6717 (2020)
10. Z. Xu, W. Liang, K. Ching Li, J. Xu, H. Jin, A blockchain-based roadside unit-assisted authentication and key agreement protocol for internet of vehicles. J. Parallel Distrib. Comput. **149**, 29–39 (2021)
11. J. Contreras-Castillo, S. Zeadally, J.A. Guerrero-Ibanez, Internet of vehicles: architecture, protocols, and security. IEEE Internet Things J. **5**(5), 3701–3709 (2018)
12. Margaret Rouse, Internet of Vehicles (IoV) (2020). https://whatis.techtarget.com/definition/Internet-of-Vehicles. Accessed on January 2020
13. Y. Li, Q. Luo, J. Liu, H. Guo, N. Kato, TSP security in intelligent and connected vehicles: challenges and solutions. IEEE Wirel. Commun. **26**(3), 125–131 (2019)
14. J. Liu, H. Guo, J. Xiong, N. Kato, J. Zhang, Y. Zhang, Smart and resilient EV charging in SDN-enhanced vehicular edge computing networks. IEEE J. Sel. Areas Commun. **38**(1), 217–228 (2020)
15. J. Wang, J. Liu, N. Kato, Networking and Communications in Autonomous Driving: a survey. IEEE Commun. Surv. Tutor. **21**(2), 1243–1274 (2019)
16. Y. Xun, J. Liu, N. Kato, Y. Fang, Y. Zhang, Automobile driver fingerprinting: a new machine learning based authentication scheme. IEEE Trans. Industr. Inform. (2019). https://doi.org/10.1109/TII.2019.2946626
17. J. Wang, C. Li, H. Li, Y. Wang, Key technologies and development status of internet of vehicles, in *9th International Conference on Measuring Technology and Mechatronics Automation (ICMTMA'17)*, (Changsha, China, 2017), pp. 29–32
18. F. Yang, S. Wang, J. Li, Z. Liu, Q. Sun, An overview of internet of vehicles. China Commun. **11**(10), 1–15 (2014)
19. S.S. Panda, B.K. Mohanta, U. Satapathy, D. Jena, D. Gountia, T.K. Patra, Study of Blockchain based decentralized consensus algorithms, in *TENCON 2019–2019 IEEE Region 10 Conference (TENCON)*, (Kochi, India, 2019), pp. 908–913
20. W. Mougayar, *The Business Blockchain: Promise, Practice, and Application of the Next Internet Technology* (Wiley, New York, United States, 2016)
21. S. Basu, U. Maulik, O. Chatterjee, Stability of consensus node orderings under imperfect network data. IEEE Trans. Computat. Soc. Syst. **3**(3), 120–131 (2016)
22. N. Chaudhry, M.M. Yousaf, Consensus algorithms in Blockchain: comparative analysis, challenges and opportunities, in *2018 12th International Conference on Open Source Systems and Technologies (ICOSST)*, (Lahore, Pakistan, 2018), pp. 54–63
23. S. Pahlajani, A. Kshirsagar, V. Pachghare, Survey on private Blockchain consensus algorithms, in *2019 1st International Conference on Innovations in Information and Communication Technology (ICIICT)*, (Chennai, India, 2019), pp. 1–6
24. G.S. Veronese, M. Correia, A.N. Bessani, L.C. Lung, P. Verissimo, Efficient byzantine fault-tolerance. IEEE Trans. Comput. **62**(1), 16–30 (2013)

25. L. Lamport, R. Shostak, M. Pease, The byzantine generals problem. ACM Trans. Program. Lang. Syst. **4**(3), 382–401 (1982)
26. L. Zhang, Q. Li, Research on consensus efficiency based on practical byzantine fault tolerance, in *2018 10th International Conference on Modelling, Identification and Control (ICMIC)*, (Guiyang, China, 2018), pp. 1–6
27. G.S. Veronese, M. Correia, A.N. Bessani, L.C. Lung, P. Verissimo, Efficient byzantine fault-tolerance. IEEE Trans. Comput. **62**(1), 16–30 (2011)
28. Z. Zheng, S. Xie, H.N. Dai, X. Chen, H. Wang, Blockchain challenges and opportunities: a survey. Int. J. Web Grid Serv. **14**(4), 352–375 (2018)
29. S. Nakamoto, Bitcoin: a peer-to-peer electronic cash system. Cryptography Mailing list at https://metzdowd.com (2009)
30. H. Dai, Z. Zheng, Y. Zhang, Blockchain for Internet of Things: a survey. IEEE Internet Things J. **6**(5), 8076–8094 (2019)
31. S. King, S. Nadal, PPCoin: Peer-to-Peer Crypto-Currency with Proof-of-Stake. 12 (2020)
32. K. Li, H. Li, H. Hou, K. Li, Y. Chen, Proof of vote: a high-performance consensus protocol based on vote mechanism consortium Blockchain, in *2017 IEEE 19th International Conference on High Performance Computing and Communications; IEEE 15th International Conference on Smart City; IEEE 3rd International Conference on Data Science and Systems (HPCC/SmartCity/DSS)*, (Bangkok, Thailand, 2017), pp. 466–473
33. P. Bagga, A. Sutrala, A.K. Das, P. Vijayakumar, Blockchain-based batch authentication protocol for internet of vehicles. J. Syst. Archit. **113**, 101877 (2020)
34. A. Vangala, B. Bera, S. Saha, A.K. Das, N. Kumar, Y. Park, Blockchain-enabled certificate-based authentication for vehicle accident detection and notification in intelligent transportation systems. IEEE Sensors J. **21**(14), 1–15 (2020)
35. M. Wazid, A.K. Das, S. Shetty, M. Jo, A tutorial and future research for building a Blockchain-based secure communication scheme for internet of intelligent things. IEEE Access **8**, 88700–88716 (2020)
36. D. Chattaraj, S. Saha, B. Bera, A.K. Das, On the design of Blockchain-based access control scheme for software defined networks, in *IEEE INFOCOM 2020 - IEEE Conference on Computer Communications Workshops (INFOCOM WKSHPS)*, (Toronto, ON, Canada, 2020), pp. 237–242
37. S. Jangirala, A.K. Das, A.V. Vasilakos, Designing secure lightweight Blockchain- enabled RFID-based authentication protocol for supply chains in 5G Mobile edge computing environment. IEEE Trans. Industr. Inform. **16**(11), 7081–7093 (2020)
38. B. Bera, S. Saha, A.K. Das, N. Kumar, P. Lorenz, M. Alazab, Blockchain-envisioned secure data delivery and collection scheme for 5G-based IoT-enabled internet of drones environment. IEEE Trans. Veh. Technol. **69**(8), 9097–9111 (2020)
39. B. Bera, D. Chattaraj, A.K. Das, Designing secure Blockchain-based access control scheme in IoT-enabled internet of drones deployment. Comput. Commun. **153**, 229–249 (2020)
40. B. Bera, S. Saha, A.K. Das, A.V. Vasilakos, Designing Blockchain-based access control protocol in IoT-enabled smart-grid system. IEEE Internet Things J. **8**(7), 5744–5761 (2020)
41. S. Son, J. Lee, M. Kim, S. Yu, A.K. Das, Y. Park, Design of secure authentication protocol for cloud-assisted telecare medical information system using Blockchain. IEEE Access **8**, 192177–192191 (2020)
42. N. Garg, M. Wazid, A.K. Das, D.P. Singh, J.J.P.C. Rodrigues, Y. Park, BAKMP- IoMT: design of Blockchain enabled authenticated key management protocol for internet of medical things deployment. IEEE Access **8**, 95956–95977 (2020)
43. M. Wazid, B. Bera, A. Mitra, A.K. Das, R. Ali, Private Blockchain-envisioned security framework for AI-enabled IoT-based drone-aided healthcare services, in *Proceedings of the 2nd ACM MobiCom Workshop on Drone Assisted Wireless Communications for 5G and beyond (DroneCom'20)*, (London, 2020), pp. 37–42
44. S. Saha, A.K. Sutrala, A.K. Das, N. Kumar, J.J.P.C. Rodrigues, On the design of Blockchain-based access control protocol for IoT-enabled healthcare applications, in *ICC 2020–2020 IEEE International Conference on Communications (ICC)*, (Dublin, Ireland, 2020), pp. 1–6

45. B. Bera, A.K. Das, M. Obaidat, P. Vijayakumar, K.F. Hsiao, Y. Park, AI-enabled Blockchain-based access control for malicious attacks detection and mitigation in IoE. IEEE Consum. Electron. Mag. **10**(5), 82–92 (2020)
46. B. Bera, A.K. Das, A.K. Sutrala, Private Blockchain-based access control mechanism for unauthorized UAV detection and mitigation in internet of drones environment. Comput. Commun. **166**, 91–109 (2021)
47. A. Vangala, A.K. Sutrala, A.K. Das, M. Jo, Smart contract-based Blockchain-envisioned authentication scheme for smart farming. IEEE Internet Things J. **8**(13), 10792–10806 (2021)
48. A. Vangala, A.K. Das, N. Kumar, M. Alazab, Smart secure sensing for IoT-based agriculture: Blockchain perspective. IEEE Sensors J. **21**(16), 17591–17607 (2020)
49. S. Banerjee, B. Bera, A.K. Das, S. Chattopadhyay, M.K. Khan, J.J.P.C. Rodrigues, Private blockchain-envisioned multi-authority CP-ABE-based user access control scheme in IIoT. Comput. Commun. **169**, 99–113 (2021)
50. S. Saha, D. Chattaraj, B. Bera, A.K. Das, Consortium blockchain-enabled access control mechanism in edge computing based generic Internet of Things environment. Trans. Emerg. Telecommun. Technol. **32**(6), e3995 (2021)
51. M. Wazid, A.K. Das, S. Shetty, J.J.P.C. Rodrigues, On the design of secure communication framework for Blockchain-based internet of intelligent battlefield things environment, in *IEEE INFOCOM 2020 - IEEE Conference on Computer Communications Workshops (IN- FOCOM WKSHPS)*, (Toronto, ON, Canada, 2020), pp. 888–893
52. F. Qu, Z. Wu, F. Wang, W. Cho, A security and privacy review of VANETs. IEEE Trans. Intell. Transp. Syst. **16**(6), 2985–2996 (2015)
53. M.A. Talib, S. Abbas, Q. Nasir, M.F. Mowakeh, Systematic literature review on internet- of-vehicles communication security. Int. J. Distrib. Sens. Netw. **14**(12), 1–21 (2018)
54. Y. Sun, L. Wu, S. Wu, S. Li, T. Zhang, L. Zhang, J. Xu, Y. Xiong, Security and privacy in the internet of vehicles, in *International Conference on Identification, Information, and Knowledge in the Internet of Things (IIKI'15)*, (Beijing, China, 2015), pp. 116–121
55. M.A. Shahid, A. Jaekel, C. Ezeife, Q. Al-Ajmi, I. Saini, Review of potential security attacks in VANET, in *Majan International Conference (MIC'18)*, (Muscat, Oman, 2018), pp. 1–4
56. N. Sharma, N. Chauhan, N. Chand, Security challenges in Internet of Vehicles (IoV) environment, in *First International Conference on Secure Cyber Computing and Communication (ICSCCC'18)*, (Jalandhar, India, 2018), pp. 203–207
57. A. Dua, N. Kumar, A.K. Das, W. Susilo, Secure message communication protocol among vehicles in Smart City. IEEE Trans. Veh. Technol. **67**(5), 4359–4373 (2018)
58. A. Samad, S. Alam, M. Shuaib, M. Bokhari, *Internet of Vehicles (IoV) Requirements, Attacks and Countermeasures* (New Delhi, India, 2018)
59. O.Y. Al-Jarrah, C. Maple, M. Dianati, D. Oxtoby, A. Mouzakitis, Intrusion detection systems for intra-vehicle networks: a review. IEEE Access **7**, 21266–21289 (2019)
60. M. Wazid, A.K. Das, K. Vivekananda Bhat, A.V. Vasilakos, LAM-CIoT: lightweight authentication mechanism in cloud-based IoT environment. J. Netw. Comput. Appl. **150**, 102496 (2020)
61. M. Wazid, A.K. Das, N. Kumar, A.V. Vasilakos, J.J.P.C. Rodrigues, Design and analysis of secure lightweight remote user authentication and key agreement scheme in internet of drones deployment. IEEE Internet Things J. **6**(2), 3572–3584 (2019)
62. Q. Jiang, S. Zeadally, J. Ma, D. He, Lightweight three-factor authentication and key agreement protocol for internet-integrated wireless sensor networks. IEEE Access **5**, 3376–3392 (2017)
63. V. Odelu, A.K. Das, A. Goswami, SEAP: Secure and efficient authentication protocol for NFC applications using pseudonyms. IEEE Trans. Consum. Electron. **62**(1), 30–38 (2016)
64. S. Chatterjee, A.K. Das, J. Sing, An enhanced access control scheme in wireless sensor networks. Ad-Hoc Sens. Wirel. Netw. **21**, 121–149 (2014)
65. D. Mishra, A.K. Das, S. Mukhopadhyay, A secure and efficient ECC-based user anonymity-preserving session initiation authentication protocol using smart card. Peer Peer Netw. Appl. **9**(1), 171–192 (2016)

66. S. Challa, A.K. Das, P. Gope, N. Kumar, F. Wu, A.V. Vasilakos, Design and analysis of authenticated key agreement scheme in cloud-assisted cyber–physical systems. Futur. Gener. Comput. Syst. **108**, 1267–1286 (2020)
67. A.K. Das, A. Sutrala, S. Kumari, V. Odelu, M. Wazid, X. Li, An efficient multi-gateway- based three-factor user authentication and key agreement scheme in hierarchical wireless sensor networks. Secur. Commun. Netw. **9**(13), 2070–2092 (2016)
68. C. Lin, D. He, N. Kumar, K.R. Choo, A. Vinel, X. Huang, Security and privacy for the internet of drones: challenges and solutions. IEEE Commun. Mag. **56**(1), 64–69 (2018)
69. M. Wazid, A.K. Das, M.K. Khan, A.A. Al-Ghaiheb, N. Kumar, A.V. Vasilakos, Secure authentication scheme for medicine anti-counterfeiting system in IoT environment. IEEE Internet Things J. **4**(5), 1634–1646 (2017)
70. M. Wazid, P. Bagga, A.K. Das, S. Shetty, J.J.P.C. Rodrigues, Y. Park, AKM-IoV: authenticated key management protocol in fog computing-based internet of vehicles deployment. IEEE Internet Things J. **6**(5), 8804–8817 (2019)
71. C.-T. Li, C.-C. Lee, C.-Y. Weng, Security and efficiency enhancement of robust ID based mutual authentication and key agreement scheme preserving user anonymity in mobile networks. J. Inf. Sci. Eng. **34**(1), 155–170 (2018)
72. J. Srinivas, A.K. Das, N. Kumar, J.J.P.C. Rodrigues, TCALAS: temporal credential- based anonymous lightweight authentication scheme for internet of drones environment. IEEE Trans. Veh. Technol. **68**(7), 6903–6916 (2019)
73. Q. Jiang, N. Zhang, J. Ni, J. Ma, X. Ma, K.K.R. Choo, Unified biometric privacy preserving three-factor authentication and key agreement for cloud-assisted autonomous vehicles. IEEE Trans. Veh. Technol. **69**(9), 9390–9401 (2020)
74. M. Wazid, A.K. Das, J.H. Lee, Authentication protocols for the internet of drones: taxonomy, analysis and future directions. J. Ambient. Intell. Humaniz. Comput. (2018). https://doi.org/10.1007/s12652-018-1006-x
75. C.T. Li, C. Chen, C. Lee, C. Weng, C. Chen, A novel three-party password-based authenticated key exchange protocol with user anonymity based on chaotic maps. Soft. Comput. **22**(8), 2495–2506 (2018)
76. Y. Zhang, D. He, L. Li, B. Chen, A lightweight authentication and key agreement scheme for internet of drones. Comput. Commun. **154**, 455–464 (2020)
77. H. Tan, I. Chung, Secure authentication and key management with Blockchain in VANETs. IEEE Access **8**, 2482–2498 (2020)
78. D. Zheng, C. Jing, R. Guo, S. Gao, L. Wang, A traceable Blockchain-based access authentication system with privacy preservation in VANETs. IEEE Access **7**, 117716–117726 (2019)
79. C. Lin, D. He, X. Huang, N. Kumar, K.-K.R. Choo, BCPPA: a Blockchain-based conditional privacy-preserving authentication protocol for vehicular ad hoc networks. IEEE Trans. Intell. Transp. Syst., 1–13 (2020)
80. C. Lee, C. Chen, P. Wu, T. Chen, Three-factor control protocol based on elliptic curve cryptosystem for universal serial bus mass storage devices. IET Comput. Digit. Tech. **7**(48–56), 1 (2013)
81. S. Challa, M. Wazid, A.K. Das, N. Kumar, A. Reddy, E.J. Yoon, Y. Kee-Young, Secure signature-based authenticated key establishment scheme for future IoT applications. IEEE Access **5**, 3028–3043 (2017)
82. D. He, S. Zeadally, B. Xu, X. Huang, An efficient identity-based conditional privacy-preserving authentication scheme for vehicular ad hoc networks. IEEE Trans. Inf. Forensics Secur. **10**, 2681–2691 (2015)

Blockchain-Based CPS and IoT in the Automotive Supply Chain

Maha Filali Rotbi, Saad Motahhir, and Abdelaziz El Ghzizal

1 Introduction

The automotive industry is built traditionally as a hierarchical model with a vertical flow of information. This model has proven to be an age-old technique and doesn't comply with the modern-day needs of the consumer. Today's consumer needs customized vehicles, cost-effective models, and more efficient systems. All these lead to complexity in the hierarchical model [1]. Also, small batch sizes and faster manufacturing cycles are not economically viable. The suppliers too are following age-old practices and sometimes fail to deliver as per expectations. The vehicle recall process is of a huge difficulty in a hierarchical model and leads to huge losses and inconvenience to the end consumers. We must keep pivoting and add technological advancements to ensure that the automotive sector is well defined and highly efficient.

The automotive supply chain is one of the most complex manufacturing processes in the world [2]. Nowadays, it is oriented more toward customized products than massive production [1].

Technological development has been happening with the Internet of things (IoT), and it has led to a specialized system for industries called the industrial Internet of things (IIoT) [3]. All efforts have led to the design and develop a revolution in the sector calling it industry 4.0 [3]. With such efforts, all the objects are getting connected and sharing data which makes the system smarter and keeps the record of all the events.

M. Filali Rotbi (✉) · A. El Ghzizal
Innovative Technologies, EST, SMBA University, Fez, Morocco
e-mail: maha.filalirotbi@usmba.ac.ma; abdelaziz.elghzizal@usmba.ac.ma

S. Motahhir
ENSA, Sidi Mohamed Ben Abdellah University, Fez, Morocco
e-mail: saad.motahhir@usmba.ac.ma

Y. Maleh et al. (eds.), *Advances in Blockchain Technology for Cyber Physical Systems*, Internet of Things, https://doi.org/10.1007/978-3-030-93646-4_7

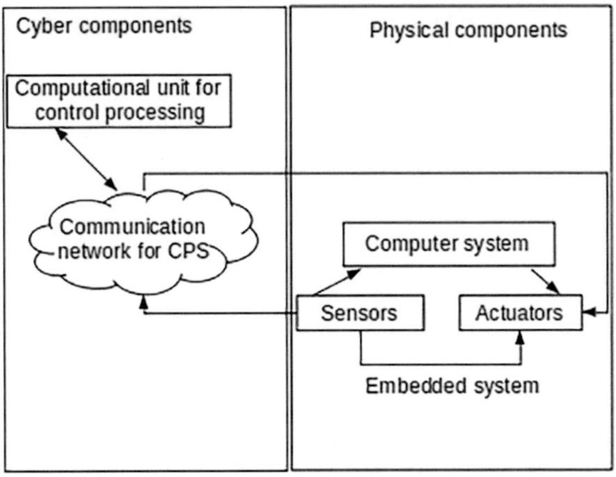

Fig. 1 Basic understanding of CPS

The connectivity is a part of a thought process called cyber-physical system (CPS), and when applied to production systems, it is termed as cyber-physical production systems (CPPS) [4]. Figure 1 showcases how physical components such as actuators and sensors can be connected through communication networks, thereby making it available for cyber-components to churn the data and take effective actions [5]. CPPSs are a core element of industry 4.0 that bring huge advantages and grass root-level changes in the day-to-day operations of the whole supply chain [6]. All the departments, including the supplier factories, need to undergo digitization to be effective for the CPPS. This leads to a new model where the hierarchical-vertical model is broken down to a decentralized-horizontal model [5]. Every single supplier becomes an implementation of CPS and shares data with the principal manufacturer. There are huge advantages of such implementation of IIoT and CPPS.

In the automotive manufacturing process, different puzzle pieces need to operate together efficiently and consistently. Connectedness, smart machines, decentralization, big data, and cybersecurity are core prerequisites for an automated and digitized SC [6].

The traditional automotive SC model is unable to meet the modern automotive manufacturing demands such as decentralization, systems connectedness, data analytics, and data traceability.

The use of CPPS and IIoT provides intelligent connectivity of the production system, high precision manufacturing, efficiency, and productivity gains [7]. However, with the increased connectivity, the large amount of data, and their sensitiveness, several challenges arise such as data integrity, immutability, and security.

In this paper, we discuss how blockchain technology, CPPS, and IIoT can be implemented together to simplify the information flow for the efficient automotive supply-chain management.

The decentralized systems in the supply chain become effective when we can make them aligned with blockchain. Blockchain is a peer-to-peer distributed network. It brings all the supply chain agents on the same platform and makes it a highly efficient decentralized-horizontal system. Modern-day demands of customization and small batch sizes can be implemented easily with no losses. With blockchain-based CPPS in the automotive sector, tracing and tracking become highly efficient, and issues can be solved well before handing out to the consumer, thereby no need for recalls. Even if needed, every part and supplier can be traced back to give faster service.

In [8], the authors present a trust model of healthcare-based Internet of things using blockchain technology, the paper presents a decentralized, interoperable trust architecture for healthcare IoHT that incorporates Blockchains. The implementation of blockchain in the IoHT (Healthcare-based Internet of Things) framework provides privacy, scalability, interoperability, availability, mutual authentication, trustworthy, and data integrity. In the context of healthcare, a study in [9] the authors propose a secure intrusion detection system for CPS in the healthcare sector using blockchain-based data transmission. The presented approach uses sensor devices to collect data and employs a deep belief network (DBN) model to detect intrusions.

An architecture design framework and a suitability application analysis flowchart for blockchain-based food traceability systems are proposed in [10], where the authors identify blockchain-based solutions for food traceability concerns and highlight the benefits and challenges of implementing blockchain-based traceability systems.

In [11], the authors propose a decentralized data management solution for secure transportation systems in smart cities using a private blockchain with Hyperledger fabric which outlines a new method for developing and deploying a decentralized platform that integrates IoT and blockchain technology for a safe, transparent, and reliable transportation system. In the context of quantum computing [12], the authors present a protocol of a blockchain framework for secure data exchanging between IoT nodes. This protocol is based on quantum-inspired quantum walks and is executable on digital computers. This paper proposes implementing quantum hash instead of regular cryptographic hashes to ensure confidentiality and integrity for IoT devices.

Blockchain technology could transform supply-chain management in many ways such as increasing product security, limiting parts counterfeiting, improving quality management, reducing the need for middlemen, and lowering the cost of supply chain transactions [13].

Blockchain turns out to be the best viable option for industry 4.0, which has CPPS as its core production technique. All these efforts make the industry smart by taking care of everything, including ways to control waste, and have the least impact on nature. Blockchain eliminates the need for intermediaries and third parties, which is a key feature of this architecture. It allows customers to track information about their products directly through the network in confidential and secured ways using cryptographic signatures.

In this work, we explored the existing literature. We analyzed the application of blockchain technology in different sectors and then discussed its implementation in the automotive supply chain. This work aims to highlight the opportunities cyber-physical production systems and industrial IoT brings to automotive supply-chain management. A blockchain-based CPPS and IIoT model is suggested to enhance the SCM efficiency. An implementation of this model in a car manufacturing factory is presented with a focus on its advantages and limitations.

Throughout the chapter, we will unravel the blockchain system, understand the CPS and CPPS with industry 4.0, discuss CPPS in automotive supply chain and its challenges, and also present how blockchain makes the whole automotive industry smarter and more efficient.

2 Revisiting the Background of Blockchain

Peer-to-Peer Distributed Networks

A system's architecture determines how the system's components are related to one another. The three main types of software systems' architectures are centralized, distributed, and decentralized. In this subsection, we will explain the difference between the architectures mentioned above and highlight the relationship between blockchain and decentralized systems. As represented in [14], Fig. 2 shows three different types of networks: Centralized, Decentralized, and Distributed.

Centralized systems are conventional systems where nodes are connected to a single central component that stores data and controls all the operations on the system. In this type of system, the failure of the authority unit causes the deficiency of the whole system.

Decentralized systems are systems where there is no central owner but instead there are several central components. Each of them has a copy of the resources that other nodes can access. In a decentralized system, the failure of one or more central nodes doesn't crash the whole system as long as at least one central owner is still running.

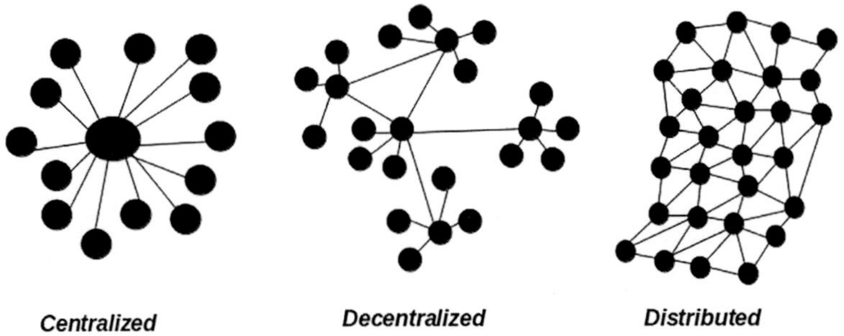

Fig. 2 Centralized, decentralized, and distributed systems

Distributed systems are systems with no central authority, and nodes are interconnected without any central control. Distributed networks eliminate centralization in a way that all nodes have equal access to resources.

Peer-to-peer distributed systems are systems where nodes make their computational resources such as processing power and storage capacity available to one another. Blockchain is a part of the implementation layer of a distributed software system that offers and maintains integrity.

Blockchain Technology

What's Blockchain?

Blockchain technology was first introduced in the context of cryptocurrency. A blockchain can be viewed as a chain of data blocks, and each block contains a number of information related to what we call transactions. Every block is linked to the previous one by means of a pointer, which makes it difficult to alter the data saved into the blocks. Blockchain is a peer-to-peer distributed ledger, and each node of the distributed network holds the final version of the ledger. The ledger is append-only, cryptographically secure and updated only after reaching a consensus among nodes.

Figure 3 represents how blockchain blocks are linked and the information stored in each block.

Generations of Blockchain

In this subsection, we will go over the main tiers of blockchain that were detailed in [15]: Blockchain 1.0, Blockchain 2.0, and Blockchain 3.0:

- Stage 1/Blockchain 1.0: this is exclusively about digital currencies (i.e., bitcoin, Litecoin, Dogecoin, etc.).

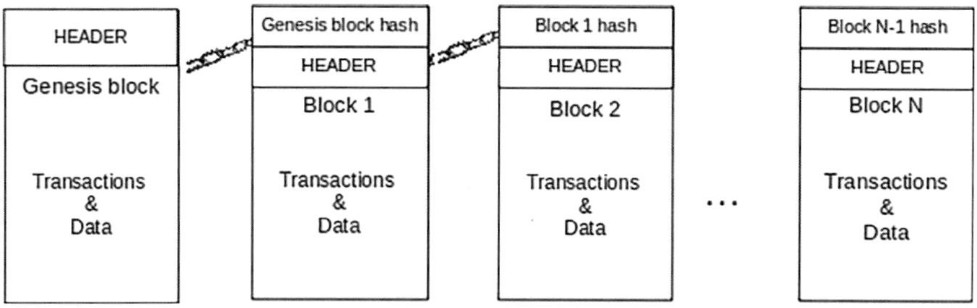

Fig. 3 Blockchain structure

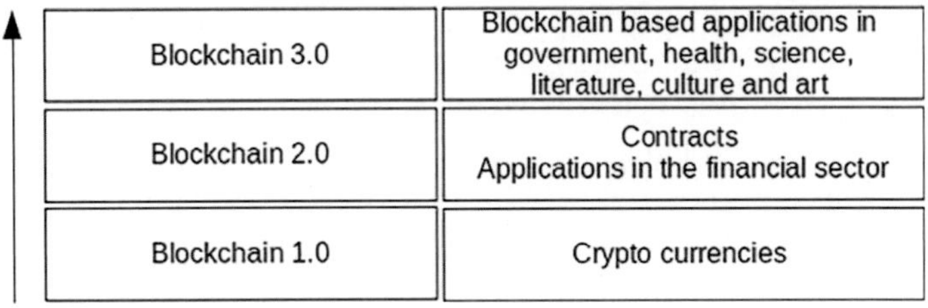

Fig. 4 Generations of blockchain

The purpose of the creation of blockchain was to manage monetary transactions between participants without requiring a third party and using cryptography to make the operation both secure and transparent.

- Stage 2/Blockchain 2.0: in this stage, the technology of blockchain took a further step, and the concepts of contracts and smart contracts were introduced. Financial services go beyond cash transactions such as derivatives, bonds, loans, etc.
- Stage 3/Blockchain 3.0: this refers to blockchain-based applications that goes beyond the financial services industry such as government services, health applications, culture, and art.

Figure 4 shows the different tiers of blockchain.

3 Model Advances in the Automotive Supply Chain

Industry 4.0

Industry 4.0 can be defined as the fourth industrial revolution. It changed the entire value chain of the life cycle of products.

Industry 4.0 is oriented toward the digitization of the manufacturing process. Its concept includes Internet of things (IoT), industrial Internet, cloud-based manufacturing, and smart manufacturing [16].

The industry 4.0 model supports the interconnection of physical components such as sensors and enterprise resources, along with the Internet [17].

The use of IoT applications has been shown to increase manufacturing productivity by 10 to 25%.

Figure 5 illustrates the technologies that revolutionized the industrial sector.

The fourth industrial revolution is considered to be a digital revolution of manufacturing industries. The purpose of industry 4.0 is to create an open, transparent, and smart manufacturing process by improving machines' performance and optimizing their maintenance. It aims to fulfill individual customer needs.

Figure 6 shows the global industry 4.0 market share.

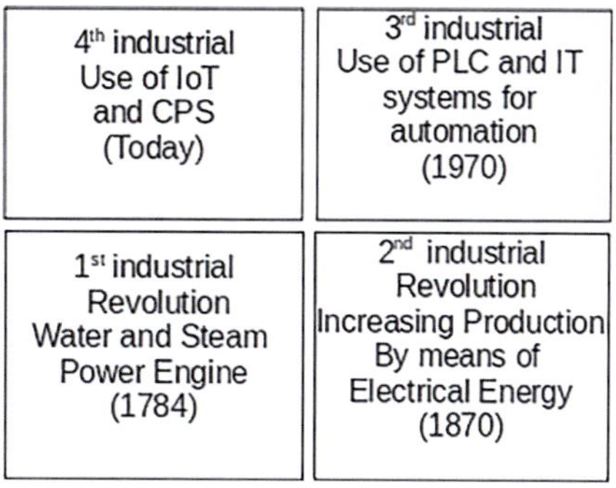

Fig. 5 Industrial revolutions [16]

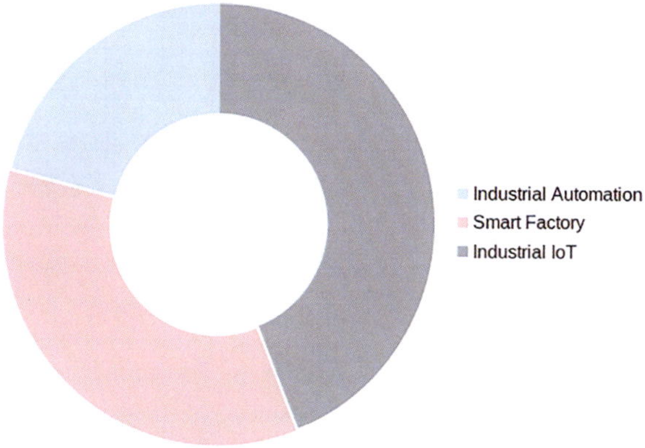

Fig. 6 Global industry 4.0 market share. (Source: Fortune Business Insights, Research report on "Global Industry 4.0 Market")

Industry 4.0 promotes the use of available information and communication networks by CPPS to automate information exchange.

It benefits from technological advances in IoT and industrial IoT to optimize production flow and automate the manufacturing process.

As shown in Fig. 7, a study expects the global industry 4.0 market to exhibit a significant growth of 16.4% in the 2021–2028 period [18].

Industry 4.0 is considered a digital revolution of the industry sector that resulted in many benefits (Fig. 8):

- Increased productivity and resources efficiency
- Real-time data for supply chains and real-time monitoring enabled by the IoT
- Advanced maintenance
- Fully automated and optimized processes
- Customized products and customer integration
- Better working conditions

The industrial Internet does not only digitize horizontal and vertical value chains but also revolutionizes a company's product and service portfolio, with the ultimate goal of better meeting customer needs.

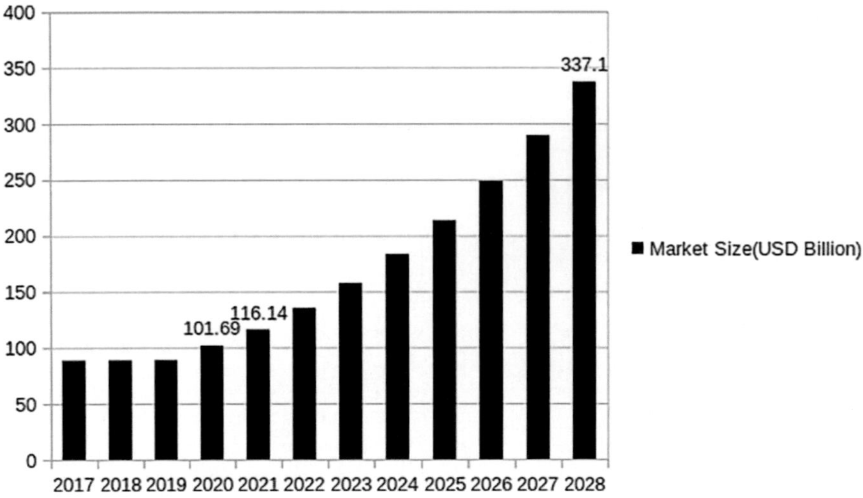

Fig. 7 Global industry 4.0 market size, 2017–2028. (Source: Fortune Business Insights, Research report on "Global Industry 4.0 Market")

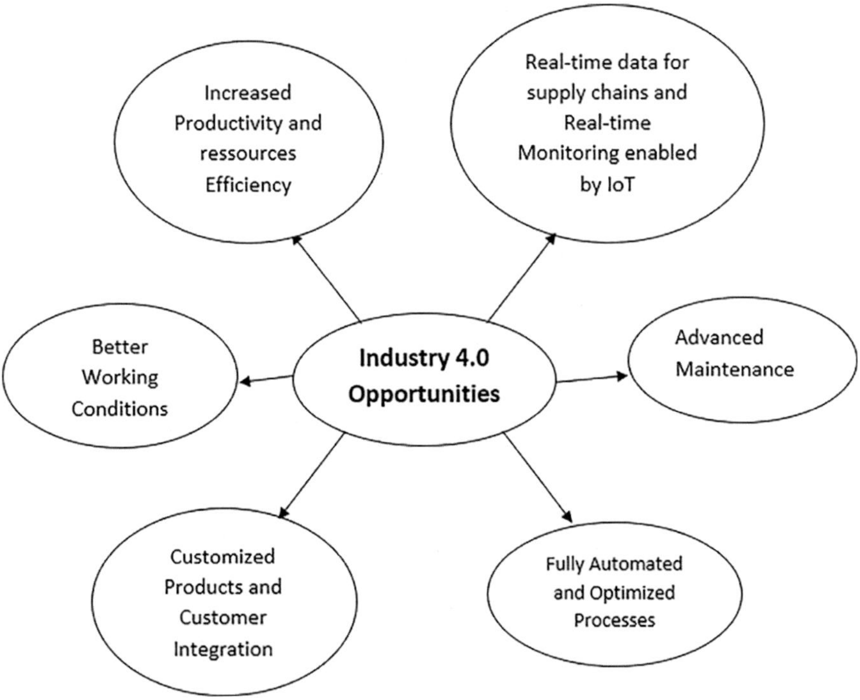

Fig. 8 Industry 4.0 opportunities

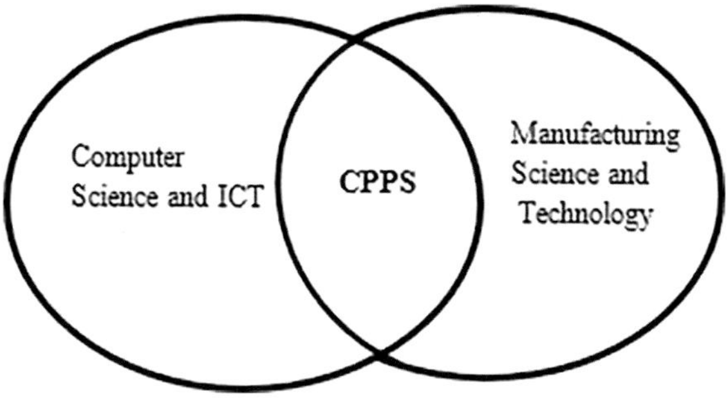

Fig. 9 Cyber-physical production systems

The industrial Internet's potential applications extend far beyond the optimization of manufacturing technology.

In the following subsection, we will focus on CPPS and IIoT and their impact on the automotive supply chain.

Cyber-physical Production Systems and Industrial IoT

Cyber-physical production systems depict the use of cyber-physical systems in a manufacturing environment. CPPSs are autonomous components and sub-systems that work in coordinated and situation-dependent ways in different phases of the production process.

As illustrated in Fig. 9, CPPS combines the technological advances in CS and ICT to provide smart systems and smart production that responds to industry 4.0 challenges.

CPPSs are interconnected and connected to the manufacturing environment. The main key elements of a CPPS are as follows:

- Self-aware: using the connected devices and sensors, CPPS measures and senses the environment where they operate.
- Self-predict: CPPS uses self-aware information to predict their issues.
- Self-evaluate: CPPS offers a synthesis of their performance to users and presents the context and background of the potential issues for diagnostics.

In the production automation domain, CPPS is used in different scenarios such as production networks, maintenance, and diagnosis.

Industrial IoT refers to a network of connected devices and objects that communicate through standard protocols in an industrial environment.

The main feature of IIoT is connectivity, and connected objects are considered smart objects that provide data related to the production process with a high precision that is a great advantage of this technology.

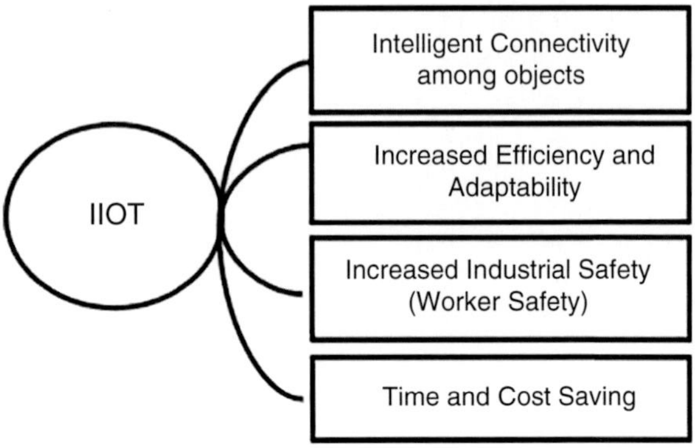

Fig. 10 Industrial Internet of things

The industrial IoT enables intelligent and automated industrial processes by facilitating machine-to-machine communication. It reduces the human interventions that lead to reducing human errors and increasing efficiency.

IIoT provides interaction with the manufacturing environment, real-time communication, and immediate response to changes which make the value chain intelligent and networked.

Research institutions and companies shifted their focus on IoT and CPS because of the flexibility and adaptability capabilities provided to the production system.

In Fig. 10, we resume the core benefits of IIoT.

Cyber-physical Production Systems in the Automotive Supply Chain

The automotive supply chain begins from basic products and components' manufacturing and is far before the automotive assembly line.

Different suppliers of various products play a crucial role in the automotive industry. For example, a small induction sensor used in the assembly line for the production of vehicles is supplied by another company and is a crucial part of the supply chain of that automotive industry [19].

CPPS gives a significant impetus to the automotive supply chain but needs operational advancements from all the participants of the supply chain (including the suppliers and their processes).

The traditional supply chain is a hierarchical vertical model and gives the least control over suppliers' products and transparency to their commitments. That leads to various issues in supply management and hampers production. With CPPS, we can overcome all these issues and provide a seamless experience for users' customization. CPPS gives customers an understanding of the real-time processes of their vehicle under production.

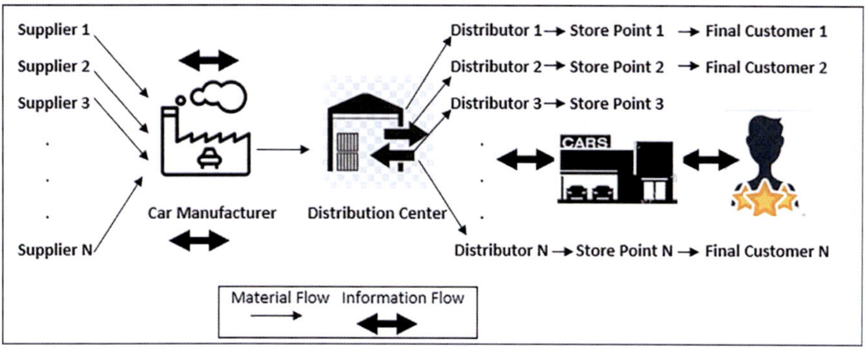

Fig. 11 Generic supply chain

As shown in Fig. 11, the generic supply chain has a simple material flow and information flow with a centralized controller and hierarchical system.

For a seamless CPPS in the automotive supply chain, there must be individual and independent CPS partnered together to fulfill the supply demands. They must create a dynamic, comprehensive, and changeable production system with a high degree of cross-linking [20].

Since it is a culmination of individual CPSs, the hierarchical-vertical supply chain model now becomes interlinked-horizontal decentralized supply chain. Such efforts lead to real-time data access at each stage making the whole automotive system intelligent and efficient. With real-time data in hand, the production hub can determine the workflow and easily monitor-predict the process outcomes and share with end users the status of the production of their vehicles. In situations of high demands, the CPS-based supplier can faster and easily share the data on the ways to fulfill the plans and make practical approaches guided by predictive analysis thanks to the CPS and vertical approach instead of a horizontal supply chain arrangement.

Figure 12 describes how IIoT plays a critical role in the implementation of CPS at each level and hence creating a better CPPS-based supply chain.

The above discussion leads to a concept of value chains which is CPPS-based automotive supply chain. The value chain will organize, optimize, and execute themselves ad hoc. The main enablers of this advancement are intelligent products and logistical objects that know and communicate their current status and location, know their target destinations within the value chain, and control the required production and logistics processes actively [21]. Thus, the digitized value network must be built at the field level. For example, in an industry 4.0 scenario, the purchasing department will be able to track inventories in the own company, as well as in the supply network in real time to keep production running and allow the customer to keep track of the status and degree of completion of his individualized product. The field of view of companies will change from the boundaries of their factories to the whole value network involving all processes and partners from the engineering, sourcing, and production up to final product delivery. For an efficient

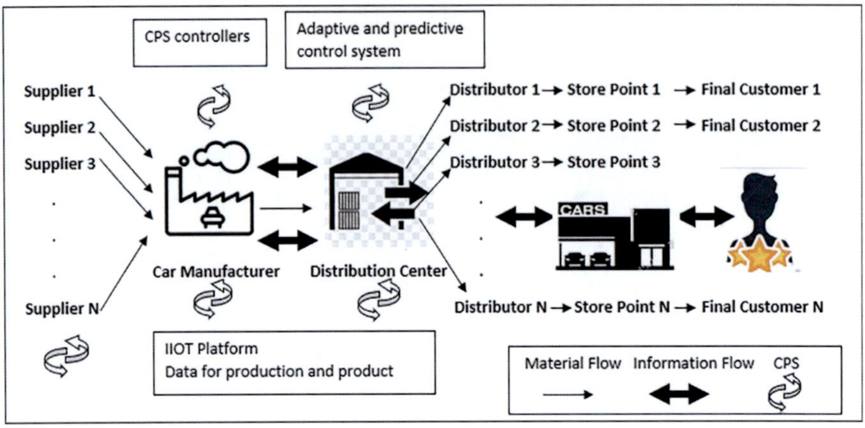

Fig. 12 CPPS-based supply chain

and dynamic exchange of information within an ad-hoc designed value chain, network standards and reference architectures are required [22]. Also, new methods and processes regarding the use of big data to identify customer needs, predictive maintenance for machines, the use of open-innovation principles, and collaborative engineering to produce products that meet the customer needs and new methods of how value chains are decentrally designed, organized, and controlled as well as costs and earnings are allocated within these dynamic value chains have to be developed [23, 24].

The new CPS establishments that will be partners in the CPPS-based value chain will require regulation and reconfiguration of material flow. The current supply chain with the centralized material flow is incapable of future requirement fulfillments of tailor-made products, smaller batch sizes, volatile procurement markets, and sales. This situation is due to centralized architectures that are rigid and unchangeable. The only way out is to have a decentralized control concept and architecture for automated flow systems. For this IIoT offers a great potential to solve weaknesses of centralized systems and create a digitized-decentralized-horizontal flow.

Challenges Facing CPPS and IIoT in the Automotive Supply Chain

Earlier in this chapter, we presented the opportunities brought to the manufacturing industry thanks to the technological advancements arising from industry 4.0.

In this subsection, we will list some challenges that are facing the successful adoption of these technologies (Fig. 13):

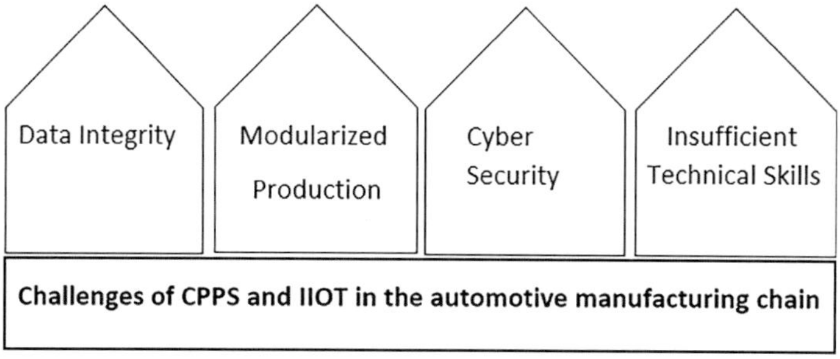

Fig. 13 Challenges of CPPS and IIoT in the automotive industry

- *Data integrity:* in the automotive supply chain, the number of operations is significant, so is the amount of data to be treated. Large amounts of data are collected, accumulated, and shared among different parts of the car manufacturing system.
  The sensitivity of the managed data makes its integrity a big challenge. It is important to ensure that the data recorded by car production systems are of high quality and integrity.
- *Modularized production:* car manufacturing is a complex process where different systems are used at different levels of the process. Equipment and sub-systems should work together and in a modularized and synchronized way to guarantee a distributed decision-making through all manufacturing phases.
- *Cybersecurity:* with the increased connectivity that the industrial Internet provides and the use of standard communications protocols that come with it, a potential need to protect systems and manufacturing data from cyberthreats has arisen. In the context of the automotive supply chain, present and emerging vulnerabilities related to the production systems are of major concern, and interoperability between digital systems expands the attack surface.
- *Insufficient technical qualification:* full automation of the automotive processes leads to a decrease in specific types of work but requires new skill sets. The understanding of the manufacturing processes and required digital tools is of high importance to the successful implementation of these new technologies. To keep up with the digital change, employees across all steps of the automotive value chain need to acquire new skills and qualifications.

4 Blockchain-Based CPS and IoT for the Automotive Supply Chain

As detailed in Sect. 2.2, blockchain is a digital distributed ledger that keeps record of financial transactions in the context of economy. With the third revolution of blockchain, its application is not limited to the financial sector anymore. Blockchain technology can be used in various domains and sectors such as government, health, art, and others.

Blockchain allows data to be shared among all nodes of the network. In this section, we will focus on the implementation of blockchain in industry 4.0 and mainly the automotive industry.

Blockchain-Based Automotive Supply Chain

Automotive manufacturing is one of the most complex sectors. Supply-chain management is critical to the success of the automotive industry. Thousands of parts from around the globe are used in this industry to deliver a high-quality final product. Globalization, changes in manufacturing processes, and customer demands are all factors that impact SCM [25].

The industrial Internet allows satisfying customers' requirements in terms of traceability of material, product, and operations data. Car manufacturers require information from their suppliers to monitor the complete life cycle of a product. With the automation level of CPPS, connectivity, and traceability offered by IIoT and transparency and security guaranteed by blockchain, we can achieve smart automotive factories.

The distributed architecture of blockchain solves the issue of a single point of failure. Participants or nodes are connected to the distributed network and have the same updated version of the ledger. With that being said, it is difficult to alter the data recorded in the blockchain.

Blockchain eliminates the need for intermediaries and third parties, which is a key feature of this architecture. It allows customers to track information about their products directly through the network in confidential and secured ways using cryptographic signatures.

In the following subsection, we present the difference between a traditional automotive supply chain and a blockchain-based supply chain.

Blockchain-Based Automotive Supply Chain

The traditional automotive supply chain model presents several gaps in terms of the relationship among intermediaries, traceability and transparency of operations, and customer ignorance of information about components and products [26].

A traditional automotive supply chain architecture comprises different actors. The main ones are suppliers, car manufacturers, distribution center, distributors, sell points, and final customers.

In a traditional automotive supply chain, information is centralized in each phase of the chain. Users cannot access information about products in different stages,

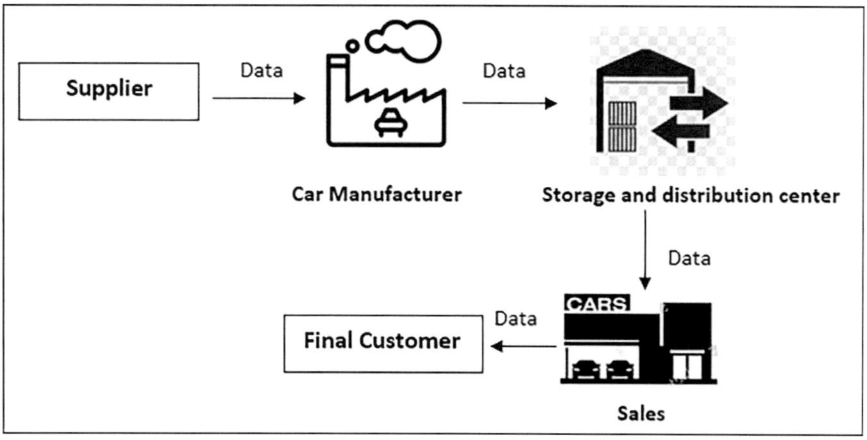

Fig. 14 Traditional supply chain

which makes it difficult to track the quality of the final product throughout the whole process.

Figure 14 presents a traditional supply chain model.

Blockchain technology reinforces supply chain reliability. It provides security, integrity of information, transparency, and traceability of operations results and data.

A blockchain-based system for SC (Fig. 15) has the following characteristics:

- *Consistency:* this ensures that all participants of the network have the same final version of data.
- *System availability:* this means that the system is functioning correctly and treating incoming requests properly at the right time.
- *Partition tolerance:* this guarantees that if a group of nodes is down due to a cyberattack or anything else, the system will still operate correctly.

Along with CPPS and IIoT, BCK (Blockchain)-based SC provides a high level of intelligence, security, transparency, and autonomy to the manufacturing process.

Blockchain Inside a Car Manufacturing Factory

The car manufacturing process is a complex process that requires hundreds of parts and a significant number of operations.

Car production is a sequence of operations and processes until the final product. Each operation comes with an impact on the car's quality. Some operations are more critical than others and require more precision.

Car manufacturing phases are (Fig. 16) as follows:

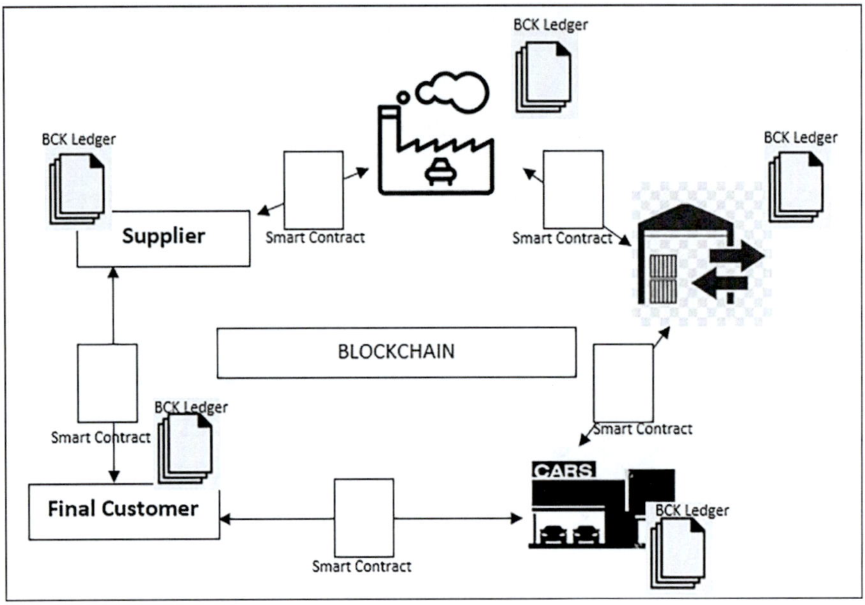

Fig. 15 Blockchain-based supply chain

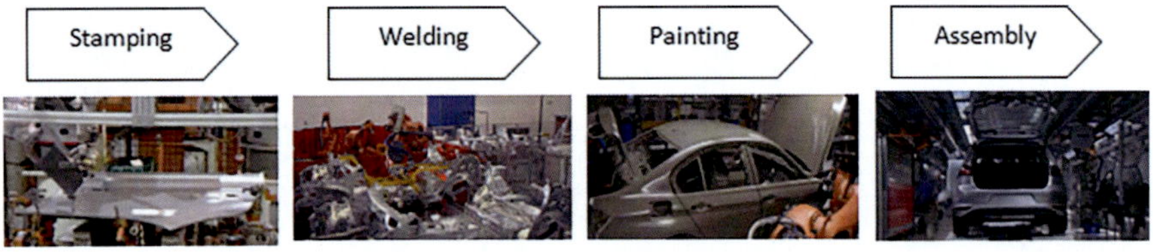

Fig. 16 Car manufacturing phases

1. *Stamping:* hoods, doors, and other body parts are made from sheet metal that has been cut and stamped.
2. *Welding:* robots weld body parts together to form the vehicle's exterior.
3. *Painting:* the welded body is washed and then painted.
4. *Assembly:* the engine, seats, tires, and all interior components of the car are attached to the painted body. Then comes the final inspection process. The output is a finished automobile.

Throughout the whole process, traceability of operation-related data is required, some operations more than others, but still, traceability of this information is essential.

Here comes the role of CPPS and IIoT that offer digitization of information. With intelligent tools and systems, we can get many data related to a given operation in the manufacturing process.

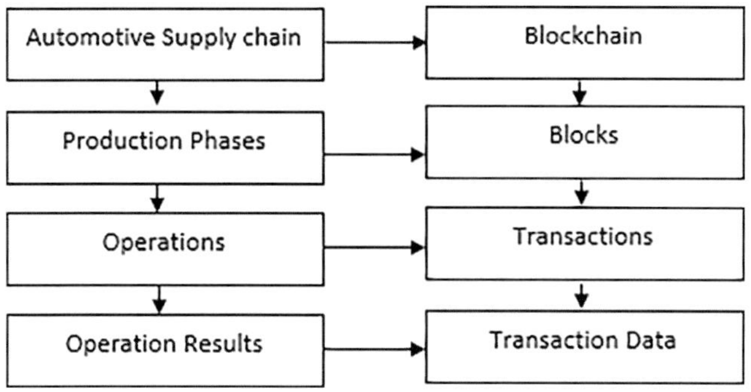

Fig. 17 Correspondence between the automotive supply chain and the blockchain

The recorded data give a history of operations results and operation status. In this way, we get to trace the final product throughout the whole manufacturing process.

The digitization of information and increased connectivity come with several challenges and threats, such as data integrity and security.

In the current supply chain model, not all data are recorded. Available data are only accessible inside the factory network to the authorized parties. To answer the above needs and make the data exchange and traceability transparent, we suggest blockchain technology.

Blockchain allows the record of data in a secure, confidential, and transparent way. Recorded data cannot be tampered or deleted.

From an architectural point of view, blockchain should be seen as a new layer in the data communication architecture as represented in Fig. 17.

Blockchain can be implemented to record operations data and allow customers to follow the production process of their products. It provides confidentiality and integrity of data.

In the context of the automotive supply chain, a private blockchain-oriented enterprise is needed to share data among participants securely, privately, and directly without the need for intermediaries.

In Fig. 18, we present an implementation for blockchain in the automotive manufacturing factory.

Blockchain Inside a Car Manufacturing Factory

- *Limiting part counterfeiting:* a vehicle has a significant number of individual parts that are either manufactured in-house or provided by a supplier. Part counterfeit is a critical problem in the automotive industry. Counterfeited components can find their way to the manufacturing line directly or indirectly. Counterfeit spare parts are untrustworthy because they frequently have degraded quality levels and often fail, causing dissatisfaction among end users and ultimately making customers lose their trust in the brand.

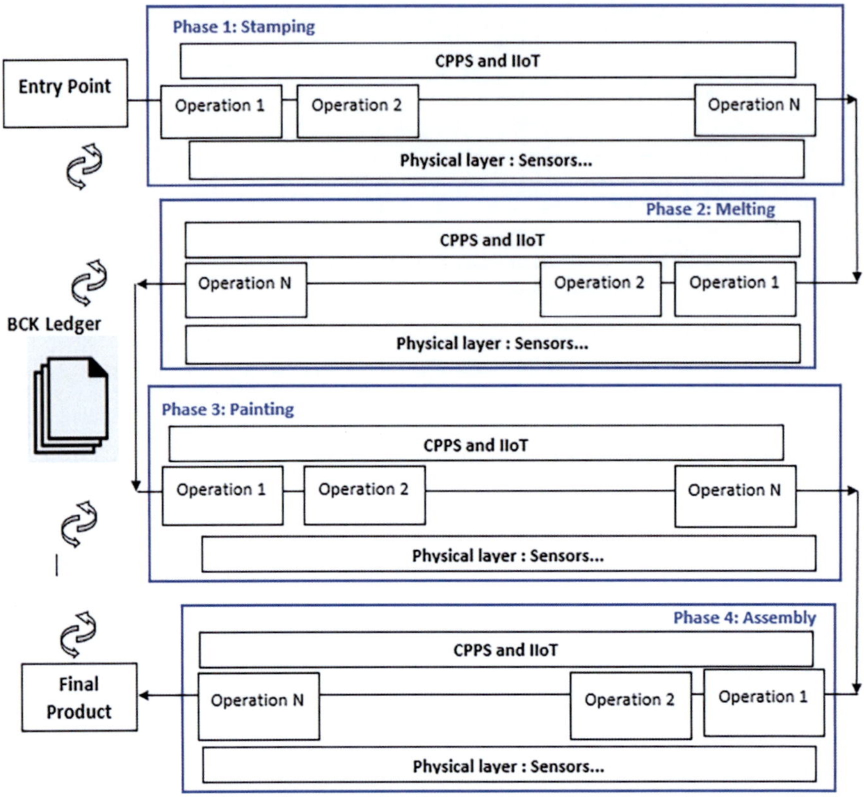

Fig. 18 Blockchain implementation in car manufacturing plant

In this context, blockchain allows to identify and represent parts digitally in a unique way, which makes the process more transparent.

- *Component tracking:* coordination among suppliers providing a significant number of components is not an easy task. Tracking components and parts is complex and prone to errors. In the traditional automotive supply chain model, participants like suppliers, distributors, and dealers do not have a common data-sharing model, which makes it difficult to exchange data related to products such as their location. Blockchain and IoT allow car manufacturers to track components everywhere, in real time and confidentially and securely.
- *Time saving and cost reduction:* in all manufacturing processes, time is money. Part tracking using blockchain prevents production disrupts.
- *Trust establishment:* as explained earlier in this chapter, blockchain offers a high level of transparency, data integrity, and consistency, which enables trust in the whole manufacturing process. The decentralized architecture eliminates the single point of failure issue and makes it complicated to alter data recorded in the blockchain.

Challenges and Limitations of Implementing Blockchain in the Automotive Supply Chain

Despite all the benefits discussed above, blockchain technology has a set of limitations that need to be taken into consideration before thinking about implementing it in the automotive supply chain, and these challenges are detailed in Table 1 as follows:

Table 1 Blockchain-based automotive supply chain challenges

Challenges of blockchain-based automotive supply chain	Explanation and examples
Transactional throughput	As the automotive supply chain is complex, the number of operations and data flow is significant. In the blockchain-based model, the number of operations represents the number of transactions to be done. Blockchain can process limited transactions per second which doesn't comply with real-life scenarios [27] For Ethereum blockchain the number of transactions 20 per second [28]
Latency	The transaction verification and approval process challenge the implementation of blockchain in many sectors [29] The average confirmation time of Ethereum transactions is 5 min [28]
Immutability	One of the core characteristics of blockchain is immutability; once an information is recorded to the blockchain, it is almost impossible to modify it or delete it
Physical limitations of the supply chain systems	The implementation of a blockchain-based model requires several changes in the existing supply chain architecture [30]
High cost of Blockchain implementation	Blockchain implementation requires a significant financial investment [13] The high cost is also due to the energy consumption. As the digital ledger needs to be updated in real time, substantial amounts of computing power are consumed [27, 29]
The gap of technical skills	The implementation and the use of blockchain technology which require a set of technical skills and also the lack of studies evaluating the application of blockchain in the supply-chain management [24] The lack of understanding of blockchain among corporate leaders, the belief that it is a fad, and the desire to wait for wider adoption before committing are all factors working against the technology's adoption [30]

- Transactional throughput: number of transactions
- Latency: time required to record data into the blockchain
- Immutability: data added to the blockchain are not erasable, which means that if any error occurs and somehow wrong information was recorded, it is very complicated to delete it from the blockchain.
- Physical limitations of the supply chain systems: blockchain operates with the IoT, IIoT, and connected systems.
- High cost of blockchain implementation
- The gap of technical skills

5 Conclusions

The process of car manufacturing is one of the most complicated production processes. Many participants such as suppliers, distributors, and dealers contribute to the automotive supply chain. With the increased need for customized products and process adaptability, the need for digitization is of crucial importance.

CPPS and IIoT are technologies that provide digitization of processes-related data and offer connectivity, efficiency, adaptability, and industrial safety. Nonetheless, the increased connectivity of objects and systems in the automotive supply chain expanded the attack surface for cyberattacks. Product tracking throughout the complete manufacturing process is a growing customer demand. Blockchain technology reinforces supply chain reliability. It provides security, integrity of information, transparency, and traceability of operations results and data.

Supply-chain management could be transformed by blockchain technology in a variety of ways, including boosting product security, minimizing counterfeit parts, improving quality management, decreasing the need for middlemen, and cutting the cost of supply chain transactions.

With the automation level of CPPS, connectivity, and traceability offered by IIoT and transparency and security guaranteed by blockchain, the manufacturing process can be intelligent, fully automated, and transparent.

In this chapter, we presented the benefits of industry 4.0 and also the benefits of implementing blockchain technology along with CPPS and IIoT in the automotive supply chain.

Blockchain's huge potential is reshaping the Internet and the entire planet.

The difficulty of blockchain's scalability, on the other hand, is the fundamental reason why this technology hasn't become mainstream yet.

References

1. E. Mavridou, D. Tzovaras, E. Beakiaris, G. Hassapis, M. Gemou, P. Spanidis, *Data Mining and Intelligent Agents for Supporting Mass Customization in the Automotive Industry* (InTech Europe Croatia and InTech, Shangai, 2011)
2. Top 4 Automotive Supply Chain Challenges and Solutions. Blume Global. [Online]. Available: https://www.blumeglobal.com/learning/automotive-supply-chain/. Accessed 04 July 2021
3. D. Serpanos, M. Wolf, Industrial Internet of Things, in *Internet-of-Things (IoT) Systems*, (Springer, Cham, 2017), pp. 37–54
4. H. Kagermann, J. Helbig, A. Hellinger, W. Wahlster, *Recommendations for Implementing the Strategic Initiative INDUSTRIE 4.0: Securing the Future of German Manufacturing Industry ; Final Report of the Industrie 4.0 Working Group* (Forschungsunion, Berlin, 2013)
5. M. Broy, M.V. Cengarle, E. Geisberger, Cyber-physical systems: imminent challenges, in *Large-Scale Complex IT Systems. Development, Operation and Management Lecture Notes in Computer Science*, (2012), pp. 1–28
6. H. Meissner, J.C. Aurich, Implications of cyber-physical production systems on integrated process planning and scheduling. Proc. Manuf **28**, 167–173 (2019)
7. S. Wiesner, K.-D. Thoben, Cyber-physical product-service systems, in *Multi-Disciplinary Engineering for Cyber-Physical Production Systems*, (2017), pp. 63–88
8. E.M. Abou-Nassar, A.M. Iliyasu, P.M. El-Kafrawy, O.-Y. Song, A.K. Bashir, A.A.A. El-Latif, DITrust chain: Towards Blockchain-based trust models for sustainable healthcare IoT systems. IEEE Access **8**, 111223–111238 (2020)
9. G.N. Nguyen, N.H.L. Viet, M. Elhoseny, K. Shankar, B. Gupta, A.A.A. El-Latif, Secure blockchain enabled Cyber–physical systems in healthcare using deep belief network with ResNet model. J. Parallel Distrib. Comput **153**, 150–160 (2021)
10. H. Feng, X. Wang, Y. Duan, J. Zhang, X. Zhang, Applying blockchain technology to improve agri-food traceability: A review of development methods, benefits and challenges. J. Clean. Prod. **260**, 121031 (2020)
11. K. Abbas, L.A.A. Tawalbeh, A. Rafiq, A. Muthanna, I.A. Elgendy, A.A.A. El-Latif, Convergence of Blockchain and IoT for secure transportation systems in smart cities. Secur. Commun. Netw **2021**, 1–13 (2021)
12. A.A.A. El-Latif, B. Abd-El-Atty, I. Mehmood, K. Muhammad, S.E. Venegas-Andraca, J. Peng, Quantum-inspired blockchain-based cybersecurity: Securing smart edge utilities in IoT-based smart cities. Inf. Process. Manag. **58**(4), 102549 (2021)
13. R. Cole, M. Stevenson, J. Aitken, Blockchain technology: Implications for operations and supply chain management. Supply Chain Manag. Int. J **24**(4), 469–483 (2019)
14. P. Baran, *On Distributed Communications: I. Introduction to Distributed Communications Networks* (Ft. Belvoir Defense Technical Information Center AUG, Fort Belvoir, 1964)
15. M. Swan, *Blockchain: Blueprint for a New Economy* (OO'Reilly Media, Inc., Sebastopol, 2015)
16. S. Vaidya, P. Ambad, S. Bhosle, Industry 4.0 – A Glimpse. Proc. Manuf **20**, 233–238 (2018)
17. K. Sipsas, K. Alexopoulos, V. Xanthakis, G. Chryssolouris, Collaborative maintenance in flow-line manufacturing environments: An industry 4.0 approach. Proc. CIRP **55**, 236–241 (2016)
18. Industry 4.0 market size, share & COVID-19 impact analysis, by application (Industrial automation, smart factory, and industrial IoT), by vertical (manufacturing, energy & utilities, automotive, oil and gas, aerospace and defense, electronics and consumer goods, and others), and regional forecast, 2021–2028. Industry 4.0 market size, growth and analysis [2021–2028]. [Online]. Available: https://www.fortunebusinessinsights.com/industry-4-0-market-102375. Accessed: 22 June 2021
19. F. Almada-Lobo, The industry 4.0 revolution and the future of Manufacturing Execution Systems (MES). J. Innov. Manag **3**(4), 16–21 (2016)

20. D. Pantförder, F. Mayer, C. Diedrich, P. Göhner, M. Weyrich, B. Vogel-Heuser, Agentenbasierte dynamische Rekonfiguration von vernetzten intelligenten Produktionsanlagen – Evolution statt Revolution, in *Industrie 4.0 in Produktion, Automatisierung und Logistik*, (2014), pp. 145–158
21. S. Dais, Industrie 4.0 – Anstoß, Vision, Vorgehen, in *Industrie 4.0 in Produktion, Automatisierung und Logistik*, (2014), pp. 625–634
22. Referenzarchitekturmodell Industrie 4.0 (RAMI4.0). VDI, 01-Apr-2015. [Online]. Available: https://www.vdi.de/ueber-uns/presse/publikationen/details/referenzarchitekturmodell-industrie-40-rami40. Accessed 25 Jun 2021
23. M. Meyer, Management control in Unternehmenskooperationen. Control. Manag **52**(5), 333–335 (2008)
24. M. Schroeck, R. Shockley, J. Smart , D. Romero Morales, P. Tufano, Analytics: the real-world use of big data: How innovative ..., Jan-2012. [Online]. Available: https://www.researchgate.net/publication/315786855_Analytics_the_real-world_use_of_big_data_How_innovative_enterprises_extract_value_from_uncertain_data_Executive_Report. Accessed 25 June 2021
25. What Are the Main Supply Chain Challenges?. Blume Global. [Online]. Available: https://www.blumeglobal.com/learning/supply-chain-challenges/. Accessed 12 July 2021
26. I.M. Ambe, J.A. Badenhorst-Weiss, An automotive supply chain model for a demand-driven environment. J. Transp. Suppl. Chain Manag **5**(1) (2011)
27. Blockchain for Industry 4.0: A comprehensive review | IEEE ... [Online]. Available: https://ieeexplore.ieee.org/document/9069885. Accessed 5 Sept 2021
28. A. Bhalla, Top cryptocurrencies with their high transaction speeds. Blockchain Certifications, 16-Apr-2021. [Online]. Available: https://www.blockchain-council.org/cryptocurrency/top-cryptocurrencies-with-their-high-transaction-speeds/. Accessed 5 Sept 2021
29. Blockchaintechnologycom, "Advantages & Disadvantages of Blockchain Technology," Blockchain Technology, 23-Nov-2016. [Online]. Available: https://blockchaintechnologycom.wordpress.com/2016/11/21/advantages-disadvantages/. Accessed 01 Sep 2021
30. S. Jabbar, H. Lloyd, M. Hammoudeh, B. Adebisi, U. Raza, Blockchain-enabled supply chain: Analysis, challenges, and future directions. Multimedia Syst **27**(4), 787–806 (2020)

BIoVN: A Novel Blockchain-Based System for Securing Internet of Vehicles Over NDN Using Bioinspired HoneyGuide

Zakaria Sabir and Aouatif Amine

1 Introduction

Researchers and manufacturers discuss and develop ITS to achieve road safety and comfortable driving (intelligent transportation systems). Current vehicles already have a set of modern technologies deployed. For instance, they can run various applications, use navigation systems, and connect to the Internet. The following purpose is achieving V2V (vehicle-to-vehicle) [1] and V2I (vehicle-to-infrastructure) [2] communications to enhance safety messages exchanged between road users [3].

Vehicular networks are characterized by dynamic topology and high connectivity, which is an issue for the current Internet architecture based on the TCP/IP model. Therefore, research communities proposed to bring NDN (named data networking) to IoV (Internet of vehicles) to tackle those issues. NDN is a future Internet architecture that is based on named content rather than IP addresses. It uses a request/reply model which doesn't need session establishment nor address allocation to exchange data.

In terms of security, NDN came with a new model that uses public and private keys to encrypt and decrypt data in the network. The main idea is to secure the data itself rather than securing the medium of transmission. However, additional security concerns still exist in NDN and require further research [4]. Thus, we thought to bring the blockchain technology to NDN-based IoV, especially to overcome the cache poisoning attack that intends to propagate bogus content to the network's nodes, and thereby to forward only safe content. Blockchain is a decentralized, distributed, and open ledger that saves transactions efficiently in a transparent and immutable way.

Z. Sabir (✉) · A. Amine
Ibn Tofail University, ENSA, Kenitra, Morocco
e-mail: za-karia.sabir@uit.ac.ma; aouatif.amine@uit.ac.ma

Y. Maleh et al. (eds.), *Advances in Blockchain Technology for Cyber Physical Systems*, Internet of Things, https://doi.org/10.1007/978-3-030-93646-4_8

This paper proposes a novel blockchain-based system for securing IoV over NDN, called BIoVN (**b**lockchain for **I**nternet **o**f **v**ehicles over **N**DN). The proposed system aims to deliver, forward, and cache only safe content over the network. To the best of our knowledge, very few publications are available in the literature that discusses the IoV over NDN in terms of blockchain technology or addresses the security of the cached and forwarded content.

We also propose a new bioinspired algorithm of HG (HoneyGuide), which we used in the BIoVN system. The principal aim of any algorithm is to discover the most outstanding couple between a collection of possible results and a goal model. This can be approached to discovering a fulfilling result in discrete search space over a reasonable execution time. Representing the problem in this manner makes it very much similar to the attitude of an optimization algorithm. Indeed, metaheuristics are considered repetitive actions that cleverly lead a subordinate heuristic to create excellent results by operating diversification and intensification methods in the space. Metaheuristics are approximate algorithms that exceed exact algorithms regarding the scope of the resolved problem. The capacity to deal with different difficulties using miniature variations usually comes from the nature inspiration. Using easy tools, different problems are resolved in nature, and a quality result is always found thanks to the attitude of living beings such as animals and insects. Therefore, we tried to discover an effective metaheuristic stimulated by the power of nature to resolve complicated problems and produce results with acceptable quality. Currently, we are in the phase of the simulation of the system. The major contributions of this work are twofold as follows:

- To prevent malicious vehicles from broadcasting poisoned content, we designed a novel reasonable blockchain-based security architecture for IoV over NDN, which ranks different nodes in the network and allows only trusted ones to exchange interest and data packets.
- We proposed a new bioinspired algorithm of the name HG, which we used in the BIoVN system.

The remainder of this paper is organized as follows: Sect. 2 presents an overview of NDN and blockchain technologies, Sect. 3 summarizes the related work, and Sect. 4 describes the proposed method. Finally, we conclude the paper.

2 Overview of NDN and Blockchain Technologies

Named Data Networking

NDN (named data networking) is an instance of ICN (information-centric networking) [5], which is recognized as a significant field of study. Among instances of ICN, NDN is proposed as a promising future Internet architecture, and it is based mainly on the content (what to send) instead of the location (where to send). Named data packets are used rather than destination and source addresses used by the current TCP/IP architecture [6]. While forwarding is done using IP address headers in IP-based routers, each packet name prefix is used by NDN-based routers to forward packets. This adoption of unique named contents allows nodes to memorize and control the state of each packet. NDN and the current Internet architecture have the same hourglass architecture except for some divergence in similar layers [7]. The named content chunks constitute the principal blocks of NDN, while in TCP/IP, the basic unit of communication is a point-to-point channel between two nodes identified by IP addresses [8, 9].

Two types of packets are engaged in NDN: interest and data [10]. They are used, respectively, by consumers and producers while communicating. Since the NDN is recipient driven, consumers express their desire for a piece of data by putting its name in an interest packet and sending it to the network. Any node which retains a copy of the desired content will play the role of the producer and reply with a data packet. This packet will then take the reverse path to come back to the consumer [11]. Each node maintains three databases [12] used in the forwarding process: the CS (Content Store) stores copies of freshly forwarded data to supply future queries and increase content distribution. The PIT (pending interest table) keeps track of forwarded interest packets that are not yet satisfied. Incoming interfaces are saved in the PIT entries, so data packets can easily retrieve the correct path to reach original consumers. And the FIB (forwarding information base) records the important forwarding information like the prefix and the next hopes. This information is used to lead the interest packet to the potential providers.

Figure 1 illustrates an example of NDN-based IoV architecture. In this example, the content "/parking/ Mimosas/P3" is desired by Consumer 1. As the forwarder has this content in its cache, it will send it directly to Consumer 1 without forwarding it to the initial producer. However, the content "/traffic/highway/A5/20" desired by Consumer 3 will be transferred to the initial producer. If Consumer 2 also expresses an interest in the same content, Consumer 3 will aggregate this request in its PIT and not forward it. Once the producer sends the data packet to the forwarder, the latter will forward it to Consumer 3, which will forward it in its turn to Consumer 2 thanks to the PIT entries.

Blockchain Technology

Blockchain technology is a distributed peer-to-peer network and an open ledger that Satoshi Nakamoto first implemented for Bitcoin [13]. In contrast to the traditional centralized ledger systems, all the participants in the network keep together a copy of the distributed ledger with the help of a consensus algorithm. Every user is authorized to add or modify data to solve a complex mathematical puzzle. Every participant in a blockchain system is recognized with a cryptographic public key shared with other users so they can interchange information. The private key, in contrast, is kept stored securely in the client's equipment. In blockchain technology,

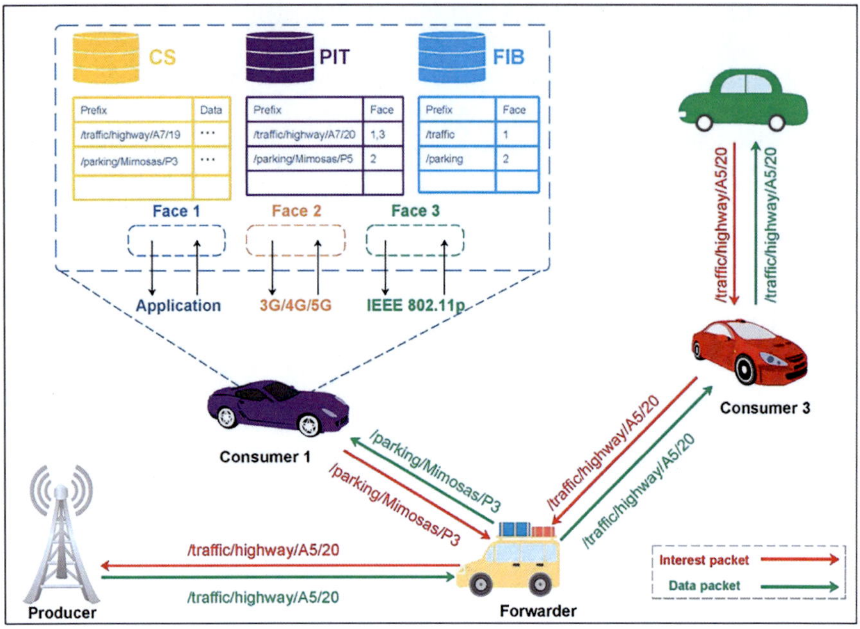

Fig. 1 Example of NDN-based IoV architecture

it is highly impossible to succeed an attack in the network system since a private key cannot be derived from a public key, according to [13] and [14]. The transactions in the blockchain are stored in a verifiable and immutable way.

The data structure of the blocks is designed in a specific way that links each block to the previous block via its unique hash value. Trustworthiness and authenticity in the network are established through a consensus algorithm which is considered an important component of the blockchain system. This process is known as "mining," and nodes that participate are called "miners." Mining is done without engaging any third party or central authority [15]. Some examples of consensus algorithms are PoW (proof of work) which is used in Bitcoin; DPoS (delegated proof of stake), which is used in Ethereum; proof of elapsed time; proof of burn; proof of space; proof of luck; and practical Byzantine fault tolerance.

Although blockchain was initially proposed for the commercial industry, it is currently revolutionizing different fields. It can support diverse applications and services [16] such as IoT (Internet of things), banking system, stock market trading, supply chain management, government record, hospitalization, voting, and property transfers. Figure 2 depicts an example of a transaction in blockchain.

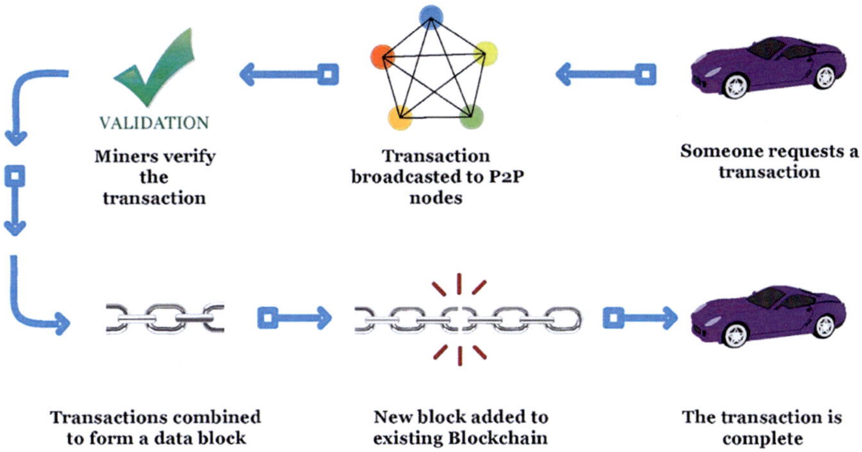

Fig. 2 Example of a transaction in the blockchain network

3 Related Work

Security in NDN-Based IoV

A huge amount of data is being exchanged in vehicular networks. The aim behind these communications is to improve road safety and enhance the driving experience. Securing the exchanged information becomes more important due to the direct impact on human life. Different researchers tried to bring the future Internet architecture NDN to IoV to improve various features, including security.

Authors in [17] proposed IFAMS (Interest Flooding Attack Mitigation Scheme), a new scheme that focuses on mitigating interest flooding attacks in VNDN (vehicular named data networking). They have edited the interest packet to create a new packet field named "consumer_id" which is a unique integer value. They have also assembled the IDs of malevolent vehicles in a table named "restricted id table" maintained by each vehicle. A randomized time value is assigned to every vehicle in the network at the time of ignition. Arsalan et al. [18] proposed a new scheme named TAP (timing attack prevention) to tackle the issue of timing attack in vehicular networks by merging NDN with SDN (software-defined networks). The proposed scheme uses the network controller to mitigate the attacker detected by a legitimate vehicle. To find out the time period in which a packet arrives from the source to the destination, a vehicle first calculates the distance using its coordinates, the sender's coordinates, and the signal propagation speed in the detection process. Then, the time period value is deducted from packet arrival time to get the last vehicle arrival time to check whether the previous vehicle added any delay in the packet or not. Authors in [19] discussed the security of VANET (vehicular ad hoc networks) using a trust function. The trust value of every vehicle interested in receiving data packets is calculated. They have also calculated the number of occasions an interest packet is shown by vehicles in a forwarding list. Sending data to neighbors is based on the number of vehicles that appear in the route. Manimaran et al. [20] proposed an adaptive IDS (intrusion detection system) for NDN named IDSNDN. Using sensor values present in the vehicle's OBU (on-board unit) and the heartbeat rate value, authors created the rules of the IDS. An additional packet named "sensor packet" is used by the IDS, which is included in the RSUs (roadside units). Once receiving a safety message, the RSU forwards it to the IDS module, verifying it based on the ruleset. If the message turns out to be fake, an alarm is returned.

Blockchain-Based IoV

IoV has become a promising research area recently. By allowing direct information exchange between vehicles, vehicular networks help in reducing traffic congestion. However, malicious vehicles may misguide the whole communication. Blockchain technology is considered a preferred technique to grant security in real-time circumstances to deal with this issue. In [21], the authors proposed a blockchain framework to address the issue of security in intelligent sensors of autonomous and connected vehicles prone to intrusion. The proposed scheme was studied based on different criteria, like a compromise of smart equipment, bogus queries of the user, authentication scenarios, etc. Shrestha et al. [22] presented a public blockchain that stocks the message and node trust in a relevant distributed ledger to solve critical data propagation issues in IoV. Authors created a new type of local blockchain useful for IoV for real-time message dissemination between vehicles up to the borderline of a country. Kudva et al. [23] studied the mitigation of attacks in IoV and proposed a blockchain-based decentralized trust score framework for the engaged vehicles to find block insider attackers. The authors proposed a detection system composed of two levels. In the first one, the trust is calculated individually by nodes, whereas trust scores for nodes are aggregated by a consortium blockchain-based system in the second one. Authors in [24] proposed a blockchain-based secure data sharing system to solve the issue of trust in IoV. This system uses blockchain to store announcement messages. Vehicles that participate either to block generation or to broadcast announcement messages are rewarded by some cryptocurrency. Kang et al. [25] addressed secure sharing and data storage in IoV by exploiting smart contract and consortium blockchain techniques which effectively prevent data distribution without permission. The authors proposed a reputation-based data sharing scheme to grant data distribution between vehicles with high quality.

Bringing blockchain to NDN-based IoV faces a lack of solutions and is still in its early research stage [26, 27]. In this paper [28], different from those studies, the primary purpose is to design a secure and credible NDN-based IoV system. This is the first work that deals with such a system using both a blockchain technology and a bioinspired algorithm to the best of our knowledge.

Bioinspired Algorithms

Metaheuristic algorithms have lately gotten a lot of interest in a variety of domains. The ACOA (ant colony optimization algorithm) [29] is inspired by ant behavior when foraging for food. They use the pheromones to spread the word to the rest of their team after they have located a nice location. A novel modelization approach is included in the BCO (bee colony optimization) [30] to demonstrate bee movement behavior while searching for food sources. For example, in the PSO (particle swarm optimization) [31], a flock of birds searching for grain works together cooperatively and intelligently to explore the surface of the target. The GA (genetic algorithm) is another well-known bioinspired algorithm [32]. Creating a species that can survive in a certain climate is similar to creating a genetic code. In the realm of computer vision, this approach has been used on a regular basis, with great success. The FA (firefly algorithm) is inspired by the firefly insect's behavior of employing light to attract the attention of other fireflies in its vicinity [33]. The BA (bat-inspired technique) [34] is another contemporary optimization algorithm. This method mimics the echolocation feature of bats, which allows them to distinguish between various prey in the shadows. Cuckoo search, according to several studies, outperforms population-based bioinspired algorithms in terms of exploration and is also suitable for large issues [35, 36]. proposes a CS-based optimization approach for extending the life span of a WSN (wireless sensor network). This approach deploys nodes in the network at random and organizes them into clusters once they have been deployed. The CS aids in cutting down on the consumption of energy. There is a proposal in [37] for improving the CS. The authors suggest a solution to the dilemma of the roadside salesperson. When compared to previous methods, the suggested algorithm performed admirably. To address the NDN-based IoV security issue, we devised a novel algorithm we dubbed HG (HoneyGuide) based on the findings from several research domains. It is possible to think of the search for valid vehicles and the detection of fraudulent ones as discrete 2-D explorations in a research area where time doesn't matter. The HG also has the benefit of having a large variety of settings. Unlike other bioinspired algorithms, the HG method only uses two parameters: the likelihood that the owner bird would discover the honeyguide's eggs and the population size (how many caves in the primary population). The suggested technique and the HG algorithm will be discussed in more detail in the next section.

4 Proposed Method

In this section, we first discuss the security of NDN-based IoV and present our proposed algorithm HG (HoneyGuide). Subsequently, we describe our proposed blockchain-based system for securing IoV over NDN.

Security of NDN-Based IoV

Enabling IoV over TCP/IP has always caused various technical issues; thus, research communities started to propose various methods to bring the future Internet architecture NDN to IoV. Since data names are used in this architecture, there is no need to establish sessions or allocate addresses to exchange data, making it a bright solution to enable effective vehicular communications. Applying NDN to IoV is very beneficial in various properties such as in-network caching, mobility and routing, and in-data security. The latter is the primary concern of this work. NDN looks at security as the main component of the data itself; it doesn't rely on the transmission medium anymore. Since all the vehicles in the network have the ability to cache content, the security attributes of the data packets have to be decoupled from their locality.

The producer uses its cryptographic key to sign each data packet by binding the content to its name while creating data. This allows any user of the network to check the integrity of the data packet. Applications in NDN must use data signatures and cannot drop security.

Nevertheless, additional security concerns still exist in NDN and require further research. Basically, two kinds of attacks need more investigation: interest flooding attack aims to send a tremendous number of interest packets to consume network resources and thus block legitimate requests. A cache poisoning attack aims to propagate fake content to the nodes of the network. Our work is considered as a contribution to resisting the last attack. We believe that blockchain can be a big revolution thanks to its decentralized nature; therefore, we propose a system that brings blockchain to NDN-based IoV to enforce security as explained later in this section.

HoneyGuide Search Algorithm

The HG is a new kind of optimization algorithm, influenced by biological principles. Processing includes using the exploring design of different biological organisms, such as insects and sharks, for its own purposes. In another scene, the HG mimics the honeyguide bird's behavior and reproduces in the same manner.

As a brood parasite (laying one egg in another species' nest), the honeyguide bird prefers hole-nesting species. Honeyguide chicks defecate on or even kill the chicks of the owner bird. They do this by sticking needles in their beaks.

There was an extra day added by the honeyguide female to ensure that her kid hatched before the host's. As a result, the honeyguide infant is constantly a step ahead of the pack in terms of development. It searches for caves in the search region to lay its eggs, and it assigns hatching probabilities to each one based on its findings.

The owner bird may be able to identify honeyguide egg generations in the future. The caverns will be replaced with new ones if this happens. New caverns are generated at random. We also feel that the area is a search area, with each cave acting as a potential vantage point from which the missing vehicle may be located.

The idea for the suggested technique came from the need to develop a robust vehicular communication algorithm that can function in a variety of challenging situations and with no prior knowledge. The HG method is convenient due to its small number of parameters. It just takes a tiny population to achieve great things. As a result, it is faster than other algorithms since its time complexity is O(n). The "HoneyGuide Search" algorithm is summarized in Algorithm 1.

To begin, the HG algorithm picks a random starting population of "n" subterranean caves at random. There is a fitness function that measures the value of these caves and ranks them according to that value. The cave with the best fitness function also doubles as a better place for eggs to hatch. As a result, there is a good chance it will be a better cave than average. After that, we will use the HG algorithm to determine the quality of the eggs in order to separate out harmful ones from the good ones and keep the good ones safe.

Algorithm 1 HoneyGuide Search

Objective function $f(x), x = (x_1,, x_d)$;
Generate an initial population of n caves x_i $(i = 1, 2,, n)$;
Generate an initial population of m BirdB; y_j $(j = 1, 2,, m)$;
Initialize the initial population of BirdB with ranks (RankB);
Set MaxIteration;
Set RankHB;
$Iteration. \leftarrow 0$
$j \leftarrow 0$
while ((j < MaxIteration) or (Stop Criterion)) **do**
Evaluate its quality/fitness F_i;
  if $f_i > f_j$ **then**
    Replace j by the new solution;
    $RankHB \leftarrow 4$
else
    $RankB\, y_j \leftarrow -1$
  end if
  Iteration ++;
  j ++;
end while
Post-process results and visualization;

Each cave (search zone) contains a number of eggs that represent vehicles in our case, as shown in Fig. 3. The HG algorithm aims to increase the rank of the HB (honeyguide baby) egg and decrease the rank of the other eggs, after a number of iterations, while this number is lower than the maximum iteration or fixed, the criterion is stopped. To increase the rank of HB (i.e., legitimate vehicle), we have to verify the trustworthiness (see Section "A Blockchain system for securing Internet of Vehicles (BIoVN)"), which is indicated by a fitness function in our case; if the

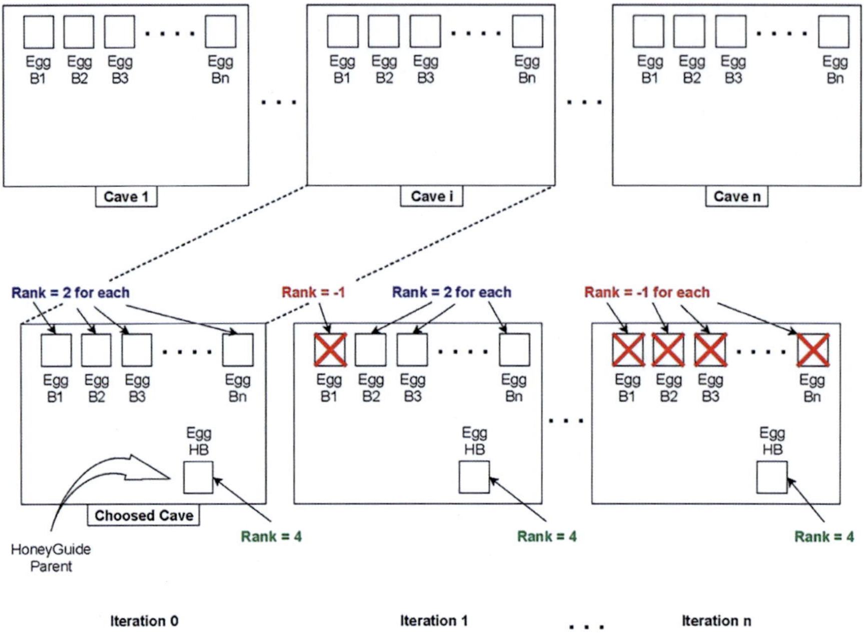

Fig. 3 HoneyGuide search algorithm

fitness function does not satisfy a fixed condition, then the rank of the other eggs (i.e., malicious vehicles) is decreased; so forth, we finally increase the overall quality of the population.

A Blockchain System for Securing Internet of Vehicles (BIoVN)

We believe that blockchain can be a big revolution thanks to its decentralized nature. Therefore, we propose a system that brings blockchain to NDN-based IoV to enforce security. The reasons behind this are summarized in three points:

- In the IoV over NDN, delivering safe content is very important, and blockchain can support this subject as it gives users of the network a new means to keep persistent and accurate databases in a decentralized way and without engaging any other authority.
- Our system belongs to the approaches whereby a rank is calculated and disseminated over a network. Such approaches can reinforce and enhance NDN-based IoV trustworthiness and security by working complementary to the existing systems.
- The design of NDN uses public and private keys to encrypt and decrypt any data in the network, which is homogeneous with the blockchain concept.

System Design

In NDN-based IoV, data privacy and security are crucial obligations. Any vehicle is allowed to cache data packets in its CS to fulfill future interest packets. To prevent malicious vehicles from broadcasting poisoned content, we describe in this part the design of our proposed blockchain-based architecture. As we mentioned before, our system belongs to approaches that calculate ranks for network nodes represented by vehicles in this case. The calculated ranks will be displayed in the blockchain network as transactions.

On the one hand, vehicles and RSUs can play different roles, from consumers to data mules to producers, and they can cache data in their cache stores. On the other hand, malicious vehicles serve their poisoned content and induce caching fake data in the network. For this reason, we suggest assigning a rank "R" to each node in the network; this rank represents the level of trust of the node and has an initial value that can increase or decrease based on the provided content by this node. A reliable content will result in raising the rank of the vehicle and, accordingly, the trust level. A fake content will result in reducing the rank of the vehicle and, accordingly, the trust level. We also propose to create a new table of name MVT (Malicious Vehicles Table), which contains the IDs of malicious vehicles detected over the network. Every node will maintain this table. System architecture

As we mentioned before, our system uses both blockchain and NDN technologies over IoV. Vehicles have different roles (i.e., consumers, data mules, or producers). Meanwhile, they can also be miners when validating transactions and running the consensus algorithm or users when generating transactions and receiving blocks. Vehicular networks are privileged from other ad hoc networks by having unlimited processing and storage capabilities; thus, we assume that vehicles don't have any difficulty dealing with transactions. Figure 4 shows the design of the blocks in the system.

The underlying characteristic of this architecture is the requirement of the peers to discover whether the intention of other nodes in the network is malevolent or honest. The purpose is to allow only legitimate vehicles to accumulate a good trust level. In this manner, non-malevolent vehicles can identify malicious ones easily and eliminate them from the transactions accordingly.

The main transaction exchanged in the BIoVN system is the one that assigns a rank (i.e., trust level) to each vehicle. Once connected to the network, the vehicle receives an initial rank value "R" which is refreshed based on its served content. The rank value raises if the vehicle replays with a safe data packet and diminished if it replays with a corrupted data packet. This transaction involves two nodes: the first one can be a consumer or a producer. It contributes to the verification of the received packet and thus modifies the rank of the sender. The second one is the packet's sender; it will have an updated rank value after the validation by the first node. The forwarding process of BIoVN is illustrated in Fig. 5. Once a node receives an interest packet, the ID of the sender is verified. If the MVT indexes it, the node drops the packet immediately. Otherwise, the process is continued according to the NDN interest forwarding process.

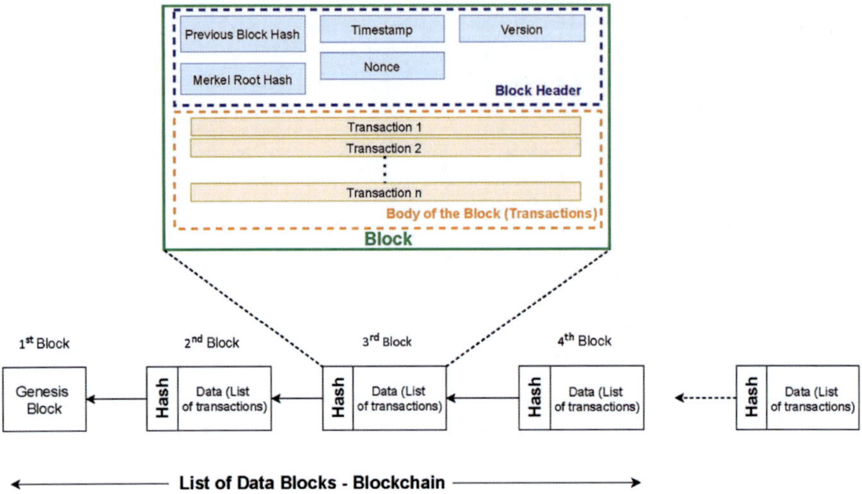

Fig. 4 Design of the blocks

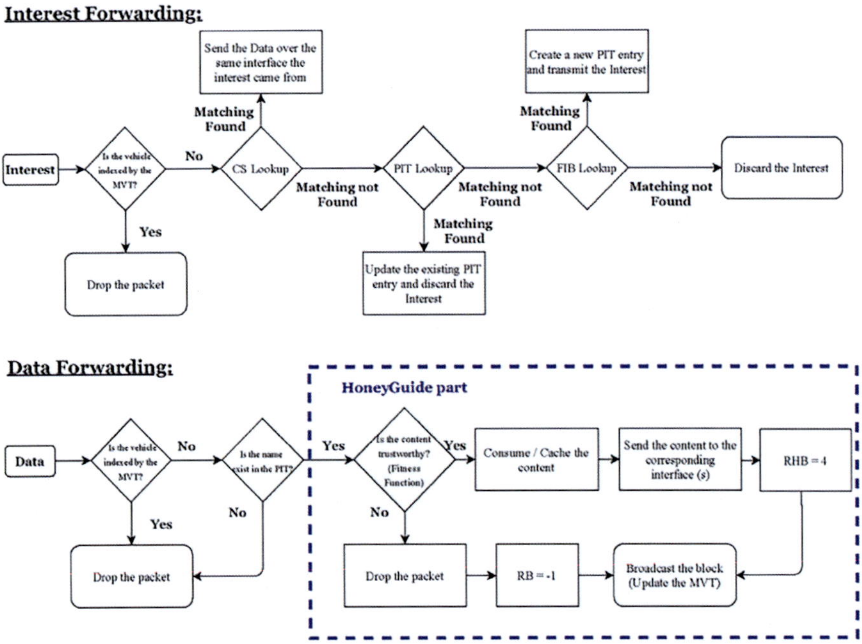

Fig. 5 Forwarding process in BIoVN

Introducing HoneyGuide to Blockchain (Data Forwarding)

The main task of the HG focuses on the update of the MVT. Our required step here is to find the most stable table among introduced nodes by the HG part. An efficient new fitness function is also introduced and used in this phase. The HG algorithm selects the most stable route via intelligent configuration of the fitness function based on the parameter rank R, which will be increased when the fitness function is satisfied and decreased otherwise (refer to Section "HoneyGuide Search Algorithm"). Following that, the table is updated in the second part.

The next step is getting information about surrounding vehicles. Once a vehicle X receives a data packet from another vehicle Y, it sends a request to the blockchain system to get Y's trust level (i.e., the rank R). If Y is found to be recorded in the MVT, X will drop its packet immediately and won't cache it in its CS. Else, if Y is legitimate, its name will not be recorded in the MVT, and X can trust it (see Fig. 5).

The process continues then, and the PIT is verified to ensure that the desired data is already recorded among packets which are not yet satisfied. If it is not the case, the packet is dropped. If it is the case, the packet enters the HoneyGuide part to verify the fitness function. The trustworthiness of the packet is verified based on the served content. If it is fake, the packet is dropped, and the consumer ranks the sender to "-1" and broadcasts the block to the network to update the MVT. If it is authentic, the consumer can consume the content if it is the original requester or caches and forward it to the corresponding interfaces. The consumer then ranks the sender to "4" and broadcasts the block to the network to update the MVT. Figure 6 depicts the basic elements of the BIoVN system.

As a preliminary result, forwarding interest packets in the original VNDN is done without validation. In contrast, in our proposed system, the interest packets are forwarded only if they are valid. In other words, BIoVN discards the packets coming from malicious vehicles, and only the interests coming from vehicles with high ranks are forwarded. It also communicates with the blockchain network about malicious vehicles and reduces their ranks. VNDN also has a significant PIT memory since each interest packet occupies an entry even if it is invalid, which leads to memory consumption compared to BIoVN, which has a miniature table. The same result also can be obtained when it comes to the memory utilization of the CS since VNDN caches all the packets and doesn't reject the invalid ones.

5 Conclusion

The main purpose behind this work was to propose a robust system to secure the Internet of vehicles. In this paper, we proposed a new system that introduces blockchain to NDN. After that, we combined a novel bioinspired algorithm HoneyGuide, with blockchain that we introduced in the data forwarding process. In conclusion, from the outcome of our investigation, we believe that bringing blockchain to the Internet of vehicles over the future Internet architecture named

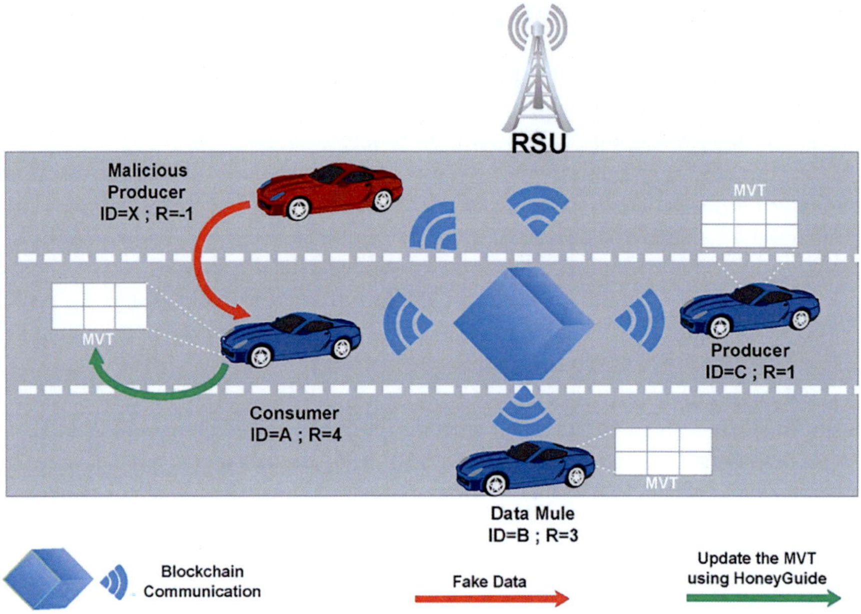

Fig. 6 Basic elements of the BIoVN system

data networking can be considered a great help against security issues faced by vehicular networks. To the best of our knowledge, this is the first work dealing with such a system using blockchain technology and a new bioinspired algorithm. The future work consists of completing the simulation and comparing our system to the original VNDN.

Acknowledgments This research work is supported by the "SafeRoad: Multiplatform for Road Safety (MRS)" Project under contract No: 24/2017, financed by the Ministry of Equipment, Transport, Logistics and Water (METLE), and the National Center for Scientific and Technical Research (CNRST).

References

1. Z. Sabir, A. Amine, A novel system based V2V communications to prevent road accidents in Morocco, in *The 2021 International Conference on Digital Technologies and Applications, Morocco*, (2021)
2. Z. Sabir, S. Dafrallah, A. Amine, A novel solution to prevent accidents using V2I in Moroccan smart cities, in *2019 International Conference on Computational Intelligence and Knowledge Economy (ICCIKE), UAE*, (2019), pp. 621–625
3. Z. Sabir, A. Amine, PrOMor: A proposed prototype of V2V and V2I for crash prevention in the Moroccan case. Adv. Sci. Technol. Eng. Syst. J **6**(1), 200–207 (2021)

4. Z. Sabir, A. Amine, Connected vehicles using NDN: Security concerns and remaining challenges, in *International Conference on Optimization and Applications (ICOA)-7th Edition, Germany*, (2021)
5. V. Jacobson, D.K. Smetters, J.D. Thornton, M.F. Plass, N.H. Briggs, R.L. Braynard, Networking named content, in *Proceedings of the 5th International Conference on Emerging Networking Experiments and Technologies, ACM, USA, CoNEXT '09*, (2009), pp. 1–12
6. E. Zhang, J. Burke, S. Thornton, T. Zhang, K. Claffy, P. Massey, W. Abdelzaher, Y. Crowley, Named Data Networking (NDN) project. CAIDA p 27. Technical Report (2010)
7. L. Zhang, A. Afanasyev, J. Burke, V. Jacobson, K. Claffy, P. Crowley, C. Papadopoulos, L. Wang, B. Zhang, Named data networking. SIGCOMM Comput. Commun. Rev **44**(3), 66–73 (2014)
8. Z. Sabir, A. Amine, NDN vs TCP/IP: Which one is the best suitable for connected vehicles? in *Recent Advances in Mathematics and Technology, Applied and Numerical Harmonic Analysis*, (Springer International Publishing, 2020), pp. 151–159
9. Z. Sabir, A. Amine, Performance of named data networking in connected vehicles, in *Proceedings of the 4th European International Conference on Industrial Engineering and Operations Management, Italy*, (2021)
10. Z. Sabir, A. Amine, Connected vehicles using NDN for intelligent transportation systems, in *The International Conference on Industrial Engineering and Operations Management, Paris, France*, vol. 2018, (2018), pp. 2433–2441
11. M.F. Bari, S.R. Chowdhury, R. Ahmed, R. Boutaba, B. Mathieu, A survey of naming and routing in information-centric networks. IEEE Commun. Mag. **50**(12), 44–53 (2012)
12. M. Amadeo, C. Campolo, A. Molinaro, Information-centric networking for connected vehicles: A survey and future perspectives. IEEE Commun. Mag. **54**(2), 98–104 (2016)
13. Nakamoto S, Bitcoin: A peer-to-peer electronic cash system (2008). White Paper URL https://bitcoin.org/bitcoin.pdf. Last accessed 2021/03/15
14. Bitcoin, What bitcoin is, and why it matters. (2011). https://www.technologyreview.com/2011/05/25/194486/what-bitcoin-is-and-why-it-matters/. Last accessed 2021/03/15
15. K. Lei, J. Fang, Q. Zhang, J. Lou, M. Du, J. Huang, J. Wang, K. Xu, Blockchain-based cache poisoning security protection and privacy-aware access control in NDN vehicular edge computing networks. J. Grid Comput **18**(4), 593–613 (2020)
16. D.B. Rawat, V. Chaudhary, R. Doku, Blockchain: Emerging applications and use cases. arXiv:190412247 [cs] (2019)
17. M.U. Sattar, R.A. Rehman, Interest flooding attack mitigation in named data networking based VANETs, in *2019 International Conference on Frontiers of Information Technology (FIT), Pakistan*, (2019), pp. 245–2454
18. A. Arsalan, R.A. Rehman, Prevention of timing attack in software de-fined named data network with VANETs, in *2018 International Conference on Frontiers of Information Technology (FIT), Pakistan*, (2018), pp. 247–252
19. V. Jain, R.S. Kushwah, R.S. Tomar, Named data network using trust function for securing vehicular Ad Hoc network, in *Soft Computing: Theories and Applications*, (Springer, Singapore, Advances in Intelligent Systems and Computing, 2019), pp. 463–471
20. P. Manimaran, P ARK, NDNIDS: An intrusion detection system for NDN based VANET, in *2020 IEEE 91st Vehicular Technology Conference (VTC2020-Spring), Belgium*, (2020), pp. 1–5
21. G. Rathee, A. Sharma, R. Iqbal, M. Aloqaily, N. Jaglan, R. Kumar, A Blockchain framework for securing connected and autonomous vehicles. Sensors **19(14)**, 3165 (2019)
22. R. Shrestha, R. Bajracharya, A.P. Shrestha, S.Y. Nam, A new type of blockchain for secure message exchange in VANET. Digit. Commun. Netw **6**(2), 177–186 (2020)
23. S. Kudva, S. Badsha, S. Sengupta, H. La, I. Khalil, M. Atiquzzaman, A scalable blockchain based trust management in VANET routing protocol. J. Parallel Distrib. Comput **152**, 144–156 (2021)
24. L. Zhang, M. Luo, J. Li, M.H. Au, K.K.R. Choo, T. Chen, S. Tian, Blockchain based secure data sharing system for Internet of vehicles: A position paper. Vehic. Commun **16**, 85–93 (2019)

25. J. Kang, R. Yu, X. Huang, M. Wu, S. Maharjan, S. Xie, Y. Zhang, Blockchain for secure and efficient data sharing in vehicular edge computing and networks. IEEE Internet Things J. **6**(3), 4660–4670 (2019)
26. D.B. Rawat, R. Doku, A. Adebayo, C. Bajracharya, C. Kamhoua, Blockchain enabled named data networking for secure vehicle-to-everything communications. IEEE Netw. **34**(5), 185–189 (2020)
27. F. Ahmad, C.A. Kerrache, F. Kurugollu, R. Hussain, Realization of Blockchain in named data networking-based internet-of-vehicles. IT Professional **21**(4), 41–47 (2019)
28. H. Khelifi, S. Luo, B. Nour, H. Moungla, S.H. Ahmed, M. Guizani, A blockchain-based architecture for secure vehicular Named Data Networks. Comput. Electr. Eng **86**, 106715 (2020)
29. M. Dorigo, G. Di Caro, Ant colony optimization: A new meta-heuristic, in *Proceedings of the 1999 Congress on Evolutionary Computation-CEC99 (Cat. No. 99TH8406), USA*, vol. 2, (1999), pp. 1470–1477
30. D. Teodorovic, P. Lucic, G. Markovic, M.D. Orco, Bee Colony optimization: Principles and applications, in *2006 8th Seminar on Neural Network Applications in Electrical Engineering, Serbia*, (2006), pp. 151–156
31. J. Kennedy, R. Eberhart, Particle swarm optimization, in *Proceedings of ICNN'95 - International Conference on Neural Networks, Australia*, vol. 4, (1995), pp. 1942–1948
32. M. Mitchell, *An Introduction to Genetic Algorithms* (MIT Press, 1998) ISBN 978-0-262-63185-3
33. X.S. Yang, Firefly algorithms for multimodal optimization, in *Stochastic Algorithms: Foundations and Applications*, (Springer, Berlin, Heidelberg., Lecture Notes in Computer Science, 2009), pp. 169–178
34. X.S. Yang, A new metaheuristic bat-inspired algorithm, in *Nature Inspired Cooperative Strategies for Optimization (NICSO 2010), Studies in Computational Intelligence*, ed. by J. R. González, D. A. Pelta, C. Cruz, G. Terrazas, N. Krasnogor, (Springer, Berlin, Heidelberg, 2010), pp. 65–74
35. M. Dhivya, M. Sundarambal, L.N. Anand, Energy efficient computation of data fusion in wireless sensor networks using Cuckoo Based Particle Approach (CBPA). Int. J. Commun. Netw. Syst. Sci **04**(04), 249 (2011)
36. E.R. Speed, Evolving a Mario agent using cuckoo search and softmax heuristics, in *2010 2nd International IEEE Consumer Electronics Society's Games Innovations Conference, China*, (2010), pp. 1–7
37. A. Ouaarab, B. Ahiod, X.S. Yang, Discrete cuckoo search algorithm for the travelling salesman problem. Neural Comput. & Applic. **24**(7), 1659–1669 (2014)

Blockchain-Based Communication for Digital Twins

Zhihan Lv, Yuxi Li, Liang Qiao, Jingyi Wu, and Anna Jia Gander

1 Introduction

Digital twins convert physical assets into virtual counterparts and create digital copies of machines/equipment or physical sites through the use of sensors. These digital assets can even be created before the assets are physically constructed. From the time the digital twins were proposed to the present, its application fields have been more focused on scenarios related to spatial architecture such as smart buildings, smart parks, and smart cities [1–3]. In fact, digital twins can be built from multiple data sources, including historical performance statistics, real-time sensor and manufacturing output, and future data provided by machine learning. It is to create the most accurate representation possible to truly understand the characteristics of the physical object and even predict its performance changes over time. With the transformation and upgrading of traditional industrial enterprises, digital twins have been extended to the field of industry 4.0 intelligent manufacturing where industrial infrastructure is increasingly integrated with general information technology [4, 5]. It collects, processes, analyzes, and interprets data through wireless connection with small devices capable of sensing, computing, and communication [6].

Integrating digital twins with industrial manufacturing is full of promise, but its introduction also brings new attack vectors. The vulnerabilities in the infrastructure system will induce different attacks, which may bring catastrophic consequences

Z. Lv (✉)
Qingdao Institute of Bioenergy and Bioprocess Technology, Chinese Academy of Sciences, Beijing, China

Y. Li · L. Qiao · J. Wu
College of Computer Science and Technology, Qingdao University, Qingdao, China

A. J. Gander
Department of Applied IT, The University of Gothenburg, Gothenburg, Sweden

Y. Maleh et al. (eds.), *Advances in Blockchain Technology for Cyber Physical Systems*, Internet of Things, https://doi.org/10.1007/978-3-030-93646-4_9

to the applications involved in the decision-making process [7–9]. Considering the involvement of many entities in the complex industrial process, to ensure the reliability of data obtained from heterogeneous databases is the first priority. Blockchain technology has made breakthroughs in product life cycle data management and data security. The use of blockchain technology allows the sharing of data on distributed ledgers to realize the traceability in case of critical failures and securely record events in an immutable and irrevocable manner [10]. The combination of digital twins and blockchains can reshape the industry. Secure data management is ensured based on blockchain technology, and then trusted data is used as input through digital twins to obtain data with guiding value to maintain the system.

In general, the integration of digital twins and blockchain ensures the safe, efficient, and credible creation of virtual models. The creation process of the digital twins usually includes four stages: design, construction, testing, and delivery. The beginning of each stage depends on the completion of the previous stage. The auxiliary application of the blockchain can ensure that the creation process of the digital twins is safe and reliable. This research focuses on the data security in the case of untrusted multiple parties sharing data, and the constraint conditions for the log storage system to achieve secure data sharing are proposed. Then, for multiple scenarios in the blockchain, an optimized resource allocation algorithm is proposed based on DRL theory, to get rid of the dependence on large data centers for data computing.

2 Related Works

The concept of digital twins was first proposed by Michael Grieves in 2003 during his product life cycle courses. In 2012, NASA formally defined the digital twins as a multidisciplinary, multi-physical, multi-scale simulation process to complete the mapping in the virtual space and listed it as one of the key technologies to drive future development. Digital twins continuously collect real data from corresponding physical entities and process them to find potentially valuable information from the data. The current creation of digital twins is mainly based on traditional methods, and this method has centralized authority to manage entity information, so it cannot ensure credibility. Raj (2021) [11] pointed out in his research that the construction of digital twins must avoid data leakage, and blockchain plays a role in this process with its unique advantages. The research of Kim and Laskowski (2018) [12] revealed the advantages of blockchain suitable for the supply chain industry, including traceability, transparency, and tamper-proof logs. These features also make the blockchain a highly secure distributed ledger, which aggregates all transaction information of stakeholders during the end-to-end creation of the digital twins and finally achieves process transparency.

Emerging blockchain technology has the potential to overcome security vulnerabilities, enabling the application and innovation of digital twins. Christidis and Devetsikiotis (2016) [13] pointed out in their research that the blockchain enables us to conduct transactions through verification through a distributed peer-to-peer network at the absence of a trusted intermediary. Blockchain allows the exchange of time-stamped events, and these events can be permanently stored securely and tamper proof in a distributed ledger. Borowski (2021) [14] believed that the establishment of a blockchain that is capable of distributed data storage, tamper proof, and immutable can achieve the acquisition of distributed data across multiple participating entities in the digital twins. Yaqoob et al. (2020) [15] envisaged the use of blockchain technology to reshape and transform the digital twins to achieve safe manufacturing while ensuring the traceability, compliance, and authenticity of data.

Above, it can be induced that based on blockchain technology, it is possible to track the creation process of digital twins safely and reliably. However, there lacks in-depth research on the constraints of data that cannot be tampered with. This research draws attention to the data security sharing mode of the blockchain from the perspective of data traceability and proposes a reliable and efficient edge computing resource allocation method.

3 Methods

Blockchain-Based Digital Twins

The developers of the digital twins have carried out in-depth research on the information foundation of the simulated physical system, which is helpful for visualization and the development of a mathematical model that simulates the physical system of the real world.

In fact, from the perspective of building objects, digital twin technology can be divided into three types, namely, things, people, and human-to-thing interaction, and it is mainly used in intelligent IoT, citizen code, and smart city [16, 17]. The creation process of the digital twins is mainly divided into four stages. In Stage 1, the engineer uses computer-aided design tools to analyze the data and directly captures the target data and converts it into a virtual copy of the entity. In Stage 2, the model will be continuously updated based on the captured data and will send the information feedback. Whether in the supply chain or any other environment that is highly dependent on data, it is necessary to ensure that the built model works well in the real environment. In Stage 3, after the digital twins are successfully built, it is tested using the test bench to eliminate logic errors or possible design defects. In Stage 4, after successful testing and verification, the digital twins can be deployed, and it is available for all owners on the blockchain. Figure 1 shows the process of creating digital twins using the blockchain as a management entity (Fig. 1).

The encryption feature of the blockchain ensures the security of data transmission. Through reliable authentication of users and data sources, the immutability of data and the security of digital twins are of high application value [18]. In the blockchain, transactions between nodes, as the basic communication utterances,

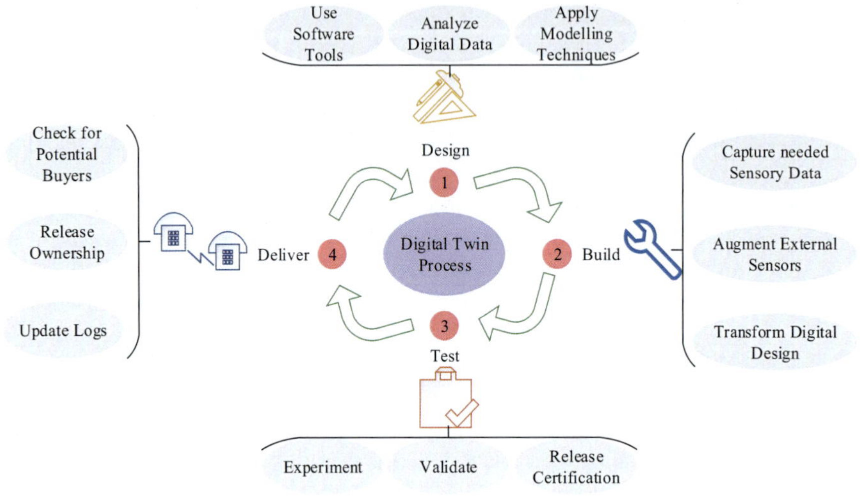

Fig. 1 The main process of creating digital twins with the blockchain as a management entity

are only recorded in the time-stamped log in the shared ledger. The blockchain stores the digital certificate, which contains all the relevant data of the product. The blockchain ensures that the data embedded in the digital certificate will not be copied, modified, or even deleted by others [19, 20]. The consensus mechanism is adopted to establish a safe and reliable network node, and then the transaction is organized into a block through the consensus algorithm and attached to the distributed ledger. Each block contains the hash value of the previous block, so it is easy to detect the tampering and destruction of the block by the attacker. Based on the consensus mechanism in blockchain technology, multiple stakeholders, partners, and users can share one umbrella, and confidential information, customer information, and corporate information can all be safely transmitted and shared. Combining blockchain technology with digital twins can expand the application scope of digital twins and provide a technical basis for building a mirrored world. Figure 2 shows the framework of digital twins based on blockchain. First, participating entities such as sensors, devices, and people are registered as authorized entities in the blockchain. Next, at the data layer, information-physical mapping is performed by monitoring, collecting, and processing the specified parameters from the physical space to the virtual space. Based on the collected data, the digital twins generate relevant knowledge and store historical data in the storage layer. The application layer further analyzes the data to schedule services or provide model calibration services, thereby completing a feedback loop. The digital twins use credible source data as input. In this process, the blockchain ensures the security of data management, and finally the data analysis is performed to predict events and evaluate related factors [14, 15, 21].

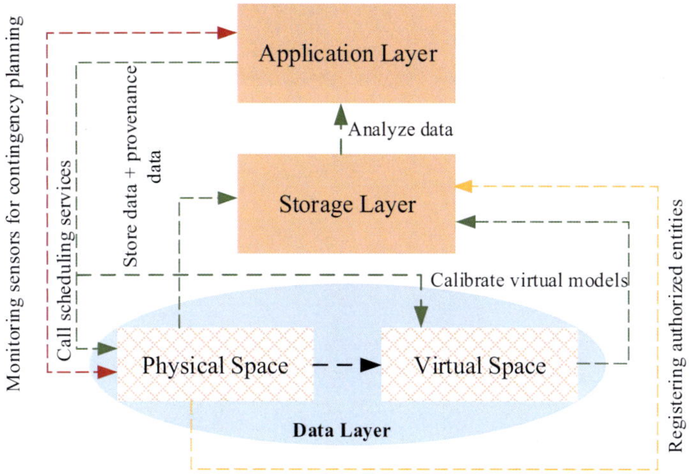

Fig. 2 Framework of the blockchain-based digital twins

Secure Data Sharing Model Based on Blockchain

The blockchain is formed by linking one block with another [22]. Each block points to the previous block, which is called a child block of the previous block. Each block has a block header, which contains the hash value generated by the parent block header, and the parent block can be found through this hash value [23–25]. When there is any change in the parent block, the hash value of the parent block also changes, and this will force the hash value of the child-block to change, and so on, the subsequent sub-sub-blocks will also be affected. If a block has many descendants, the recalculation of all descendants of this block requires much effort, so a longer blockchain is more difficult to be changed. Now that new transactions are generated every second, to achieve tampering, the results must be quickly calculated, while new transactions are not generated, whereas under the current computer capabilities, the attack takes much longer than normal transaction time, so it is impossible to tamper [26, 27].

The log storage system of the blockchain is mainly used as the interface for writing and reading data operation logs. In this research, a hybrid storage strategy for data is proposed, as shown in Fig. 3. Only the hash of the log data will be stored in the log storage system, and the log data itself will be stored in other storage media with lower overhead. Data is usually broken down into three parts of data shards, data index information, and check hash. Although data shards and data index information are stored on an untrusted medium, the data is still credible because the data hash is kept unchanged on the storage medium.

The validity of the log data needs the signature of both the system and the user at the same time, and a single-party signature is invalid. For example, in a scenario, there are three roles of the user P_1, system P_2, and log storage system P; P_1 and P_2

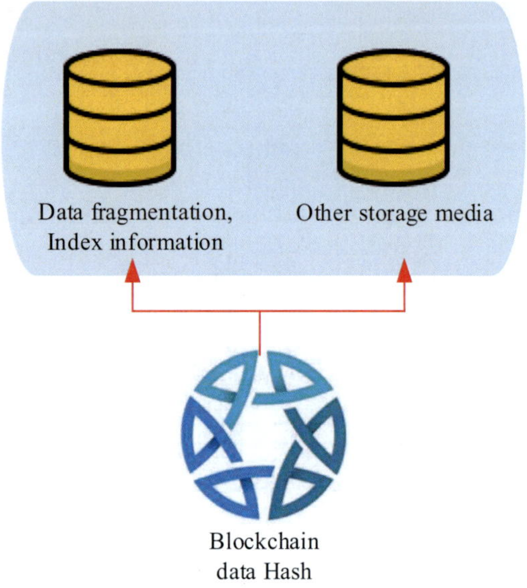

Fig. 3 Hybrid storage strategy for data

can encrypt data in their own way, but they can all be decrypted by the log storage system P. During the key generation process, the log storage system P selects an odd integer $p \leftarrow [2^{\eta-1}, 2^{\eta}] \cap (2Z+1)$ of η bit as the private key sk. $q_0, q_1 \dots q_\tau \leftarrow Z \cap \left[0, \frac{2\gamma}{p}\right)$ is chosen to maximize q_0. $r_0, r_1 \dots r_\tau \leftarrow Z \cap [-2^{\rho}, 2^{\rho}]$ is chosen, and then the equation is obtained as follows:

$$x_0 \leftarrow q_0 p + 2r_0, x_i \leftarrow [q_i p + 2r_i]_{x_0} \tag{1}$$

At this time, the public key can be expressed as follows:

$$pk = \langle x_0, x_1 \dots x_\tau \rangle \tag{2}$$

In the encryption process, a set $S_i \subseteq \{1, 2, \dots \tau_i\}$ and an integer $t_i \leftarrow \{-2^{\rho}, 2^{\rho_i}\}$ are randomly selected, and the output ciphertext can be expressed as follows:

$$c_i = \left[m_i + 2t_i + \sum_{j \in S}^{x_{i,j}} \right]_{x_{i,0}} \tag{3}$$

During the decryption process, the log storage system P can decrypt to obtain the plaintext $m_i \leftarrow [[c_i]_p]_2$ according to the key $sk = p$.

To ensure the reliability of the data source, this research uses the Flume structure to complete the collection, aggregation, and transmission of data. Figure 4 shows the Flume logic architecture. The Flume event is defined as a data flow unit with byte payload and optional string attributes [28]. The core of Flume is an agent. This

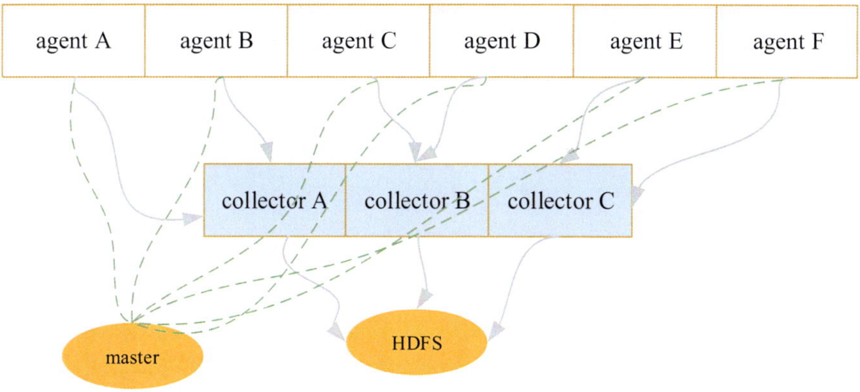

Fig. 4 Flume logic architecture

agent has two external interaction places, one is to receive data input-source, and the other is data output-sink [29]. After the source receives the data, it sends the data to the channel. The channel acts as a data buffer to temporarily store the data, and then the sink sends the data in the channel to the designated place. When the log storage system receives a request for data reading, Influence Chain will first read the data index information on the untrusted low-overhead storage medium according to the request and then obtain the data information and verify the data by reading the data hash. Finally, the data is returned to the user.

Optimized Resource Allocation in Blockchain

Aimed at the trusted resource allocation in the decentralized blockchain network, this research proposes an optimal resource allocation strategy for each user in the multiuser and multiserver scenario (Fig. 5). In the scenario where there are several edge service providers in the blockchain network, the set of the edge server and the set of the user are defined as N and M, respectively. To prevent the system from directly offloading the computing task submitted by the user each time to the edge server [30–32], first, the task needs to be submitted to the system, and then a more reasonable calculation offloading and resource allocation strategy is implemented, with user satisfaction as the standard.

When the user i offloads the task to the edge computing server j, the corresponding upload rate is as follows:

$$\tau_{i,j} = B * \log_2 \left(1 + \frac{p_i * g_{i,j}}{B * N_0} \right) \quad (4)$$

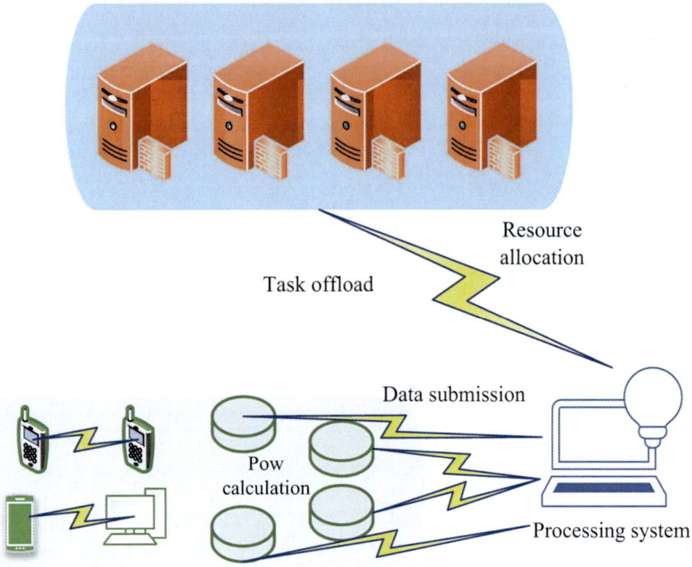

Fig. 5 Multiuser and multiserver scenario

where B represents bandwidth and N_0 refers to the variance of Gaussian white noise. p_i is the transmission rate that needs to be achieved when the terminal device i uploads data, and $g_{i,j}$ represents the channel gain between the device i and the edge computing server j.

The transmission delay for users to offload tasks to the edge server can be expressed as follows:

$$t_0 = \frac{s_i}{\tau_{i,j}}\alpha_{i,j} \tag{5}$$

where s_i represents the amount of transactions that users can monitor in the blockchain network and $\alpha_{i,j}$ is the offloading decision vector.

The delay of edge server processing user computing tasks can be expressed as follows:

$$t_1 = \frac{s_i * d}{f_j} \tag{6}$$

where f_j represents the CPU frequency of the server.

Taken together, the overall delay of task offloading is as follows:

$$T_{i,j} = t_0 + t_1 = \frac{s_i}{\tau_{i,j}}\alpha_{i,j} + \frac{s_i * d}{f_j} \tag{7}$$

To allow more users to participate in the blockchain network, the user satisfaction St is used as an effective indicator of resource allocation to maximize the results. After the user submits the task to the system, the system will make a resource allocation decision $\beta_{i,j}$. The definition of St is based on the system delay, and they are inversely proportional to each other. Then, the equation is obtained as follows:

$$St_i = \frac{pure_i}{T_i} \tag{8}$$

The target expression of the system can be derived as follows:

$$\max \frac{1}{M} \sum_{i \in M} \frac{pure_i}{T_i} \tag{9}$$

$$s.t. \begin{cases} \alpha_{i,j} \in \{0, 1\} \\ \sum\limits_{j \in N} \alpha_{i,j} = 1 \\ \sum\limits_{i \in M} \beta_{i,j} < V \\ V \geq v_{i,j} > 0 \\ V \geq \beta_{i,j} > 0 \\ \beta_{i,j} = 0 if\ \alpha_{i,j} = 0 \end{cases} \tag{10}$$

The optimal resource allocation vector $\beta_{i,j}$ can be obtained by solving the above target expression, but it is not a convex problem, and the complexity of the problem will increase with the increase of the number of users. In this regard, DRL is introduced. The deep Q network uses four adjacent frames in time as the original image input. After process by the deep CNN and the fully connected neural network, the state-action Q function is output to realize end-to-end learning control.

$S = \{s(t)\}$ is used to represent the network state space, and then the network state $s(t)$ at moment t can be expressed as follows:

$$S = \{s(t)\}\, s(t) = \left(v_j(t), s_i(t), d(t), \Gamma_{i,j}(t)\right) \tag{11}$$

where $\Gamma_{i,j}(t)$ represents the signal-to-noise ratio between i and j at time t.

Each resource allocation in the blockchain network includes user unloading decisions and resource allocation decisions, and then the actions of the network are defined as follows:

$$a(t) = \left(\alpha_{1,1}(t) * \beta_{1,1}(t), \ldots, \alpha_{M,N}(t) * \beta_{M,N}(t)\right) \tag{12}$$

With the user satisfaction used as the reward in the reinforcement learning process, the income of user i at the moment t can be expressed as follows:

$$I_i^t = \frac{\sum_j \alpha_{i,j}^t * \beta_{i,j}^t}{\sum_i \sum_j \alpha_{i,j}^t * \beta_{i,j}^t} * \left(T + rs_i^t\right) * e^{-\frac{1}{\lambda}\mu s_i^t} \tag{13}$$

The cost of user *i* can be expressed as follows:

$$\text{cost}_i = \gamma d * s_i^t \sum_j \frac{f_j}{v_j} * \alpha_{i,j} \tag{14}$$

Finally, the reward at the moment *t* is obtained:

$$r(t) = \sum_i \frac{pure_i^t}{T_i^t} = \sum_i \frac{I_i^t - \text{cost}_i}{T_i^t} \tag{15}$$

The above expression reflects the reward brought by each action of DRL, and parameters need to be supplemented on this basis to get the overall reward.

$$R(t) = \sum_{t'=t}^{T} \in^{t'-t} r(t) \tag{16}$$

where $\in$ represents the rate of loss of subsequent earnings. If $\in$ approaches 0, the system will pay more attention to the current profit at this time.

The resource allocation algorithm based on DRL can be divided into two solving processes, namely, forward transmission and reverse training. In the forward transmission process, a deep neural network is constructed with the purpose to maximize user satisfaction, while in the reverse training process, the deep neural network is trained in the reverse direction by minimizing the time difference error.

Simulation Experiment and System Setting

For the blockchain-based network data sharing system, the system test environment is as follows. The host machine CPU is Intel® Core™ i7, with 16 GB running memory and Windows 10 operating system. The client and server are Ubuntu operating system. The server is used to simulate the test environment, and the client is responsible for initiating the test. The blockchain benchmark environment is the Ethereum blockchain system; the client environment adopts a server with JDK 8. It is proved that the introduction of the log storage system can meet the constraint requirements of reading and writing and will not affect the original performance of the system.

To prove the effectiveness of the algorithm proposed in this research in the edge computing resource allocation of blockchain network, five other algorithms are selected for comparative experiments, namely, genetic algorithm, random allocation algorithm, preference allocation, ant colony, and Q-learning algorithm. Q-Learning is a value-based reinforcement learning algorithm. The Q function is recorded by the state-action table. The Bellman equation can be used to solve the optimal strategy for the Markov process.

A simulation platform is built based on Python, and the number of users in the blockchain network is set to 10, 20, 30, 40, and 50; the number of edge service areas is 2, 3, 4, 5, and 6, and the server processing frequency is 1, 2, 3, 4, and 5. The channel bandwidth is 180 kHz, the noise is 90 dBm, the upper limit of the number of resources of a single edge server is 100, and the channel gain is in the following range [1, 10].

4 Results and Discussion

Evaluation of Data Security Sharing Strategy

In this research, the blockchain benchmark environment is used as a control, and the query request from the client is used as the workload. The workload acts on the base environment and the main chain environment, respectively. To ensure the accuracy of the test results, three repeated experiments are carried out, and finally, the average delay under different load conditions is calculated. Figure 6 shows the experimental results. It is noted that in the three experiments, the difference between the blockchain benchmark environment and the main chain environment is very subtle in terms of average latency. This also suggests that the introduction of the log storage system in the blockchain network will not have a significant impact on

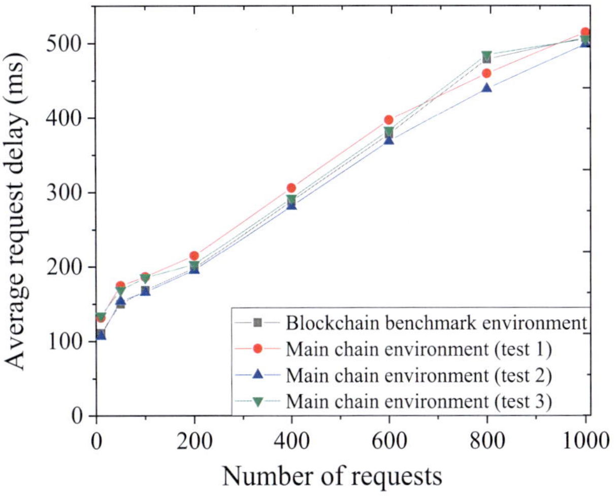

Fig. 6 Comparison of request delays under different loads

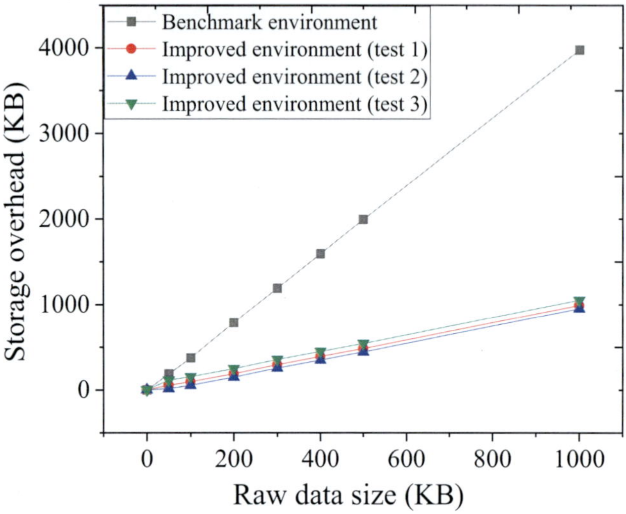

Fig. 7 Comparison of data storage overhead under different source data volumes

the performance of the reference environment to read data, and it can also meet the constraint requirements in terms of effectiveness. Hence, this solution is feasible. Figure 7 shows the storage overhead of the system when the file data size is between 1 KB and 1 MB. It is noted in Fig. 7 that the improved environment after the introduction of the log storage system reduces the storage overhead of the system by about 75% compared with the blockchain benchmark environment. Although different storage modes will have varying degrees of impact on the experimental results, as long as the storage overhead of the local file system less than the blockchain network, the solution is considered effective.

Figure 8 shows the corresponding execution time on the endorser node when the client simultaneously initiates different numbers of data write requests. It is noted that there is not much difference in the calculation time of the endorser node between the cache enhancement strategy and the non-cache enhancement strategy. Figure 9 shows the calculation amount corresponding to the endorser node when the client initiates 100 data write requests at the same time. It is noted that the calculation amount of the system under the cache enhancement strategy is smaller, and the calculation efficiency is about 1–2 times that under the non-cache enhancement strategy.

Performance Analysis of Resource Allocation Algorithms

Figure 10 shows the user satisfaction under different numbers of users. The increase in the number of users means that limited computing resources will be shared by

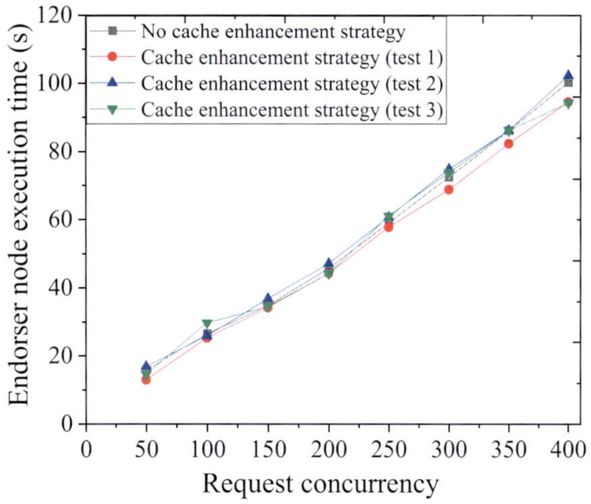

Fig. 8 Execution time results under different numbers of requests

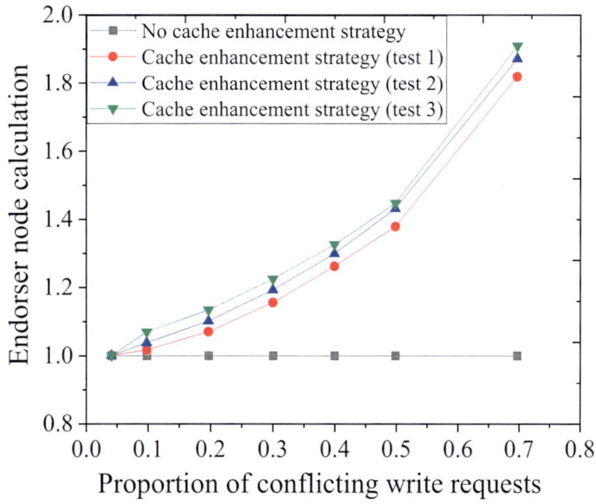

Fig. 9 Calculated results under different proportions of conflicting write requests

more users, and thus the average user satisfaction will decline. However, compared with other resource allocation algorithms, the allocation method based on DRL significantly improves user satisfaction. The Q-learning algorithm also has good performance. In the case of a small number of users, the needs of most users can be met, and user satisfaction is high. Figure 11 shows the user satisfaction under different edge server performance. With the increase in the frequency of edge servers, the system will process data faster, so user satisfaction will increase linearly. Figure 12 shows the user satisfaction under different numbers of servers. The increase in the number of servers means that more parallel tasks can be supported and the user satisfaction will increase accordingly. The advantages of the resource allocation algorithm based on DRL theory are not obvious under a small number of servers. When the number of servers is greater than 5, user satisfaction has been

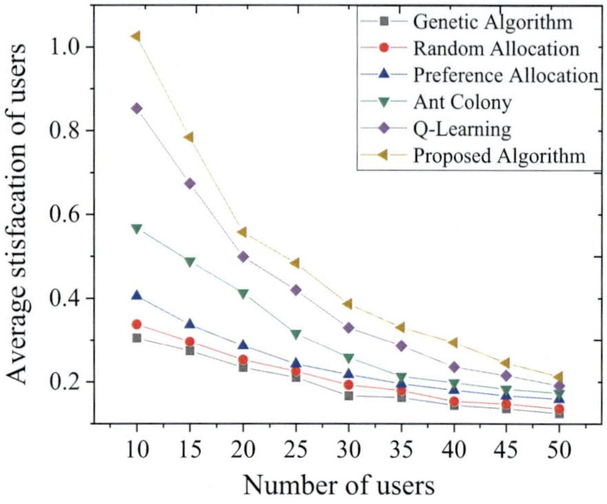

Fig. 10 The relationship between the number of users and the user satisfaction

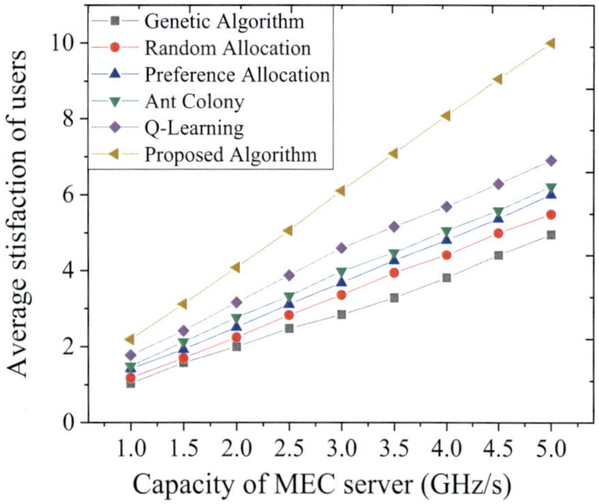

Fig. 11 The relationship between edge server performance and user satisfaction

significantly improved compared with the other five resource allocation algorithms, and compared with the Q-learning algorithm, its user satisfaction has increased by about 15%.

Figure 13 shows the running time. The running time of the algorithm will increase as the number of users increases. When the number of users is less than 30, the running time of the random allocation algorithm is the shortest, and the resource allocation algorithm based on DRL has the longest running time among the six because of the addition of the link of training the deep network. When the number of users increases, the number of two-dimensional tables stored by the Q-learning algorithm also increases, and the running time of the algorithm increases almost exponentially. Figure 14 shows the time delay. It is noted that the preference allocation algorithm adopts a greedy strategy and preferentially allocates servers with low delay to users, so its time delay is the shortest among the six. The Q-

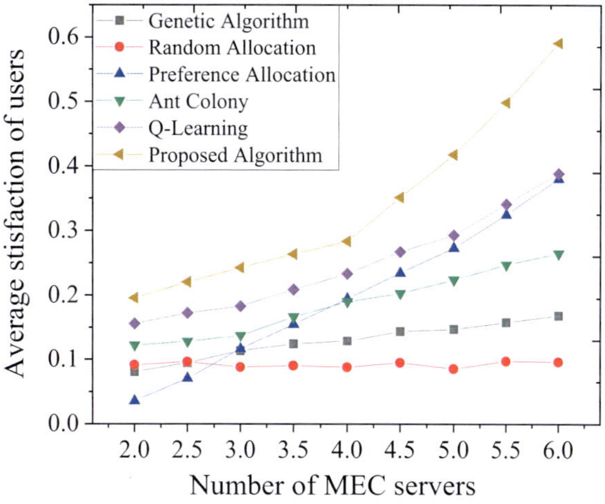

Fig. 12 The relationship between the number of servers and the user satisfaction

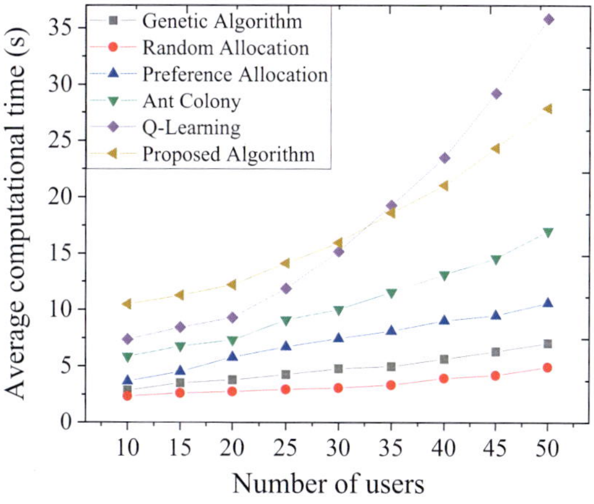

Fig. 13 The relationship between the number of users and the running time

learning algorithm is found to have a serious impact on the system delay when the number of users is large. Figure 15 shows the changes in the revenue. It is noted that as the number of users increases, the Q-learning algorithm and the one based on DRL can obtain more user benefits, and the overall growth is exponential.

5 Conclusion

With the vigorous development of interactive Internet of things devices, a large amount of multi-structured data is generated, transmitted, and stored in local and remote storage systems. By building digital twins for IoT devices, it is possible to have a deeper understanding of their structural characteristics and behaviors

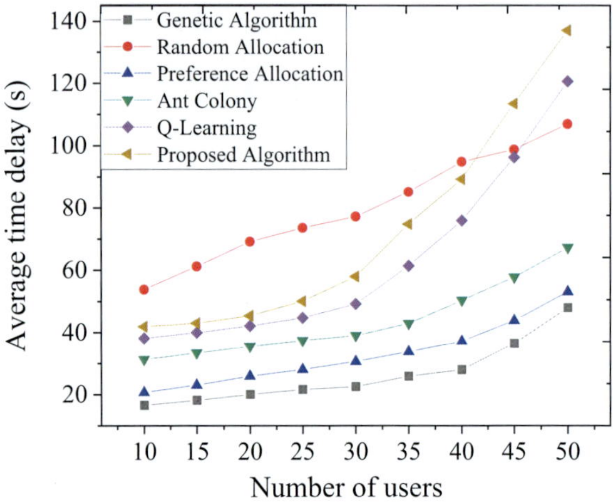

Fig. 14 The relationship between the number of users and the algorithm delay

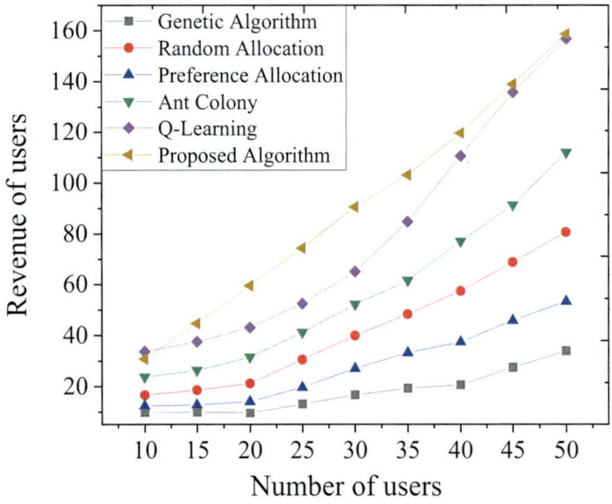

Fig. 15 The relationship between the number of users and the revenue

in different environments. Digital twins can even simulate the working status of physical entities based on changing conditions, but at the same time, hackers can remotely penetrate IoT devices and cause irreparable damage to IoT systems and applications. This will result in the digital twins may not be accessible, and if the tampered data is submitted to the digital twin model, it will lead to wrong decisions. In this regard, in this article, possible connections between the digital twins and the blockchain are explored to defend against security threats. The decentralization of blockchain technology allows nodes that do not trust each other to trust the stored data.

A resource allocation method based on DRL theory is proposed to maximize the user satisfaction. First, the operation delay and user benefits of the blockchain are taken into account, and then the optimization problem is transformed into a DRL model to obtain a reasonable network state. Finally, the optimal resource allocation plan is obtained after the network training. With the optimal user satisfaction as the strategy, simulation experiments have further confirmed that the allocation method based on DRL significantly improves user satisfaction. Moreover, the algorithm can obtain more user benefits, and the overall growth is exponential. This article provides ideas for the data maintenance, prevention of tampering, and data traceability during the construction of digital twins under decentralization. However, there are still some shortcomings. Because the experiment is carried out on a simulation platform, it fails to consider the deployment and execution methods in the real network environment. Therefore, in the follow-up, the system throughput should be expanded to enhance the practicability of the plan.

References

1. A. Fuller, Z. Fan, C. Day, et al., Digital twin: Enabling technologies, challenges and open research. IEEE Access **8**, 108952–108971 (2020)
2. A. Francisco, N. Mohammadi, J.E. Taylor, Smart city digital twin-enabled energy management: Toward real-time urban building energy benchmarking. J. Manag. Eng. **36**(2), 04019045 (2020)
3. E. Shahat, C.T. Hyun, C. Yeom, City digital twin potentials: A review and research agenda. Sustainability **13**(6), 3386 (2021)
4. D.A. Ivanov, M.A. Ivanova, B.V. Sokolov, Analysis of transformation trends in enterprise management principles in the era of industry 4.0 technology. Inform. Autom. **60**, 97–127 (2018)
5. M. Bortolini, E. Ferrari, M. Gamberi, et al., Assembly system design in the industry 4.0 era: A general framework. IFAC-PapersOnLine **50**(1), 5700–5705 (2017)
6. F. Tao, Q. Qi, L. Wang, et al., Digital twins and cyber-physical systems toward smart manufacturing and industry 4.0: Correlation and comparison. Engineering **5**(4), 653–661 (2019)
7. S. Wang, L. Hong, M. Ouyang, et al., Vulnerability analysis of interdependent infrastructure systems under edge attack strategies. Saf. Sci. **51**(1), 328–337 (2013)
8. M. Ouyang, Critical location identification and vulnerability analysis of interdependent infrastructure systems under spatially localized attacks. Reliab. Eng. Syst. Saf. **154**, 106–116 (2016)
9. M. Ouyang, F. Tao, S. Huang, et al., Vulnerability mitigation of multiple spatially localized attacks on critical infrastructure systems. Comput. Aided Civ. Inf. Eng. **33**(7), 585–601 (2018)
10. M. Niranjanamurthy, B.N. Nithya, S. Jagannatha, Analysis of Blockchain technology: Pros, cons and SWOT. Clust. Comput. **22**(6), 14743–14757 (2019)
11. P. Raj, Empowering digital twins with blockchain. Adv. Comput. **121**, 267 (2021)
12. H.M. Kim, M. Laskowski, Toward an ontology-driven blockchain design for supply-chain provenance. Intell. Syst. Account. Finance Manag. **25**(1), 18–27 (2018)
13. K. Christidis, M. Devetsikiotis, Blockchains and smart contracts for the internet of things. IEEE Access **4**, 2292–2303 (2016)
14. P.F. Borowski, Digitization, digital twins, blockchain, and industry 4.0 as elements of management process in enterprises in the energy sector. Energies **14**(7), 1885 (2021)
15. I. Yaqoob, K. Salah, M. Uddin, et al., Blockchain for digital twins: Recent advances and future research challenges. IEEE Netw. **34**(5), 290–298 (2020)
16. B. He, K.J. Bai, Digital twin-based sustainable intelligent manufacturing: A review. Adv. Manuf. **9**(1), 1–21 (2021)

17. Z. Jiang, Y. Guo, Z. Wang, Digital twin to improve the virtual-real integration of industrial IoT. J. Ind. Inf. Integr. **22**, 100196 (2021)
18. Y. Lu, X. Huang, K. Zhang, et al., Communication-efficient federated learning and permissioned blockchain for digital twin edge networks. IEEE Internet Things J. **8**(4), 2276–2288 (2020)
19. B. Putz, M. Dietz, P. Empl, et al., Ethertwin: Blockchain-based secure digital twin information management. Inf. Process. Manag. **58**(1), 102425 (2021)
20. C. Zhang, G. Zhou, H. Li, et al., Manufacturing blockchain of things for the configuration of a data-and knowledge-driven digital twin manufacturing cell. IEEE Internet Things J. **7**(12), 11884–11894 (2020)
21. C. Mandolla, A.M. Petruzzelli, G. Percoco, et al., Building a digital twin for additive manufacturing through the exploitation of blockchain: A case analysis of the aircraft industry. Comput. Ind. **109**, 134–152 (2019)
22. I. Makhdoom, I. Zhou, M. Abolhasan, et al., PrivySharing: A blockchain-based framework for privacy-preserving and secure data sharing in smart cities. Comput. Secur. **88**, 101653 (2020)
23. C. Feng, K. Yu, A.K. Bashir, et al., Efficient and secure data sharing for 5G flying drones: A blockchain-enabled approach. IEEE Netw. **35**(1), 130–137 (2021)
24. Y. Lu, X. Huang, K. Zhang, et al., Blockchain empowered asynchronous federated learning for secure data sharing in internet of vehicles. IEEE Trans. Veh. Technol. **69**(4), 4298–4311 (2020)
25. K. Fan, S. Wang, Y. Ren, et al., Medblock: Efficient and secure medical data sharing via blockchain. J. Med. Syst. **42**(8), 1–11 (2018)
26. I.A. Ansari, M. Pant, C.W. Ahn, SVD based fragile watermarking scheme for tamper localization and self-recovery. Int. J. Mach. Learn. Cybern. **7**(6), 1225–1239 (2016)
27. X. Ling, J. Wang, T. Bouchoucha, et al., Blockchain radio access network (B-RAN), pp. towards decentralized secure radio access paradigm. IEEE Access **7**, 9714–9723 (2019)
28. H.B. Ribeiro, A.L.A. Simões, L.D. da Luz, et al., Stability of solids in stepped flume nappe flows: Subsidies for human stability in flows. J. Appl. Fluid Mech. **14**(3), 681–690 (2021)
29. E.L. Lydia, M.B. Swarup, Big data analysis using hadoop components like flume, mapreduce, pig and hive. Int. J. Sci. Eng. Comp. Technol. **5**(11), 390 (2015)
30. Z. Xiong, Y. Zhang, D. Niyato, et al., When mobile blockchain meets edge computing. IEEE Commun. Mag. **56**(8), 33–39 (2018)
31. M.U. Javed, M. Rehman, N. Javaid, et al., Blockchain-based secure data storage for distributed vehicular networks. Appl. Sci. **10**(6), 2011 (2020)
32. X. Luo, P. Yang, W. Wang, et al., S-PoDL: A two-stage computational-efficient consensus mechanism for blockchain-enabled multi-access edge computing. Phys. Commun. **46**, 101338 (2021)

The Role of Blockchain Technology in Enhancing Security Management in the Supply Chain

Zakariya Chabani and Widad Chabani

1 Introduction

Generally, all the supply-chain activities are associated with the flow of materials and information, which raises several security-related concerns that may compromise the entire supply-chain management [1, 2]. Several organizations from all over the world have reported various forms of challenges involving security breaches. The security concerns associated with supply-chain management come in potential damage to the transported materials and information security risks. Failure to control such barriers often triggers perceived insecurity, hindering an organization's physical and economic performance [3]. Due to these reasons, the concept of security management has always been integrated into supply-chain management to help curb the associated dangers [3].

Cybercrime is the main threat to information management in nearly all organizations, regardless of the industries in which they operate. It is defined as any crime facilitated or undertaken using a computer, a network, or an associated hardware device [4]. The computer or network involved in cybercrime can be the criminal's target, facilitator, or agent during the time of the attack. The rates of cybercrime attacks continue to rise, a condition that suggests increased levels of information insecurity. For instance, Symantec's study suggests an 81% increase in malicious attacks between 2010 and 2011 [4]. According to experts, such attacks substantially negatively impact communities and industries [4, 5]. Cyberattacks are also reasonable for other damaging consequences like brand image loss [4].

Z. Chabani (✉)
Canadian University Dubai, Dubai, UAE
e-mail: zakariya.chabani@cud.ac.ae

W. Chabani
HEC Alger, Tipasa, Algeria

Y. Maleh et al. (eds.), *Advances in Blockchain Technology for Cyber Physical Systems*, Internet of Things, https://doi.org/10.1007/978-3-030-93646-4_10

Despite companies' financial and reputational damages, most of them continue to make the same mistakes considered the causes of such problems. This situation suggests a general lack of information and awareness of the causes and damages associated with poor security management. The studies also fail to provide the most basic information regarding the causes and solutions to the problems emanating from poor security and privacy standards in the supply chain. This study tries to analyze the role of blockchain technology in enhancing security management in supply-chain management. It is evident, from statistics, that most companies are ignorant of the benefits of blockchain technology, especially its role in enhancing supply-chain security. Institutions that have employed this technology in the past also seem to be aware of the security loopholes. Thus, the project intends to enlighten organizations and individuals on the benefits of embracing security measures when undertaking supply-chain activities.

This study intends to achieve several vital objectives for enhancing security standards in typical supply-chain situations. The first objective is to bridge the gap in current literature and general information about supply-chain management concepts since empirical studies about blockchain technologies are almost absent [6, 7]. It also attempts to popularize blockchain technology to expand the extents to which it is employed in various organizations. Since blockchain is a relatively emerging concept, it is evident that most institutions are not entirely familiar with its application. Consequently, there are several cases in which this technology's application becomes prone to security issues due to a lack of adequate knowledge. The paper proceeds to provide a more detailed analysis of the blockchain approach to supply-chain management, which is expected to improve the technology's general knowledge and application. The paper also analyzes the general security issues while offering the most recommendable approaches to counteracting them. Overall, the paper is expected to change the public's understanding of blockchain technology related to security and privacy, two of the essential aspects of supply-chain management in contemporary corporate society.

The rest of this paper is organized as follows: the second section is designed to study the previous papers related to the subject. The third section is dedicated to explaining the research design, the hypotheses, and the methodology. The fourth section explains the data and the results, followed by a discussion (Sect. 5). Section 6 is the conclusion of the paper.

2 Literature Review

One way to think about supply-chain management is to think of it as an organized and systematic network between a business, its suppliers, and the end client to lower expenses and be competitive on the market [8]. Diverse processes, data flows, people, and other resources make up the system. All the phases and organizations involved in transporting a product from its initial condition to the end consumer are included in the supply chain. Blockchains are used to enable agreements, payments, and delivery in contemporary supply-chain management systems. Effective management is required to improve supply-chain operations and achieve lower costs and a faster production cycle. The article gives information on how vital a blockchain is to the supply-chain system. In the current world, security is one of the critical elements one should consider before doing any transaction within the market. Through the use of the blockchain technique, several activities have been achieved in the market. The technological idea generated by using the blockchain system enables the customer to believe and trust the supply-chain system. Through blockchain technology, most of the activities within the supply-chain environment can highly be monitored and protected against malicious individuals.

Blockchain technology has emerged as one of the most discussed concepts in the business sector. The technology is associated with limited movement of information among individuals involved in an organization. It offers a relatively more effective way of ensuring that the activities and information involved in an organization's supply chain are only accessible to the persons involved.

As a distributed database system, blockchain records are stored in a series of digital blocks linked together in an impenetrable and immutable way. Hash codes are produced using a one-way cryptographic algorithm utilizing block data to identify each block uniquely [8]. We cannot use a blockchain to handle the whole supply chain because of its complexity. Partial blockchain adoption can improve inventory control because of its security, transparency, and efficiency. Besides, blockchain adoption is used by most current supply-chain systems [9]. In addition to reducing supply-chain costs, Blockchain-Supply Chain Management (BC-SCM) improves consumer, supplier, and retailer confidence by enabling them to trace goods from their origin to their final destination, according to a recent study. In that way, they can check for a product or material authenticity and avoid product fraud. This article will review how the blockchain system improves the security of the supply chain and creates confidence in the market in the transaction field.

Joshi et al. performed a survey highlighting the blockchain technology structure concerning varying consensus algorithms. These researchers also emphasize their study on other aspects of the strategy, such as challenges and opportunities from the perspectives of existing users and individuals with adequate knowledge about it. This study's principal areas of focus were the privacy and security of the data circulating within a given blockchain setup. It proceeds to forecast the technology's expected trends shortly while highlighting the expected evolutionary changes in security and privacy concepts. The study also focuses on the impacts of such factors in the various contexts in which they are employed [10].

Blockchain security can be achieved through various ways depending on the security and privacy objectives of a given blockchain and the nature of data traffic involved. Joshi et al. hypothesize five different perspectives from which blockchain security can be attained. The main principles of blockchain security enhancement include defense in penetration, minimum privilege, vulnerability management, risk management, and patches management [10]. Each of these techniques aims to protect the information belonging to transacting parties in a blockchain from theft or any other damages that come with unauthorized access to confidential information.

The study by Joshi et al. offers a clear insight into blockchain security based on the information gathered from various respondents during their study. It highlights the possible sources of security threats and the recommended measures that should be taken to protect servers from the resulting vulnerabilities. However, this study has a weakness that limits its complete application. It discusses the idea of security in a far too public domain to be considered applicable in this paper's topic. Instead of focusing primarily on the security concerns in supply-chain blockchain, the study generalizes the entire topic of cybersecurity, which is only partially relevant to this paper. Overall, these researchers presented information that helps create a basic understanding of privacy and security concepts in network or server-based blockchain transactions [10].

A typical blockchain's overall security functions three major elements: confidentiality, integrity, and availability. The confidentiality of a network or server refers mainly to the level of privacy of the data traffic within such a system. According to these researchers, adequate confidentiality refers to only the people involved in blockchain transactions can access the associated data or communication messages. Thus, this strategy helps to protect data from unauthorized access by malicious individuals [11].

The principal interests of Dorri et al. study are key components of information security and privacy in a blockchain and how they can be exploited to ensure security and privacy are attained. Using their blockchain-based model, they present what they believe to be the most common sources of security and privacy compromises in blockchain technology. They proceed to describe how network administrators can counteract such threats. Their model is used to illustrate such systems' vulnerabilities and how they can be counteracted. Under this model, blockchain systems can be subjected to either the DDoS or linking attacks [11].

In admitting that DDoS attack is one of the critical threats to handling, storage, and traffic of information in a network or server-based transactions, the blockchain offers practical methods of counteracting such problems. Their blockchain-based model is equipped with various hierarchical systems that make it resilient to DDoS attacks [11]. The model designed is characterized by the property where all transactions are checked by the system miners, a condition that makes it impossible for hackers to install malware on the system directly [11]. The first hierarchy of preventing such acts, based on this model, is by making the component devices inaccessible. The following two hierarchies of this defense setup consist of specially designed configurations managed by network administrators. While the first defense layer prevents the DDoS attackers from making any director alterations to the critical components of a network such as codes and firmware, the following configurations are intended to ensure that the network administrators are in control of any activities within the network or server, such that the permission to access the information within the network is granted only to authorized persons.

Another attack is the linking attack. It refers to the situation where the connection between one social media user and others can be inferred with high precision levels. It aims to reveal the connections that a specific user attempts to hide as a means of maintaining confidentiality [12]. Linking attacks can take forms such as links reconstruction, which often result in the exposure of information that is meant to be confidential to third parties [12]. The criminals who use this technique have been highly enhanced by the emergence of public media platforms such as Facebook, Twitter, and Instagram. Users from various parts of the world often use vital information that describes their relations and other essential properties to hackers by presenting them on their social media pages. Many users' attempts to keep specific categories of information help the attackers identify valuable enough information to be their specific targets while orchestrating link reconstruction attacks [12].

A significant percentage of the data shared on social media sites often comes in personal information such as names of the high schools and colleges attended, smartphone numbers, dates of birth, and email addresses [12]. Conventionally, sites such as Facebook, which provide online social media services, attempt to offer security measures to protect their users' personal information. However, personal data can still be accessed even when a user takes precautionary measures, such as revealing none of their data except links to the social network [12]. This issue arises from the fact that friends can still infer a user's information using even the subtlest clues available to them via the shared links.

The Fire et al. research uses an evidence-based approach to investigate the primary sources of security threats in blockchain technology. The components of system security, as discussed by these scholars, are similar to the ones hypothesized by Joshi et al. This study's first role is that it highlights the potential causes of security issues in a blockchain. It mentions issues such as the DDoS and linking attacks. It proceeds to claim that it is easier for network administrators whose operations and transactions are designed as blockchain to control these issues using various techniques. These discussions and the associated recommendations attempt to reveal the blockchain's superiority, especially regarding its role in providing security and privacy in activities such as the supply chain characterized by the traffic of precious information that should be kept confidential [12].

Despite its benefits in reinforcing the ideas of security, society breaches, and recommended measures to counteract such challenges, the study fails to provide information that can be generalized to serve large target audiences. The scholars narrow their discussions to blockchain technology as employed in managing security systems in smart homes. They also provide their approach to developing a smart home configuration that is not vulnerable to the stated security threats. This approach is highly discriminative concerning the size of the audience to which it is essential. Discussing the causes and solutions to security and privacy concerns within a blockchain in a general form would have made the study more generalizable and useful to a broader audience.

For blockchain technology to be considered relevant to supply-chain management in any sector, it has to offer all constituent components of data security. Halpin and Piekarska mainly focus on security and privacy as the core features that make blockchains popular. According to these authors, the technology enhances data security by overcoming the typical database's weaknesses, not offering any cryptographic advantages. Thus, many people and organizations, especially those operating in the advertisement environments, are always enthusiastic about the blockchain as it offers the cryptographic advantages necessary for secure transactions [13].

Halpin and Piekarska's study offers a deep insight into both privacy and security concepts. These scholars' main security element is restricted access to blockchains and their contents. Blockchains are characterized by permissionless access, which raises security concerns. This feature is expected to open avenues for malicious individuals to inject more users into the blockchain while remaining in remote control of such users. The researchers agree that using the proof-of-work method solves this problem by ensuring that only authorized persons can access blockchains and their contents. The study proceeds to discuss privacy in blockchains using the bitcoin situation as a prominent example. Bitcoins, which are typical examples of blockchains, offer high anonymity and pseudonymity levels. Since pseudonymity and anonymity are two of the most fundamental privacy components, Halpin and Piekarska successfully persuade their audience about blockchain technology's effectiveness in maintaining user privacy.

Despite the strengths identified in this reference, it lacks specific properties considered for the current study. The paper's principal focus is to investigate blockchain technology concerning its roles in enhancing or threatening security and privacy in the supply chain. However, the study mainly focuses on general security and privacy features that relate to blockchain technology. Most of the ideas presented here are purely theoretical and are not supported by research-based evidence. Thus, it is evident that this resource is likely to be characterized by limited generalizability.

Kopyto et al. conducted an interdisciplinary Delphi survey. They analyzed long-term judgments from 108 selected experts from academia, industry, and politics/associations with various context-related backgrounds (blockchain, SCM, hybrid functions). The results reveal the Supply Chain Management (SCM) related obstacles that should be solved in order to have a successful blockchain application by 2035. One key finding indicates that active trust management between supply-chain partners will still be needed to successfully transfer data to the SCM. Furthermore, the study identifies data availability and data authenticity as two major SCM-specific barriers that could prevent the exploitation of potential benefits [14].

Blockchain provides product provenance and traceability. The authors claim that it is crucial to identify the origin of a product in sectors like food and machinery spare parts by documenting all its journey traces. These initiatives are primarily aimed at ensuring product traceability via blockchain by providing traceable evidence of product movement. Kouhizadeh et al. say that suppliers and retailers may use Everledger to create a proof of product origin for their goods to gain the confidence of conscientious customers. Such information is available to consumers on any device [15].

Everledger is used to help build a more trustworthy supply chain for the diamond industry and its trade. Because it is built on blockchain technology, the platform can authenticate services using ISO27001-compliant standard-based methods. In this way, stakeholders engaged in the diamond-trade supply chain may input and extract secure diamond-related data such as characteristics and pictures and videos, certifications, and other vital documents. Suppliers that invest in ethical and sustainable business practices may be rewarded via the platform.

A retailer's brand reputation is protected by the platform's capacity to track the diamond's trip through the supply chain, allowing retailers to create a compliance and sustainability profile that matches their purchasing criteria. A new layer of value is added by adding compliance, nation, mining, and polishing to the platform, enabling merchants to move beyond only the 4Cs. As a result of the platform, they can communicate with their suppliers better. At a glance, customers may view the total inventory, sustainability indicators, as well as diamond compliance. Using an audit trail that cannot be altered may also credibly assert certificate claims to other parties. Since the blockchain is being used, we know that product provenance can be tracked on the platform. As a result, it assumes compliance with ISO27001, which regrettably does not offer any objective criteria that the machine can immediately verify when anything goes wrong with it. In other words, it is impossible to automatically identify frauds caused by subjective compliance with the authentication methods. Besides, effective management must improve supply-chain operations and achieve lower costs and a faster production cycle. The article gives information on how vital a blockchain is to the supply-chain system [16].

Several industrial tools focus on the dependability and control of digital trade systems based on blockchain technology, particularly in the freight and financial trading industries. Besides, in using a safe and traceable payment mechanism, these solutions comply with requirements for achieving traceability. Parties may anonymously strive for fair financial commerce by creating transparent and traceable interactions based on blockchain technology. Moosavi et al. stress a blockchain-based technology called OriginTrail, which offers reliable data sharing in supply chains. Supply-chain management systems may be easily linked with the application level. Supply-chain management, banking, insurance, and other industries benefit from decentralized applications on top of OriginTrail. Scalability is also enabled by the OriginTrail independent network layer, a network that is not part of the blockchain; this is particularly true for supply chains focused on data management and accessibility [17].

The OriginTrail Decentralized Network (ODN) data layer offers a decentralized graph database, which connects datasets across supply chains. Through support for worldwide data-exchange standards, this layer promotes interoperability, while protecting sensitive data using zero-knowledge privacy lower-layer GS1 standards for master data, transaction data, visibility data, as well as IoT and comply data are used to create interoperability [18]; this means that the protocol can fully use the relational nature of supply-chain data. Consensus procedures and data verification may take place once the data has been matched throughout the supply chain. Besides, through blockchain technology, most of the activities within the supply-chain environment can be highly monitored and protected against malicious individuals. As a distributed database system, blockchain records are stored in a series of digital blocks linked together in an impenetrable and immutable way. Hash codes are produced using a one-way cryptographic algorithm utilizing block data to identify each block uniquely.

It is worth noting that OriginTrail can verify the amount of a complete supply chain by using encrypted data exchanged among stakeholders in a single supply chain. OriginTrail may also be utilized with various blockchains because of its virtualization of the blockchain layer. The virtualization guarantees the protocol's flexibility and long-term viability. OriginTrail exclusively provides data security and privacy at the network level [19]. It cannot be used directly to ensure the confidentiality and security of data from beginning to finish as it travels through the supply chain. In addition, many additional layers must be considered, such as the software-specific application layer, which uses a variety of security and privacy-protection methods.

In using blockchain technology, industry professionals could smoothly interoperate different cross-border systems and infrastructures. As a result of such solutions, parties can build confidence and trust with one another across borders. Therefore, given that cross-border communication requires dispute resolution, some experts have called for the development of systems that automatically resolve digital disputes arising from cross-border businesses by documenting all interactions between participants in a blockchain. IBM's blockchain has created such a repeatable solution that changes the process of resolving disputes between many supply chain actors. Specifically, it is used by IBM as a single source of truth accessible only to authorized parties. Smart contracts automatically implement agreements and business rules. In theory, stakeholders transmit processed data directly from their recording systems to the blockchain, giving selected/permitted participants visibility. In a multi-participant environment, privacy is protected by not allowing others to see it. Data input mistakes are also avoided. Following that, the solution's business logic detects inconsistencies between data components and documents to identify the underlying cause of any disputes. A supply-chain disagreement may arise because of a faulty measurement unit, incorrect delivery location, or incorrect amount supplied. In taking the telecom industry as an example, everyone must agree on data, SMS, voice use, and prices to be paid. As new data becomes available, all comparisons are made in near real time. The consequence is a significant reduction in the time it takes to resolve conflicts. Disagreement data is synthesized using IBM's platform's automatic dispute settlement rules [20].

Xu et al. are also making strides to enhance the regulation of supply-chain networks via blockchain in various fascinating application areas such as payment and healthcare. A limited number of payment systems' security systems have been created following the agreed-upon strategy and directive to ensure traceability. Many additional solutions have been created to protect intellectual property by creating transparent and traceable ownership rights based on blockchain technology [19]. On the contrary, one system looked at mechanisms for cross-border coordination that were visible and traceable. The activity was done to aid in developing the design and analysis of coordination policies [19]. For example, the Gcoin blockchain is utilized to make transactions irreversible, consensus driven, and transparent in the medication supply chain. It also aims to enable a surveillance net by extending the management structure of the medication supply chain. Through data mining, government agencies establish a risk threshold to sustain the surveillance net. Later, if a drug stakeholder's/transaction company's behavior does not meet the criteria, the smart contract may raise an alert and request an inspection. Whenever a product has software, tracking the dynamic characteristics of the product becomes more difficult.

To ensure that all participants are aware of the consensus decision and the rules that have been implemented, the decision is made publicly available. Therefore, final consensus choices are transmitted back to the system of record for documentation. The blockchain distributed ledger stores all data, inconsistencies, and consequential decisions, creating a complete and immutable audit history. When cross-chain and cross-border contacts are permitted, IBM's technology has a significant flaw: it cannot settle conflicts. In addition, it is unable to resolve disputes arising from procedures that are not designed using smart contracts. First of all, the blockchain is decentralized. The ledger verification is decided by agreement among network members. Besides, to avoid misuse, most blockchains require users to do complicated computations repeatedly, resulting in high costs associated with this need. Similarly, verifiable data audit does not rely on a blockchain and does not waste energy resources. Another advantage is that it has a treelike structure instead of a chain. Blockchain is comparable in many respects to the verified data audit, though. Upon adding an entry to a ledger, a "cryptographic hash" value is generated. This value summarizes not just the current entry but also all of the initial values in the ledger. An entry's hash value will be affected and that of the entire tree due to this impact [18].

Etemadi et al.'s solutions to enable more realistic business models are of great interest to many sectors in question. In this respect, industries are primarily concerned with securing cryptocurrency operations and activities by ensuring that various stakeholders comply. IBM's IoT platform, for example, ensures that IoT-enabled supply-chain activities are compliant in their operations with one another. As a result of supply-chain management, goods and their components have a remarkable history, including key events influencing their lives or planned maintenance. Suppliers, OEMs, and regulators may securely access this data. The use of decentralized edge computing allows third-party devices to process tasks, such as analytics, securely. Services are paid for using micropayments. By using distributed role control and micropayments via micro-services, dispersed devices may request and pay for services [21].

The industry has created techniques for identifying vulnerabilities in payment systems and their virtual execution environment and developing tools for securing cryptocurrency payments and operations. While being parsed, the intermediate language has additional information about the program. The compiler produces a parse tree when a program is created, reflecting its functionality [21]. The compiler may further enhance an element from the control flow by adding information such as taint data, the location of the source, and other things that might have affected it. Besides, languages such as Solidity provide inheritance, allowing procedures and methods to be created beyond the bounds of a particular contract. The ability to manipulate variables is crucial to discovering more sophisticated vulnerabilities from a static posture. Information is propagated iteratively from function arguments to program state, capturing control flow information over possibly many transactions. By enriching the information and statically verifying the presence of common vulnerabilities that may be reached under specific suspicious circumstances, Slither enhances contracts' security [17].

3 Research Design and Methodology

Research Hypotheses

The study intends to provide answers to two basic categories of questions that often trouble the administrators of small and large corporations. The first class of questions revolves around the source of the security challenges that paralyze most managers' attempts to attain successful supply-chain activities. This part of the study question is mainly intended to be a fact-finding tool for the stated concerns. The second part of the question dwells on searching for the perfect response to these concerns to evade the damaging impacts associated with them. The blockchain, the most contemporary concept in supply-chain management, has been selected as the main avenue through which solutions to these issues can be obtained. The research's specific study questions are as follows:

What are the most common forms of security issues in supply-chain management?
What causes security compromises in supply-chain management?
How does blockchain technology affect people's management in a typical supply-chain scenario?
How does blockchain technology affect asset and facility management in a typical supply chain?
How does the implementation of blockchain technology affect the overall security performance of a typical supply chain?

In order to answer the study's questions, and based on the literature review, the suggested hypotheses are as follows:

- H1 – the most common security issues in the supply chain include physical damages of products, supplies, and information.
- H2 – the most common causes of supply-chain security compromises include the affected companies' poor management of people, assets, facilities, and information.
- H3 – blockchain technology improves the efficiency of managing the people involved in supply-chain activities.
- H4 – the blockchain technology improves the efficiency of managing assets and facilities employed in supply-chain activities.
- H5 – the blockchain technology improves the overall security performance in the supply chain.

Methodology

Due to the lack of adequate databases from which reliable information could be obtained, several information sources were used, including company websites, case studies, existing literature, past news reports, and conference papers. The main objective of using as many data sources was to ensure that enough information was available to facilitate further analysis to answer the study's central questions. The gathered information was grouped into four basic categories. The first category consisted of all the security complaints presented by companies between years 2000 and 2018.

The second category of information consisted of all the privacy violation complaints from various companies in ten different industries within the United Arab Emirates. The selected industries were agriculture, engineering, finance service, consultancy, healthcare, manufacturing, transport, automobiles, consumer electronics, and education. Twenty companies, including small and medium businesses, were selected from each industry, and the data was arranged appropriately.

The final category came in the form of a percentage of supply-chain operations that were undertaken using the blockchain technology in different organizations. A total of 200 UAE companies were selected for this section. Meta-analysis was used to sort the data into two sets of variables in which security and privacy were regarded as the primary independent variable. The number of firms using the blockchain and the percentage of supply-chain activities undertaken using the supply chain were regarded as dependent variables.

The sorted data was then presented for analysis to establish the most crucial relationships among selected sets of variables. The analysis was done using the EViews software, which offers a simple approach to undertaking this process. Three principal tests were conducted to establish the required relationships. The selected tests were correlation, cross-correlation, and Granger's causality tests. Each of these tests was aimed at realizing specific objectives. For instance, correlation analysis was intended to determine the possible existence of any form of relationship between the dependent and the independent variables in each section. The cross-correlation aimed to investigate a more detailed insight into the correlations established among variables. On the other hand, Granger's causality test was selected to enable the researchers to determine if any of the two variables cause the other. No ethical considerations characterized this study since there was no need for any permissions or actual participants whose privacies needed to be protected.

4 Results and Data Analysis

Descriptive Statistics

Various properties characterized the data corresponding to the privacy of blockchain technology. The average percentage of operations that the selected firms undertake using blockchain technology is 51%. On the other hand, the average number of security-based complaints encountered by these organizations is 49.54%. The first conclusion that can be deducted from this data is that firms only use blockchain technology to an average extent. This information can also justify scholars' arguments claiming that this concept is still new and has not been embraced by many firms. An average number of security-based complaints also imply that a relatively sizeable number of complaints characterized the period under investigation. The standard deviations corresponding to the percentage of operations done using blockchain technology and the number of security-related complaints are 29.15 and 36.75, respectively. These figures represent high standard deviations, which imply that most data points are widely scattered away from the stated means (see Table 1).

Relatively similar trends characterized the data corresponding to both security and privacy issues that were investigated during the research. The mean values for the number of firms using the blockchain method and the corresponding number of privacy-related lawsuits or complaints were 10.60 and 11.40, respectively. These values imply that there were a relatively low number of organizations using the blockchain technology. However, the mean for the number of complaints or lawsuits is only 11.40, which implies a relatively high number of complaints in each selected industry, especially considering that only a few firms represented each of the selected industries. Just like in the case of security data, this data was characterized by high standard deviations of 13.01 and 6.74 for the number of operations done using the blockchain technology and the number of privacy-based lawsuits and

Table 1 Descriptive statistics for the blockchain security data

	% of operations done using blockchain	No. of security complains
Mean	51.00000	49.54000
Median	51.00000	44.50000
Maximum	100.0000	137.0000
Minimum	2.000000	0.000000
Standard deviation	29.15476	36.74746
Skewness	1.03E-17	0.534709
Kurtosis	1.799040	2.265195
Jarque-Bera	3.004804	3.507482
Probability	0.222595	0.173125
Sum	2550.000	2477.000
Sum square deviation	41650.00	66168.42
Observations	50	50

Table 2 Descriptive statistics for the privacy data

	Number of firms that use blockchain	Number of privacy-based lawsuits
Mean	10.60000	11.40000
Median	7.000000	11.00000
Maximum	45.00000	21.00000
Minimum	1.000000	1.000000
Standard deviation	13.01452	6.736303
Skewness	2.046725	−0.050852
Kurtosis	6.123299	1.735472
Jarque-Bera	11.04639	0.670573
Probability	0.003993	0.715133
Sum	106.0000	114.0000
Sum square deviation	1524.400	408.4000
Observations	10	10

complaints, respectively. This trend implies that most of the data points used for this analysis were widely distributed away from the mean (see Table 2).

Data Analysis

This study's principal objective was to investigate the relationship between firms' decisions to employ blockchain technology and the security or privacy issues at organizational levels. Practical conclusions regarding this investigation can only be reached after performing several tests to provide the statistical relationships among variables used. The selected tests for this study included correlation, Granger's causality, and cross-correlation. Each test produced results that can be used to provide a detailed insight into the data and the principal themes being investigated.

Correlation Analysis

The test was conducted to evaluate the possible relationship between the blockchain technology in undertaking various activities within an organization and the overall security of privacy trends within the industries studied. Pearson's correlation coefficient (*P* value) was used as the primary metric for determining and quantifying these two variables' relationships. The *P* value describing the relationship between the extent to which organizations employed blockchain technology in company operations and the resulting number of complaints encountered was −0.980536. This figure can be interpreted as a strong negative correlation between two variables (see Table 3).

For privacy data, the results obtained were relatively similar to those corresponding to security, especially about the trends shown by the analysis results.

Table 3 Correlation analysis for security data

	No. of blockchain operations	No. of security complains
No. of blockchain operations	1.000000	−0.980536
No. of security complains	−0.980536	1.000000

Table 4 Correlation analysis tests for the blockchain privacy data

	No. of blockchain operations	No. of security-based complaints
No. of blockchain operations	1.000000	−0.752065
No of security-based complaints	−0.752065	1.000000

Pearson's coefficient describing the relationship between the number of operations organizations conducted using the blockchain technology and the number of privacy-related complaints and lawsuits is −0.752065. This number indicates a strong negative correlation between the number of supply-chain operations undertaken using blockchain technology and the number of privacy concerns within each analyzed industry. It also implies that the number of complaints is higher when there are low rates of applying this technology (see Table 4).

Granger's Causality Tests

Correlation is a useful method of determining the relationship between sets of variables. However, this type of analysis informs the analyst about the existence or nonexistence of relationships among data. It also tells more about this relationship's degree without offering any additional information about such a relationship. Granger's causality test is regarded as an additional test that enables data analysis to learn about the relationships investigated. It mainly indicates if one of the variables causes the other while also offering reliable information about the direction of this causality.

The P values of each of the possible causality directions were used to determine the possible causality between two sets of variables in the security data. The decision was guided by the principle that the null hypothesis can be rejected when the P value is lower than 0.05, while it was not rejected in the cases where the P values were higher than 0.05. This reasoning system also made it possible to determine the overall direction of causality in case it was identified. The P value associated with the ability of security-based complaints to affect the extent to which firms employed the blockchain technology was 0.008, while the P value associated with the ability of blockchain operations to affect the number of security-based complaints is 0.0925. Since 0.008 is lower than 0.05, the null hypothesis number of security-based complaints causes the degree of application of the blockchain technology in undertaking supply chain within organizations to be rejected.

Table 5 Granger's causality test results for blockchain security data

Pairwise Granger causality tests			
Date: 10/09/19 Time: 15:53			
Sample: 1 50			
Lags: 2			
Null hypothesis	Obs	F-Statistic	Prob.
NO_OF_SECURITY_BASED_COM does not Granger Cause NO_OF_BLOCKCHAIN_OPERATI	48	5.29235	0.0088
NO_OF_BLOCKCHAIN_OPERATI does not Granger Cause NO_OF_SECURITY_BASED_COM		2.51679	0.0925

Table 6 Granger's causality test results for the blockchain privacy data

Pairwise Granger causality tests			
Date: 10/09/19 Time: 15:30			
Sample: 1 10			
Lags: 2			
Null hypothesis	Obs	F-Statistic	Prob.
PRIVACY_BASED_LAWSUITS does not Granger Cause BLOCKCHAIN_APPLICATION	8	0.29347	0.7649
BOCCHAIN_APPLICATION does not Granger Cause PRIVACY_BASED_LAWSUITS		1.77636	0.3098

On the other hand, the hypothesis that the extent to which firms use blockchain technology cannot be rejected since the associated *P* value is 0.0925 higher than the 0.05 limit. Consequently, it is worth concluding that there is a unidirectional causality between these variables (see Table 5).

The same test was conducted on the private data to determine any form of causality between the two variables used here. The *P* values for each of the null hypotheses (privacy-based lawsuits do not Granger-cause the application of blockchain, and the blockchain application in organizations does not Granger-cause privacy-related complaints) were 0.7649 and 0.398, which are higher than the 0.05 limit. Consequently, none of these hypotheses could be rejected. This analysis implies that there is no causality between these two sets of variables in any direction. Thus, neither the application of the blockchain technology nor the number of privacy-related lawsuits and complaints Granger-causes the other (see Table 6).

Cross-correlation Tests

The final test performed on the collected data is cross-correlation analysis. This test's main objective was to assess the two variable sets' movement relative to each other. The principal rationale behind the test is that if the number of blockchain operations in various companies within the industry affects the number of security-related complaints in the same industry, then the values representing one variable will increase as the values representing the other increase. In essence, cross-correlation is employed when dealing with time-series information. The numbers representing the correlation range between -1 and $+1$ such that high cross-correlation among given sets of data is indicated by cross-correlation values close to 1. According to the given data, there is a gradual increase in the values corresponding to the number of complaints from -0.9805 to -0.0162. A similar trend is observed in the dataset corresponding to the organizations' blockchain technology level within the sampled industry. These trends imply a negative cross-correlation between the level of application of blockchain technology in supply-chain operations and the number of complaints among industrial firms. Consequently, this data suggests a steady increase in the number of complaints as the level of application of the blockchain technology continues to decrease (see Table 7).

The data corresponding to the relationship between privacy and level of application of the blockchain technology does not offer much information regarding the series' trends. There are high levels of irregularities characterizing the relationships between these two sets of data. However, close observation of individual values reveals that the number of firms complaining about privacy issues through lawsuits is higher whenever the extent of application of the blockchain technology is low (see Table 8).

5 Discussions

The study's primary approach to representing security and privacy is through complaints related to each of these concepts. On the other hand, the blockchain technology application in organizations was measured using two principal approaches. The first approach involves using the percentage of company supply-chain operations undertaken using blockchain principles. This strategy enables the researcher to differentiate between security status within organizations when the blockchain technology is employed and the same trends in cases where the technology is not applied. The second method of quantifying blockchain technology's widespread application is the data describing the number of companies that used blockchain as their main approach to handling transactions and data security in the supply chain. Overall, two variables were represented with reasonable activities that would help make it possible to perform quantitative analysis.

The correlation test results revealed a correlation between the blockchain technology's application in organizations and the corresponding privacy and security trends at both organizational and industry levels. Such statistics indicate that blockchain technology has become an integral aspect of both security and supply-chain management information. While this test offered a clear link between these two variables, it failed to offer more details. The cross-correlation test analysis provided additional information that was missing in the correlation test results. Data

Table 7 Cross-correlation analysis for security data

<table>
<tr><td colspan="5">Date: 10/08/19 Time: 18:35
Sample: 1
50Included observations: 50 Correlations are asymptotically consistent approximations</td></tr>
<tr><th>BLOCKCHAIN OPERATIONS_NUMBER OF COMPLAINS (−i)</th><th>NUMBER OF COM-PLAINTS_BLOCKCHAIN OPERATIONS (+i)</th><th>i</th><th>lag</th><th>lead</th></tr>
<tr><td>**********| . |</td><td>**********| . |</td><td>0</td><td>−0.9805</td><td>−0.9805</td></tr>
<tr><td>*********| . |</td><td>*********| . |</td><td>1</td><td>−0.9324</td><td>−0.8956</td></tr>
<tr><td>*********| . |</td><td>********| . |</td><td>2</td><td>−0.8834</td><td>−0.8218</td></tr>
<tr><td>********| . |</td><td>********| . |</td><td>3</td><td>−0.8354</td><td>−0.7502</td></tr>
<tr><td>********| . |</td><td>*******| . |</td><td>4</td><td>−0.7857</td><td>−0.6828</td></tr>
<tr><td>*******| . |</td><td>******| . |</td><td>5</td><td>−0.7371</td><td>−0.6170</td></tr>
<tr><td>*******| . |</td><td>******| . |</td><td>6</td><td>−0.6860</td><td>−0.5549</td></tr>
<tr><td>******| . |</td><td>*****| . |</td><td>7</td><td>−0.6352</td><td>−0.4927</td></tr>
<tr><td>******| . |</td><td>****| . |</td><td>8</td><td>−0.5847</td><td>−0.4327</td></tr>
<tr><td>*****| . |</td><td>****| . |</td><td>9</td><td>−0.5346</td><td>−0.3738</td></tr>
<tr><td>*****| . |</td><td>***| . |</td><td>10</td><td>−0.4812</td><td>−0.3182</td></tr>
<tr><td>****| . |</td><td>***| . |</td><td>11</td><td>−0.4321</td><td>−0.2642</td></tr>
<tr><td>****| . |</td><td>.**| . |</td><td>12</td><td>−0.3818</td><td>−0.2088</td></tr>
<tr><td>***| . |</td><td>.**| . |</td><td>13</td><td>−0.3330</td><td>−0.1561</td></tr>
<tr><td>***| . |</td><td>. *| . |</td><td>14</td><td>−0.2851</td><td>−0.1062</td></tr>
<tr><td>.**| . |</td><td>. *| . |</td><td>15</td><td>−0.2331</td><td>−0.0611</td></tr>
<tr><td>.**| . |</td><td>. | . |</td><td>16</td><td>−0.1858</td><td>−0.0162</td></tr>
<tr><td>. *| . |</td><td>. | . |</td><td>17</td><td>−0.1375</td><td>0.0264</td></tr>
<tr><td>. *| . |</td><td>. |* . |</td><td>18</td><td>−0.0911</td><td>0.0648</td></tr>
<tr><td>. *| . |</td><td>. |* . |</td><td>19</td><td>−0.0458</td><td>0.1007</td></tr>
<tr><td>. | . |</td><td>. |* . |</td><td>20</td><td>−0.0037</td><td>0.1339</td></tr>
<tr><td>. | . |</td><td>. |**. |</td><td>21</td><td>0.0372</td><td>0.1644</td></tr>
<tr><td>. |* . |</td><td>. |**. |</td><td>22</td><td>0.0757</td><td>0.1940</td></tr>
<tr><td>. |* . |</td><td>. |**. |</td><td>23</td><td>0.1146</td><td>0.2207</td></tr>
<tr><td>. |* . |</td><td>. |**.|</td><td>24</td><td>0.1511</td><td>0.2453</td></tr>
</table>

analysis findings suggest that the rate of security and privacy-related complaints decreases with an increase in this technology's application. The final observation was that security concerns at both organizational and industrial contexts Granger-cause the blockchain technology level in the supply chain.

Cases of security complaints and lawsuits about privacy violations become less when companies decide to use the blockchain. This observation's rationale is that technology offers better security than traditional approaches to supply-chain operations. It suggests that the data describing documents and transactions associated with supply-chain activities are always safer when organizations opt to use the blockchain method in their supply-chain activities. The causality analysis findings can be interpreted to mean that companies who have previously fallen victims to data security and privacy violations often ensure that such situations

Table 8 Cross-correlation analysis for the blockchain privacy data

Date: 10/09/19 Time: 15:24 Sample: 1 10
Included observations: 10

Correlations are asymptotically consistent approximations

<table>
<tr><th>NUMBER_OF_FIRMS_THAT_USE
NUMBER_OF_PRIVACY_BASED_(−i)</th><th>NUMBER_OF_FIRMS_THAT_USE
NUMBER_OF_PRIVACY_BASED_(−i)</th><th>i</th><th>lag</th><th>lead</th></tr>
<tr><td>********| . |</td><td>********| . |</td><td>0</td><td>−0.7521</td><td>−0.7521</td></tr>
<tr><td>. |* . |</td><td>. |** . |</td><td>1</td><td>0.1196</td><td>0.1586</td></tr>
<tr><td>. |* . |</td><td>. | . |</td><td>2</td><td>0.0950</td><td>−0.0346</td></tr>
<tr><td>. |**** . |</td><td>. | . |</td><td>3</td><td>0.3529</td><td>0.0265</td></tr>
<tr><td>. *| . |</td><td>. **| . |</td><td>4</td><td>−0.1064</td><td>−0.1741</td></tr>
<tr><td>. |*** . |</td><td>. |* . |</td><td>5</td><td>0.3389</td><td>0.0535</td></tr>
<tr><td>. **| . |</td><td>. | . |</td><td>6</td><td>−0.2003</td><td>−0.0196</td></tr>
<tr><td>. |*** . |</td><td>. |* . |</td><td>7</td><td>0.3330</td><td>0.1079</td></tr>
<tr><td>. | . |</td><td>. | . |</td><td>8</td><td>−0.0179</td><td>0.0338</td></tr>
</table>

do not recur. Therefore, they try to implement corrective mechanisms such as introducing the blockchain concept into their supply-chain operations. Therefore, from the data analysis results, it is evident that the rate of application of this encrypted data approach to the supply chain becomes more applicable whenever there are increasing trends in the number of security violation complaints. Overall, blockchain technology is becoming more popular among organizations in different sectors, attributed to the increasing cases of security and privacy violations during supply-chain operations.

6 Conclusion

Blockchain technology remains one of the most exciting areas of research in contemporary society. This technology is applied in various sectors, such as the financial service sector, in which cryptocurrencies are used to provide anonymity, security, and confidentiality. According to the information provided by other scholars who have performed research in this field, blockchain technology is a practical approach to attaining and sustaining information security in any field of operation. Activities such as the proof-of-work equip this concept with the ability to control the identities of the people who have access to confidential information. The analysis of primary data reveals that security and privacy levels are always enhanced when data eruption strategies are used in the supply chain. The data also reveal that more organizations are becoming attracted to this strategy as they try to react to security and privacy violations. As revealed by the data, the most discouraging trend is that most organizations are not aware of this concept and its importance to their operations. Consequently, the cases of data insecurity remain high at both organization and industry levels.

These findings can be very crucial in opening avenues for future research in this field. The primary role that the research plays in academics is that it offers a critical and analytical insight into the blockchain concept, especially concerning vital issues such as security and privacy.

Future researchers can use this study's findings to pioneer innovation into other fields such as the business opportunities associated with the technology and have not been exploited by society. The findings also guide the future society on the best ways through which the technology can be implemented in various sectors to ensure that privacy and data security issues in the supply chain are resolved. Another critical field that is open for further research is the idea of instances where the technology is considered to have weaknesses regarding its ability to offer absolute security and privacy.

The main limitation of this study comes in the form of a lack of readily available data. Several data sources had to be used, including existing literature and statistics from reputed sites like Statista, to ensure that sufficient information was collected to successfully facilitate the study's key objectives. Generally, this study's objectives were successfully attained as the data analysis results were entirely consistent with the hypotheses.

References

1. D.M. Lambert, M.G. Enz, Issues in supply chain management: Progress and potential. Ind. Mark. Manag. **62**, 1–16 (2017)
2. N. Dao, J. Daniel, S. Hutchinson, M. Naderpour, Logistics and supply chain management investigation: A case study, in *Australian Symposium on Service Research and Innovation*, (2018), pp. 1–15
3. J. Hintsa, P. Wieser, X. Gutierrez, A.-P. Hameri, Supply chain security management: An overview. Int. J. Logist. Syst. Manag., 1–10 (2009)
4. L. Urciuoli, T. Männistö, J. Hintsa, T. Khan, Supply chain cyber security – Potential threats. Inf. Secur. Int. J. **29**, 51–68 (2013)
5. J.E. Gould, C. Macharis, H.-D. Haasis, Emergence of security in supply chain management literature. J. Transp. Secur. **3**, 287–302 (2010)
6. B. Müßigmann, H. von der Gracht, E. Hartmann, Blockchain technology in logistics and supply chain management—A bibliometric literature review from 2016 to January 2020. IEEE Trans. Eng. Manag. **67**, 988–1007 (2020)
7. S.F. Wamba, M.M. Queiroz, Blockchain in the operations and supply chain management: Benefits, challenges and future research opportunities. Int. J. Inf. Manag **52**, 102064 (2020)
8. E.M. Abou-Nassar, A.M. Iliyasu, P.M. El-Kafrawy, O.Y. Song, A.K. Bashir, A.A. Abd El-Latif, DITrust chain: Towards blockchain-based trust models for sustainable healthcare IoT systems. IEEE Access **8**, 111223–111238 (2020)
9. S.E. Chang, Y. Chen, When blockchain meets supply chain: A systematic literature review on current development and potential applications. IEEE Access **8**, 62478–62494 (2020)
10. A.P. Joshi, M. Han, Y. Wang, A survey on security and privacy issues of blockchain technology. Math. Found. Comput. **1**, 121–147 (2018)
11. A. Dorri, S.S. Kanhere, R. Jurdak, P. Gauravaram, Blockchain for IoT security and privacy: The case study of a smart home, in *2017 IEEE International Conference on Pervasive Computing and Communications Workshops (PerCom Workshops)*, (IEEE, Kona, 2017), pp. 618–623
12. M. Fire, G. Katz, L. Rokach, Y. Elovici, Links reconstruction attack, in *Security and Privacy in Social Networks*, ed. by Y. Altshuler, Y. Elovici, B. Cremers, N. Aharony, A. Pentland, (Springer, New York, 2013), pp. 181–196
13. H. Halpin, M. Piekarska, Introduction to security and privacy on the blockchain, in *2017 IEEE European Symposium on Security and Privacy Workshops (EuroS&PW)*, (IEEE, Paris, 2017), pp. 1–3
14. M. Kopyto, S. Lechler, H.A. von der Gracht, E. Hartmann, Potentials of blockchain technology in supply chain management: Long-term judgments of an international expert panel. Technol. Forecast. Soc. Chang., 1–13 (2020). https://doi.org/10.1016/j.techfore.2020.120330
15. M. Kouhizadeh, S. Saberi, J. Sarkis, Blockchain technology and the sustainable supply chain: Theoretically exploring adoption barriers. Int. J. Prod. Econ. **231**, 107831 (2021)
16. P. Dutta, T.M. Choi, S. Somani, R. Butala, Blockchain technology in supply chain operations: Applications, challenges, and research opportunities. Transp. Res. Part e: Logist. Transp. Rev. **142**, 102067 (2020)
17. A. Shahid, A. Almogren, N. Javaid, F.A. Al-Zahrani, M. Zuair, M. Alam, Blockchain-based agri-food supply chain: A complete solution. IEEE Access **8**, 69230 (2020)
18. J. Moosavi, L.M. Naeni, A.M. Fathollahi-Fard, U. Fiore, Blockchain in supply chain management: A review, bibliometric, and network analysis. Environ. Sci. Pollut. Res., 1–15 (2021)

19. P. Xu, J. Lee, J.R. Barth, R.G. Richey, Blockchain as supply chain technology: considering transparency and security. Int. J. Phys. Distrib. Logist. Manag. **51**, 305–324 (2021)
20. X. Zhang, P. Sun, J. Xu, X. Wang, J. Yu, Z. Zhao, Y. Dong, Blockchain-based safety management system for the grain supply chain. IEEE Access **8**, 36398–36410 (2020)
21. N. Etemadi, Y. Borbon-Galvez, F. Strozzi, T. Etemadi, Supply chain disruption risk management with blockchain: A dynamic literature review. Information **12**(2), 70 (2021)

Using Hyperledger Fabric Blockchain to Improve Information Assurance of IoT Devices for AI Model Development

Anthony Kendall, Arijit Das, Bruce Nagy, Bonnie Johnson, and Avantika Ghosh

1 Introduction

Artificial intelligence (AI) software exercises a high degree of control over a particular system function (e.g., movement/guidance of a drone or manned aircraft). If the function creates a hazard, this can cause a mishap that has a consequence of a catastrophic and critical event, resulting in death or resources destroyed. There is no redo, reboot, or retraining of an AI function that fails in this scenario. Software safety engineering and test and evaluation efforts to ensure fidelity include data-related elements such as process flow, code level, and data structure analysis. Cybersecurity plays a role in ensuring cyberattacks do not compromise the integrity of the data elements such as AI training sets.

These flows and interaction with data and software, for example, are similar to the use of blockchain (BC) in supply chains used in our previous research, the Navy supply chain process, which we believe can be adapted for system safety and software integrity and play a role in cybersecurity especially in complex systems relying more and more on IoT (Internet of things) sensors on "the edge."

Complex systems go beyond IoT sensors supporting "ships at sea" or deployed resources, but Abbas [1] notes that the rise of smart cities depends on data streams from many sensors. His article introduces prior knowledge using a Hyperledger Fabric-based data architecture that is a secure and trusted smart transportation system. Smart cities can also be vulnerable to cyberattacks on smart electrical grids, and therefore various blockchain efforts may provide countermeasures.

A. Kendall (✉) · A. Das · B. Nagy · B. Johnson
Naval Postgraduate School, Monterey, CA, USA
e-mail: bruce.nagy@navy.mil; adas@nps.edu; bruce.nagy@navy.mil; bwjohnso@nps.edu

A. Ghosh
University of California, Berkeley, Berkeley, CA, USA
e-mail: ghoshavantika_123@berkeley.edu

Y. Maleh et al. (eds.), *Advances in Blockchain Technology for Cyber Physical Systems*, Internet of Things, https://doi.org/10.1007/978-3-030-93646-4_11

Abd El-Latif [2] proposes a quantum-inspired blockchain framework to secure smart edge utilities in IoT-based smart cities, and this proposed framework would have the ability to withstand the probable attacks from both digital and quantum computers. Their authentication and encryption protocol is based on quantum-inspired quantum walks (QIQW) to secure data transmission among IoT devices. QIQW is employed for linking blocks of the chain.

Software and data integrity of course apply to the IoT supporting healthcare decision-making as we see the increasing use of smart thermometers, oximeters, and other biometrics collected by IoT devices.

Abou-Nassar [3] proposes a decentralized interoperable trust (DIT) framework where smart contracts guarantee the authentication of health budgets and Indirect Trust Inference System (ITIS) to reduce semantic gaps and enhances the trustworthy factor (TF) estimation via the network nodes and edges. He notes, importantly, the simultaneous parallel creation of IoT systems-based ontology has not produced a common communication protocol or universal coding language for all IoT devices present and, in the future, an important feature of blockchain as a uniter.

Other work by Nguyen [4] demonstrates the flexibility of blockchain smart contracts to improve information assurance of important assets such as healthcare data acquisition process using sensor devices, and intrusion detection takes place using deep belief network (DBN) model. Blockchain technology is applied for secure data transmission to the cloud server, which executes the residual network (ResNet)-based classification model.

AI/machine learning (ML), the training sets, algorithms, and associated software supporting critical systems such as electrical grids or defense systems are targets for increasingly sophisticated adversarial machine learning attacks, which attempt to fool models through malicious input into the system such as AI poisoning and other attacks. Athalye et al. [5] showed that it is even possible to fool an AI into having it identify a turtle wrongly as a rifle and such attacks are likely to become more sophisticated. BC could be used as a countermeasure to prevent such poisoning as well as safeguard system integrity. BC, specifically Hyperledger Fabric (HLF), is a tamper-resistant decentralized trusted ledger that provides proof of transaction where trust is implemented through distributed consensus to ensure that only authorized people can modify the code base, AI algorithm, or training set and that the modification is traceable and transparent. Distributed ledgers provide system safety through BC provenance and policy enforcement through a feature called smart contracts, which imbed logical code. Data in support of AI and software development can also suffer not only from deliberate sabotage or ruse but also from human error.

Machine learning increasingly requires complex data sources from repositories and sensors down to the edge for training sets supporting AI development. Getting the right and accurate data can be a complex process, and error or intentional manipulation is always a concern. The number of sensors and IoT devices, such as smart thermometers/oximeters to track COVID-19, has caused an explosion of data generation but not an increase in safeguards to ensure system safety if these edge devices are used to control machines or make life-critical decisions.

Centralized security and authentication controlling IoT devices could lead to a single point of failure, a new target for a cyberattack, and cause a bottleneck and high latency [6]. Typically, an ML project may require diverse data sources and modalities. One example may be drones flying over an urban area, which requires its ML training set data on the region, including crime rate, weather, and road conditions/constraints. For just this simple example, data needed may include war-gaming, tabletop exercises, lessons learned, product performance specifications, contractor specifications, test evaluation results, a diverse set of sensors, IoT devices, and so on. Once an AI is trained, BC can be used to ensure the integrity of the data during operations. BC can be used to find the right data, what is in it, who owns it, and how to get it with quick authorization. Data scientists have long recognized that just getting the right data and permission to use it can be an arduous and long process.

2 Cybersecurity Through System Safety

AI has the potential of creating a technological leap [7]. That potential leap, especially when dealing with critical systems controlled is partially controlled by AI, needs scrutiny. This scrutiny focuses on the specificity of the composition and size of the training data algorithm. This research describes how an HLF architecture can be used to increase safety and confidence in the deployment of AI functions. There must be confidence in the data and training sets and the algorithms, and there must be confidence that they are tamper proof and free from anomalies, intentional or by accident. Acquisition communities cannot identify and certify operational constraints of an ML algorithm for deployment without having confidence in the training data quality, including any negative side effects [8] that might result from the training process.

The *system safety* concept calls for a risk management strategy based on identification, analysis of hazards, and application of remedial controls using a systems-based approach. This is different from traditional safety strategies [9].

AI or autonomous systems safety issues deployed at sea or other challenging environments usually have not included consideration of adversarial attacks that might affect functional performance. AI adversarial network attacks using techniques like deepfakes, putting an image/video into another image/video for miscategorization [10], will be considered within our BC discussion.

In assessing safety in the Department of Defense, the goal is to identify anything that might be safety critical. Safety critical is "a term applied to a condition, event, operation, process or item whose mishap severity consequence is either catastrophic or critical (e.g., safety-critical function, safety-critical path, and safety-critical component)" [11]. Specifically, the publication MIL-STD-882E [11] helps software engineers determine the level of rigor (LOR), which specifies the depth and breadth of software analysis and verification activities necessary to provide a sufficient level of confidence that a safety-critical or safety-related software function will perform as required.

ML/AI usually falls into the system safety two highest software control categories: Level 1 (autonomous) and Level 2 (semiautonomous). We contend that BC could contribute to the analysis and verification of software activities by ensuring data integrity and better accessibility to the data.

Our previous research used the Hyperledger Fabric (HLF) Blockchain to generate three general use cases for naval logistics, including financial and inventory transaction audit trails, serial number tracking, and maintenance log integrity. We believe the BC network derived from these three use cases could be adapted for system safety and cybersecurity purposes since all our previous demonstrations dealt with the integrity of the data supporting work processes and events. BC tracks food/part items as assets recorded on ledgers, and training data are assets and are also created with similar work processes and events. With HLF, you can control who, what, and when and identify those who have access to the logistics data representing assets as well through an immutable ledger containing logistics data that cannot be tampered with. These HLF attributes are similar to what is needed to curate data assets. HLF is as transparent as needed but can hide data from those without a need to know.

The data source flows of data and training sets supporting data scientists are similar to previous BC research on naval supply chains to improve transparency and the safety of the related supply chain data and transactions, but there is a higher level of risk since they are often at Level 1 or Level 2 autonomous systems. In a sense, training sets and analytical data are like the tracking of parts and food since they point to resources represented by the information that needs to be protected and distributed in a friction-free manner. Control of these sources during the integration process to create training data and general analysis is vital to ensure the training sets and AI algorithms are transparent to those who need them and are controlled and their validity supported by an audit trail that BC provides. Training set alterations could go unnoticed within the AI function build process but revealed during operation in hazards affecting unwanted human death or resource destruction. Our previous research demonstrated how BC can provide a needed data management technology through a tamper-resistant decentralized trusted ledger that provides proof of transaction where trust is implemented through distributed consensus.

Only authorized people can modify the code base, AI algorithms, or training set modifications that are detectable, traceable, and transparent. Distributed ledgers provide system safety through BC provenance and policy enforcement through smart contracts.

HLF is a consensus-based network that a large organization can control and has no "proof of work" protocol, which is a wasteful use of computer resources. HLF uses channels to control who can see what data and through consensus; a large organization can control what is allowed to be put on the BC ledger. Such technologies can not only be used in naval supply and logistics to streamline and improve effectiveness in terms of how workflow can be improved to provide more rapid and secure distribution of material and two-way financial transactions but can also be used on data transactions such as datasets requested by data scientists. Data scientists have long recognized that obtaining "clean data" and the permission to use it has been hampered by administrative friction, which can be caused by data owner's requirements, trust issues from generated data source transactions, and other administrative processes.

The benefits of BC technology described in this paper support system safety in terms of providing objective quality evidence about data integrity, as well as test and evaluation teams in terms of data management control. We believe elements of BC, such as smart contracts, could contribute to all acquisition groups involved. We will discuss our previous logistic use case as well as new use cases specifically for software safety.

3 The Hyperledger Fabric Blockchain Solution

HLF provides proof of transaction where trust is implemented through distributed consensus and not centralized policy enforcement. The specific version of BC we used is HLF, which is open-source from the Linux Foundation. HLF is a permissioned, distributed ledger that works on the consensus model that is an integral component of the "trust system" in the BC. Essentially, the fabric environment provides the "common logging" and service management components on the platform, and the containerized infrastructure allows developers to build a BC network where data is recorded on distributed ledgers where the data written can be trusted and transactions are immutable and tamper proof. Smart contracts can embed legal knowledge, laws, and regulations and enforce data policy. BC/HLF can also provide "provenance" of an item, such as food or a part, and trace back to the source of that part or food item in case of contamination or counterfeit/defective parts as well as other times such as blocks of data in support of AI.

Of course, BC can be used for cryptocurrency such as bitcoin, but cryptocurrency is not a part of this study, and a semiprivate BC in support of data integrity needs a specific set of BC features other than Everledger or Ethereum, which uses an inefficient way to verify blocks called proof of work (PoW) instead of the more efficient consensus algorithm such as proof of stake.

With our previous research questions—could BC simplify and enable access and identity management for navy supply and logistics systems in a cost-effective manner to reduce this friction? How could BC improve Navy logistics to the last tactical mile?—we demonstrated the feasibility in previous research of using IBM and Oracle versions of HLF to track assets such as food items. Tracking and moving assets could be applied to data assets and adapted for software safety use because in both cases we care about the integrity of the data generated. There have been planned pilot projects in the DoD, usually supply chain scenarios [12].

Although HLF is a Linux open-source project, several software companies have adapted HLF as its core BC enterprise solutions and have added additional value through add-ons, cloud support, and company expertise that goes beyond the plain vanilla HLF. This is common with open-source products as you pay for more capability and support. We compared enterprise versions of HLF, IBM, and Oracle HLF BC platforms and evaluated their ability to maintain an efficient, streamlined, and accurate ledger of all shipment transactions during transportation. Additionally, the team developed a ledger serialization function in the smart contracts for synchronized connection on ships and bases to the HLF framework. The characteristics of enterprise BCs include the following:

- Permissioned architecture.
- Highly modular.
- Pluggable consensus.
- Open smart contract model—flexibility to implement any desired solution model.
- Low latency of finality/confirmation.
- Flexible approach to data privacy—data isolation using “channels” or share private data on a need-to-know basis using private data collections.
- Multilanguage smart contract support—Go, Java, and JavaScript.
- Designed for continuous operations, including rolling upgrades and asymmetric version support.
- Governance and versioning of smart contracts.
- Flexible endorsement model for achieving consensus across required organizations.
- Queryable data (key-based queries and JSON queries).
- The use of X.509 public key infrastructure (PKI), which is quite familiar to the DoD for a signed data structure that binds a public key to a person, computer, or organization. Certificates are issued by certification authorities (CAs).
- Cloud support and SaaS (software as a service).

Figure 1 is an example of a very simple generic BC ordering network. A1, A2, and A3 are different “off-chain” applications that could be on IoT devices or Web browsers on computers or smartphones. These applications connect the

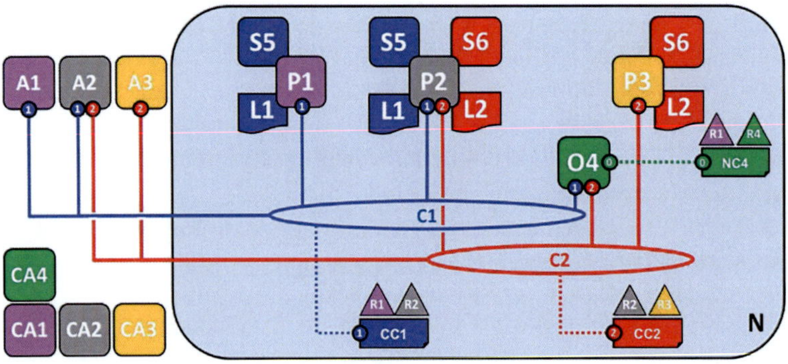

Fig. 1 Generic HLF BC network

on-chain world with the BC network/database. These client applications represent the "last mile" and could include legacy programs pre-BC. The blue-shaded background represents the BC logical infrastructure layer—not whatever physical layer infrastructures might be used, such as satellite or fiber optics. O4 is an ordering service. Network configuration (NC4) gives administrative rights to organizations R1 and R4. At the network level, certificate authority CA4 is used to dispense identities to the administrators and network nodes of the R1 and R4 organizations. Certification authorities CA1 and CA4 provide entity validation, and other CAs are shown in the diagram. In this example, there are two consortiums (common interest parties), represented by R1 and R4 entities who set network configuration policies, seen CC1 and CC4 which set up channels. Channels are ways to decide who gets to see what ledgers. There are three peers: P1, P2, and P3. On the left, P1 has S5, which is a smart contract that provides the rules for the ledger L1. Only those who have access to Channel 1 (C1) have access to the ledger L1. You see that if you have access to A1 or A2, you have access to C1, but the A2 application has access to both C1 and C2 and, therefore, access to ledgers L1 and L2, which are set by configuration control (CCL).

4 Methodology

Our methodology involves two sets of use cases which we implemented as basic demonstrations on various Hyperledger Fabric platforms. The first set (original cases using Oracle and IBM versions of HLF) was used in our previous research in Navy logistics, which we believe can also demonstrate BC use for system safety if modified, as both sets of use cases track assets—one tracks food items and the other tracks datasets as assets. The key for repurposing a supply chain for use in software safety support is through the addition of off-chain application programming interface (API), such as representational state transfer (REST, or many others), which provides an interface between the BC and the outside world and to what is called "the last mile," which in most of our use cases is a Web client. In our first set of use cases (the original use cases), we built two demos (Oracle and IBM cloud versions) illustrating the Navy logistics/supply chain. We demonstrated how BC can document and authenticate transactions along the supply chain, which would be similar to a data supply chain used for data system safety. We worked with both Oracle and IBM enterprise BCs to demonstrate the first set of use cases. In a work in progress, we have an additional set of use cases (labeled new cases) specifically for use with system safety using the open-source version of HLF (https://www.hyperledger.org/).

Blockchain Use Case Examples for the Naval Logistics/Supply Chain

We looked at three general use cases to apply BC technology using both cloud versions of IBM and Oracle BC platforms: (1) financial and inventory transaction audit trails, (2) serial number tracking, and (3) maintenance log integrity. Maintenance log integrity involves the same issues as AI dataset integrity. The three examples are as follows:

- *Original case 1*—financial and inventory transaction audit trials. An investigatory inventory and financial transactions via audit trails can be a costly and timely process, and the audit trails could encompass different systems throughout a vast network in such an organization as the Navy. The questions to be answered might include what, where, and who—where a distributed ledger would be able to track "what" through immutable data blocks that make up the ledger. One of the BC's strengths is identity verification and management, which would be able to verify and track the "who" in any financial and inventory transactions on the BC.
- *Original case 2*—serial number tracking/BC tracking can also be applied to the tracking of specific items in the supply chain, such as serial numbers. Also, the tracking could include a visual identification of the item by an individual, which would automatically be identified as a trusted agent to make that verification along with the where and the when.
- *Original case 3*—maintenance log integrity/maintenance repairs, such as on naval aircraft, ground, or ship systems, typically generate data on various transactional databases, which in turn may be sourced to other databases or repositories such as data warehouses, enterprise resource planning (ERP) systems. Our past research on aviation and ground maintenance systems databases shows that there are errors in the databases and often information is not updated. At the tactical and operational levels, this could have an impact on the effective efforts to ensure maximum mission readiness. Smart contracts, which are integral to HLF, are codes that can check, enforce, or flag bad data. Certainly, relational databases can have triggers to check for illogical data entries, but it isn't always being done, and typically several databases and sources may be involved in a maintenance information system to make such error checking costly or not practical. While some minor errors may be acceptable in transactional databases, these errors could have an impact on data analysis and ML/AI if the data in these systems are used as training datasets. BC could use smart contracts to flag errors over a diverse set of data sources and provide basic provenance.

Blockchain Use Case Examples for System Safety

In our second set of use cases (using Hyperledger Fabric open-source), we specifically address three software system safety use cases applied to the open-source HLF:

- *New case 1*—a researcher/data scientist needs to manage data or training sets for research or ML to process text or binaries (images, RFI signals), structured and unstructured.
- *New case 2*—a data scientist needs to derive metrics on a dataset but is not allowed to see raw data.
- *New case 3*—BC is used as a database for relatively small source code.

Figure 2 is a simplified HLF BC network that could support our three scenarios for software safety in the blue background square on the right (see https://www.hyperledger.org/). This is the BC. This BC is supported by a physical network that could be cloud based and supported by the Internet. The "off-chain" applications, IoT, and storage are shown outside of the square. These are applications developed in a normal way and not a new technology. The applications use standard APIs such as REST to interface between the user, databases, and the outside world to connect to the BC. They are called off-chain because while they interface with the BC, they are not part of the BC. From left to right are the identify certificates—CAs such as CA1, CA2, and CA3 in our example to identify those who have access. BC is good at leveraging existing technologies, and CA is old technology using X.509 public key infrastructure (PKI), which is used to encrypt and sign email. A1, A2, A3, and so on are off-chain client applications that have access to various ledgers (our database) which are controlled through CC1 and CC2 (CCL), which set up channels and their access. P1 and P2 are peer nodes that in the example host ledgers L1 and L2 for P1 and L3 for P2. Each ledger is supported by smart contracts (S5, S6, S7) that determine the business rules and logic of how the ledger is to be written and who can write on it. C1 and C2 are channels to determine what applications or entities are allowed to see what ledger, which makes Hyperledger very powerful as

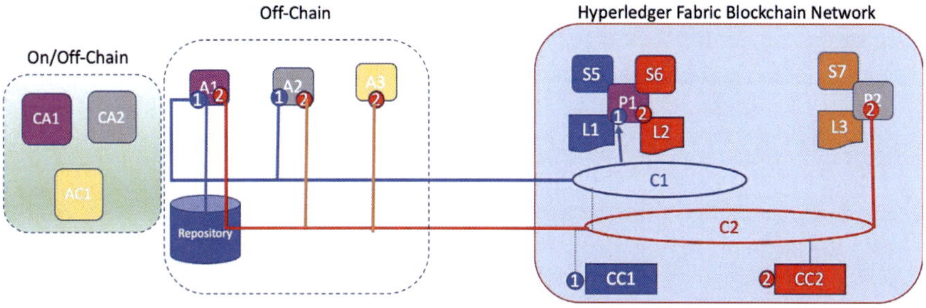

Fig. 2 HLF scenarios (new use cases)

you can control who sees and changes what—such as organization personnel and contractors having access to different data.

Off-chain A1 is an application that administers access to the repository and writes to the ledger, which records the metadata in each dataset and provides a digital signature/hash value. CCL provides access to Channels 1 and 2 and, as shown, access to all ledgers. For structured data in the repository (maybe more than the one shown in the diagram), Al would post/write the metadata of a dataset of interest including, if practical, all of the data fields and DTG and record a hash value or signature. This would be entered in either L1, L2, L3, or other ledgers created. It is not practical to record/post large datasets on a BC ledger, but metadata and pointer/anchors to the data could be provided through URLs. It is possible that through the administrator interfacing with a peer node, the BC could store some small datasets through CouchDB, which would provide the current information/state of an asset such as a dataset.

New Case 1 *New Case* 1 is about cybersecurity and trust in the data used for analytics and the building of AI models. Figure 2 shows application A2, which could be a customer/client such as a data scientist that is interested in datasets or training sets for an AI project. This customer per the diagram (set up by CC) has access to Channels 1 and 2, which means he can view Ledgers 1 through 3, which would be information about various datasets that can be accessed. In one scenario, the person using A2, the Web application, for example, could search for a specific dataset or topic and then request that dataset through the application, which would check the smart contract—let's say for L2—to see if the system allows read access to the repository.

Existing off-chain software would complete the task and send an anchor or link (URL) to retrieve that dataset. The customer could later check back and see if the data have changed/been tampered with or if the data were given to another user. Also, the client would be provided the provenance and metadata and even points of contact, including subject matter experts and the owner of the data. The client can check to see if the dataset has changed and who changed it, since any changes to the repository would be recorded in the appropriate ledger as to who, when, and what. Smart contracts could also provide some prefiltering through smart contracts to reduce unintentional errors. In the past, this has been done pre-analysis, but by using smart contracts, this would only need to be done once and not by each researcher or customer. This AI system safety idea is similar to the IBM concept [13], where the authors sought a trusted AI environment through provenance with a BC library exposed by REST or Python APIs that provided support for "immutable recording of the AI process, querying for traceability and audit, fair value attribution, etc." We take it a further step to suggest that BC can be part of a smart repository solution that allows clients to search and find trusted datasets and safeguard them. A variation of this use case is a federated learning (FL) scenario that uses a collaborative ML technique whereby the devices collectively train and update a shared ML model while preserving their datasets. Even in a trusted military network using a private BC, some devices on the edge may prove untrustworthy, and ur Rehman et al. [14]

propose a reputation-aware FL that enables trust through BC consensus and trust algorithms through BC smart contracts.

Related to FL are the existence and rise of highly complex problems that require solutions that can adapt to changing events and environments. An engineered solution must address highly complex problems through adaptive architectures and the embedding of constituent systems with the intelligence to learn, self-organize, collaborate, and evolve to achieve desired adaptable emergent behavior. Advances in information and computational technologies and the emergence of IoT devices that perceive their internal and external environments to enable the potential development of complex, adaptive, and intelligent capabilities needed to engineer a complex adaptive system of system solutions. Blockchain learning coupled with a federated learning approach could be key enablers for engineered solutions that address highly complex problems through self-organization and desired emergent behavior. A blockchain approach can ensure trust among IoT devices (intelligent constituent systems) exchanging information and collaborating in a Federated Learning architecture. Johnson [15] notes that system information assurance is critical in an adaptive complex system as communication issues could take the form of a cyberattack, injecting unauthenticated and/or false data into the system or causing denials of service. Blockchain could monitor IoT devices and prefilter out any unreliable data sources or unauthenticated IoT devices to ensure trusted collaborations using trust algorithms embedded in blockchain smart contracts.

New Case 2 A user wants to compile metrics but is not allowed access to the raw data because of security or cross-domain restrictions. Lampropoulos et al. [16] proposed a similar scenario, where one Telco A holds private datasets and internally processes a data request by another Telco B and Telco A only returns the results to Telco B and not the raw sensitive data. The whole process is performed with transparency, ensuring the quality of the results and the privacy of the processed data. A3 in Fig. 2 is an application that only has access to Channel 2. The user then picks the dataset to use and looks at the metadata and fields; then the smart contract (S7) executes the query through A1 and posts the results in the ledger L3. This use case could also be used for a cross-domain solution setting up rules when a user could have access to a different domain, the raw data, or just the results. Channels that are associated with one specific ledger are the means to control who sees what ledgers.

New Case 3 Our last scenario is the data are not stored off-chain but in the BC itself.

HLF has the option of using CouchDB that can use standard JSON queries to get the "world" or current state of an asset (like a dataset). Perhaps this use case would apply to IoT devices where you want real-time data from sensors but still want to ensure software safety. The data would be immutable but replicated throughout the network.

Figure 3 summarizes the flow in our simple scenarios. First, the "customer"—a data scientist or developer—wants to access data such as for training sets in ML, or a developer wants access to code. The customer wants to find the right data quickly, know who owns it, and know that it can be reasonably trusted. In our example, this data resides in a repository that may include both structured data (relational databases) and semi-structured and unstructured data such as in the form of .JSON files, text, or graphics. The customer starts a request for the data, and an answer comes back with the metadata, data fields, date-time group, and hash value of the set. This information is in a ledger in addition to an encrypted link to access the dataset.

The customer can also see the complete history of changes to the data and can verify that the training set, data, or code has not been tampered through the hash code both in the metadata and the ledger on the BC. Only those authorized can add to the chain, and it is immutable.

5 Use Cases Using Three Hyperledger Fabric Versions

We discuss our results using the IBM, Oracle, and Linux Foundation versions of HLF and their application to system safety scenarios. Figure 3 provides a simplistic view of the system safety scenario where the data scientist is looking for training sets or related data.

The data scientist (the client) uses a Web browser, enabled by REST API or other development interfaces, and searches for a dataset or training set through the BC which, through certificates (x.509) and smart contracts, knows who the client is. Based on governance, the BC and smart contract will decide if that data scientist has the authority to retrieve the data. If so, the client will be sent a link to access the repository or even an IoT device or a BC repository with frequently used datasets,

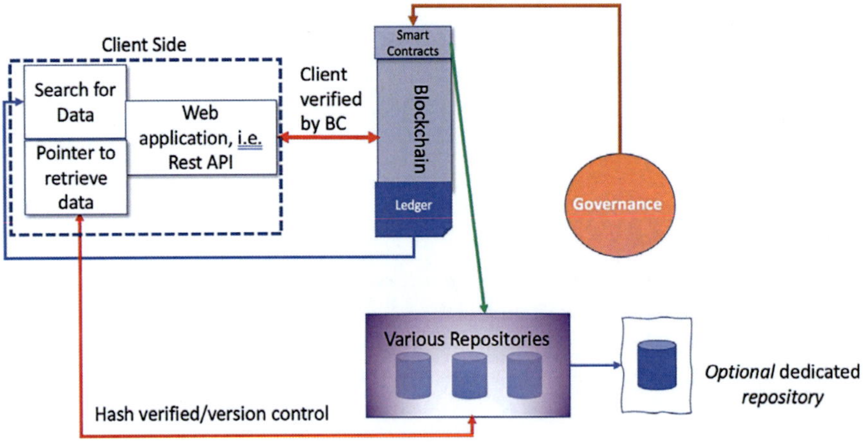

Fig. 3 Data scientist use case example and smart repositories

through x.509 certificates, which will have the identity verified by the BC, as well as the sending to the client hash value making sure the dataset hasn't been tampered with. The smart contract may do some initial cleaning up and filtering of the data. What normally takes months to get the data may only take a day and comes with the assurance that the data had not been tampered through an immutable BC. Large organizations may require a myriad of forms or approvals to get the data to the data scientists, and smart contracts that embed rules and authorizations could streamline that burdensome process.

Successful Applications of Blockchain for Naval Supply Chain Tracking

As discussed, our previous research investigated how BC could simplify and enable access and identity management for the Navy supply and logistics systems in a cost-effective manner to reduce administrative friction and how BC could improve Navy logistics to the last tactical mile. In our scenario, the first destination transportation (FDT) refers to the movement and cost of moving shipments from free on board (FOB) points of origin to the location at which the shipment is first received for use or storage. As naval regulations apply, the first checkpoint of where a shipment is received, whether within the United States (CONUS) or outside (OCONUS), begins with a supplier outside of the DoD supply system or industrial activity that creates the shipment. The labor and transportation charges, including freight drayage, cartage, port handling, and other in-transit costs, are processed at the FDT. Freight cartage refers to any inland transit of cargo between locations, which serves as the "checkpoints" in the BC network.

When a location is assigned responsibility for "cartage of consignments" to land-based activities, ships, or other transport units, the charges of transportation are given to the location of assigned responsibility, which acts as a peer node checkpoint in the network. At this point, the initial entry in the ledger may be created and committed by the peer node belonging to the FDT and the orderers. It is important to note that FDT does not only include shipments of equipment but also the initial transportation of Navy-owned materials that are provided to a contractor for research. This indicates that the charges of a shipment from a contractor's facility to its final destination point are paid by the government. However, to maintain the legitimacy of a decentralized ledger in this research study, the network for which the ledger is maintained consists of only contractors, supply facilities, and final base destinations. Essentially, tracking responsibility is passed down from supplier to checkpoint. The checkpoint managers responsible for the charges in a shipment delivery may create and commit the transaction over the BC network, and the next checkpoint manager may agree or disagree about the condition and extraneous details of the shipment that the previous manager signed. Currently, the Department of Navy (DON) uses service-wide transport (SWT) as a clearinghouse, which is a centralized operation and maintenance manager created to provide transportation funds for naval shipments and mail. Since naval cargo and the movement of mail to bases are not responsibilities of a destination location, the SWT was created to pay for the movement of material, such as aircraft engines, mission module packages, catapult and arresting gear, propellers, shafts, civil engineering support equipment, safety equipment, drones, overseas mail, and Navy Exchange Service Command (NEXCOM) merchandise shipped from the United States to international locations.

For disconnected operations, to maintain an accurate ledger with the consensus algorithm, the peer nodes must be connected to the fabric environment unless the peer node decides to save the ledger as a .JSON file and re-upload the ledger as a .CSV file once back online. The ledger is automatically updated after the node reconnects following disconnections due to shipboard communications. The fabric environment will make BC technology a more viable option for all naval transportation activities.

The Navy requires a multifunctional and secure platform that enables personnel to track multiple shipments from production facilities to bases and a secure ledger of inventory that can only be modified with either an undisputed consensus or an access to the smart contract. Once a peer node administrator or user in the network has access to their smart contract, they can modify the transaction protocol that occurs on transactions in the network. However, the network will not instantiate a new version until there is an agreement with the channel creator or the majority of the channel members.

In this simplified logistics BC network, the smart contract contains six methods that carry out the protocol for each transaction on the ledger: foodAssetExists, createFoodAsset, readFoodAsset, updateFoodAsset, trackFoodAsset, and delete-FoodAsset. The method of using names indicates that each shipment is checked to verify if it already exists at a location denoted by a string. After checking for duplication, the asset is created in the ledger using a key-value pair, such as "001: a shipment of supplies." Once the asset is created, it is always a good practice to read the asset's details into the ledger so that users further down the network have a detailed understanding of what a package is supposed to contain. Also, if a shipment is changed—say, a package is redirected to a base that requires supplies urgently—the shipment's location is updated within the ledger and deleted once the shipment arrives.

A multifunctional and secure platform that enables personnel to track multiple shipments from production facilities to bases or ships in transactions involving money, items, material, and history should be trusted, transparent, and traceable back to the origin of the item. These transactions involving information, money, or physical items such as food or parts usually involve the enforcement of the policy, technical, or legal requirements that require the enforcement of business rules. BC can maintain a secure ledger of inventory (or transactions involving data or information) that can only be modified with either an undisputed consensus or an access to the smart contract, which can enforce business rules and flag "violations."

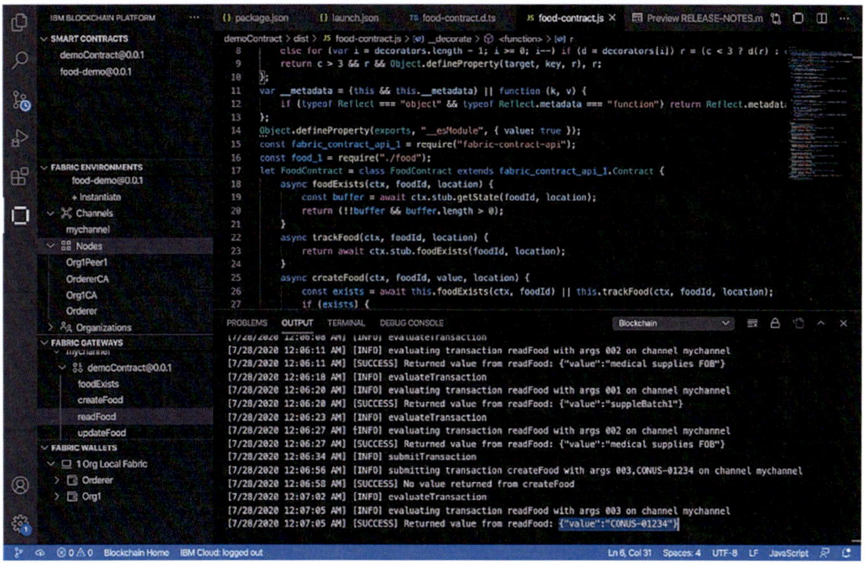

Fig. 4 Sample ledger of shipments that are added and updated (contents/location)

Once a peer node administrator or user in the network has access to their smart contract, they can modify the transaction protocol that occurs on transactions in the network. However, the network will not instantiate a new version until there is an agreement with the channel creator or the majority of the channel members.

Based on the above process, we showed how a food or item tracking scenario would work using both IBM and Oracle cloud versions of HLF (see Fig. 6). In these BC networks we set up, the smart contracts (Fig. 4) contain six methods that carry out the protocol for each transaction on the ledger: foodAssetExists, createFoodAsset, readFoodAsset, updateFoodAsset, trackFoodAsset, and deleteFoodAsset. In our food/item tracking scenario, the method of using names indicates that each shipment is checked to verify if it already exists at a location denoted by a string. After checking for duplication, the asset is created in the ledger using a key-value pair, such as "001: a shipment of supplies." Once the asset is created, it is always a good practice to read the asset's details into the ledger, so that users further down the network have a detailed understanding of what a package is supposed to contain. Also, if a shipment is changed, say, a package is redirected to a base that requires supplies urgently, the shipment's location is updated within the ledger and deleted once the shipment arrives.

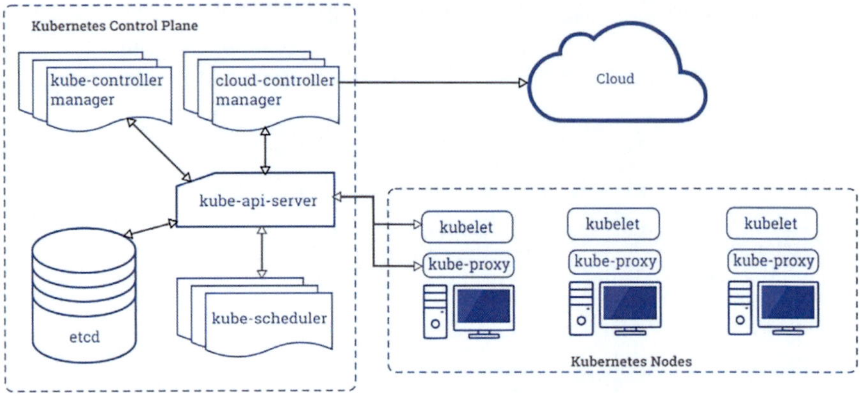

Fig. 5 Visual representation of the interaction between Kubernetes and Cloud

Blockchain Use Case Examples for the Navy Logistics/Supply Chain

Using IBM BC Platform™ To use the IBM BC Platform, users are required to install four vital components: (1) the Visual Studio Code environment, (2) Node.js, (3) Docker, and (4) Kubernetes. The Virtual Studio Code environment is the off-line integrated development environment (IDE), where developers create smart contracts using the open-source programming language Typescript, which was developed by Microsoft (see Fig. 4).

Smart contracts serve as the fundamental basis of all enterprise BCs because they give certified users the ability to create new transactions and assets, as well as other functions specific to a project. In this project, the team's main goal was to create a consensus network that has the power to create food shipment assets, update or delete them from the ledger when required, and track their location using the "foodId" string, which may be replaced by radio-frequency identification (RFID).

The HLF (from the Linux Foundation) is the basis of both IBM and Oracle platforms. Its components are created in a Kubernetes cluster usually within the IBM Cloud. A Kubernetes cluster contains a set of working machines (nodes) that run containerized applications. The nodes within the cluster host the components of the application workload. Within the cluster, the control plane manages the nodes and workloads that run across multiple machines, as shown in Fig. 5.

Figure 6 illustrates the ordering service and integration of security nodes on a blockchain channel. When the fabric environment is running, you can create the ordering service. The ordering service is a group of orderers that accepts approved transactions endorsed by the peer nodes based on the smart contracts and organizes the transactions in the appropriate order in the ledger blocks based on the consensus algorithm.

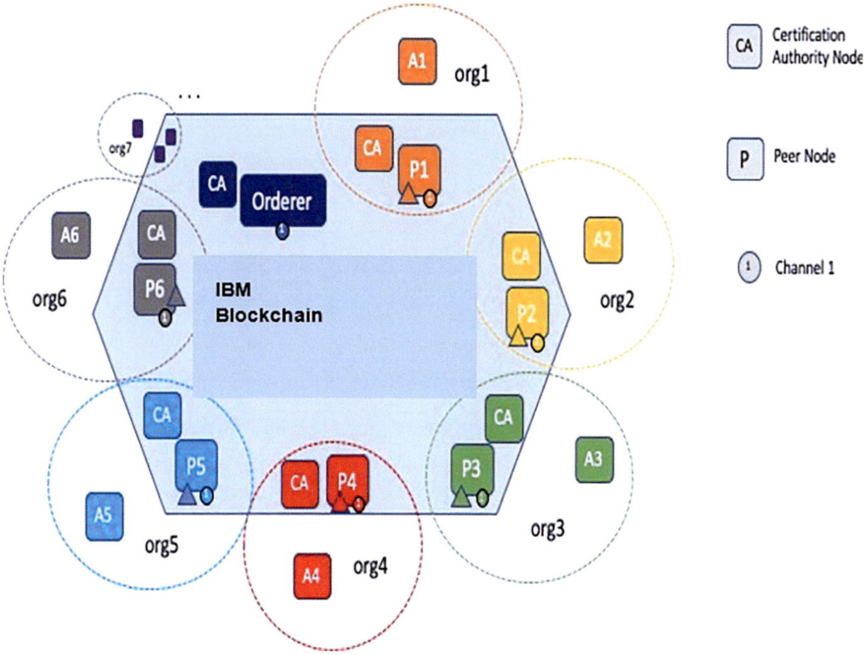

Fig. 6 Ordering service and the integration of security and nodes in a BC channel

The peer nodes host ledgers and smart contracts—the backbone of the BC network.

The smart contract—the transaction protocol—automatically executes, controls, and documents transactions or events occurring on the network.

Like all BC frameworks, the network's integrity is upheld by the consensus algorithm.

Each node in the network reviews the entire BC and checks that all previous blocks are valid so that a new transaction may be initiated into the network. However, alternatively, in a permissionless public BC, the consensus algorithm is replaced by the PoW, which creates a hash system of all of the transactions.

In a PoW system, miners constantly attempt to solve the algorithm so that they may mine new blocks and be the first to extend their BC. HLF doesn't use the wasteful PoW but uses a system closer to the "proof of stake" as a consensus mechanism. Essentially, decisions are authorized by users who are permitted to join the system and specific channel, as not everyone can join the network. Unlike PoW, computational power is not required, since there are no puzzles needed to obtain "currency." In a "proof of stake" system, "validators" are discouraged from creating faulty empty blocks because they have the motivation to incorporate a maximum number of transactions for gains.

To ensure security, the hash must be solved by all peer nodes in the network so that new transactions may be approved for the network. While this alternate

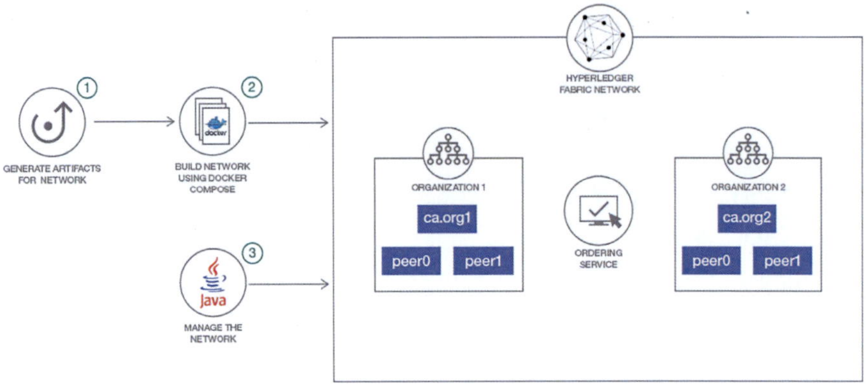

Fig. 7 Interaction between the external software and the HLF environment

approach is viable, it is also time-consuming because ensuring that the ledger is tamper-free requires each ledger copy in the nodes to be changed and hashes to be solved.

Developers should install Node.js and Docker unless the developer exports both items into a .JSON file and re-uploads both files onto the peer nodes as a .CSV file. Docker serves as an OS-level platform to package containers and bundled software, libraries, and configuration files.

Figure 7 shows that in using well-defined channels within the software, these containers communicate with each other to allow the user to connect to the fabric environment and add to or change the ledger. Finally, the Kubernetes system, which was designed by Google and maintained by the Cloud Native Computing Foundation, is the main system that allows the IBM BC Platform to package, install, deploy, and manage multiple peer nodes in the platform.

Figure 8 provides an overview of how you would manage the off-chain (UI[2]) and the actual BC network consisting of three fabric components: CA(4), peer nodes, and ordering service.

Using the Oracle Blockchain Platform On both Oracle and IBM platforms, we were able to set up a BC network with peer nodes (stakeholders) with smart contracts that set up the rules for transferring and tracking assets such as food items discussed previously. More work needs to be done on enhancing the network to accurately represent this aspect of the supply chain.

The team set up the network using an Oracle cloud with four peer nodes set up over a single channel and used Oracle Identity Management for role-based access. Separate roles are required for adding users to a role with BC provisioning entitlement, which requires tenancy admin. Additionally, the cloud platform was used instead of the software package due to the amount of storage memory required to host the software appliance VM packages on a local computer. However, the

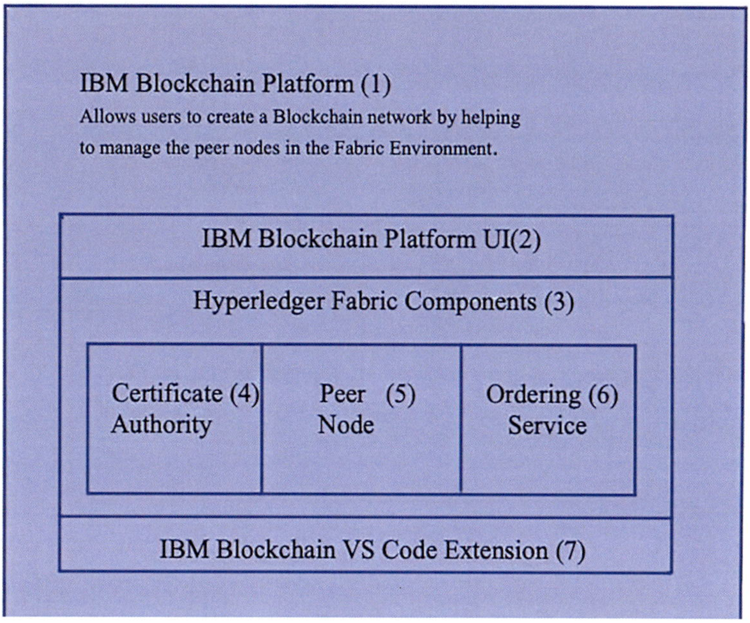

Fig. 8 High-level representation of IBM blockchain platform architecture

fundamental concepts of using an HLF environment and consensus algorithm remained the same for both platforms to build a BC network.

Oracle Blockchain Platform also provides wizards to simplify joining multiple instances to the network, creating new channels, and deploying chaincodes. Implementation of smart contracts is through Typescript (see Fig. 9). These and other DevOps functions are also available via extensive REST APIs for off-chain applications to interface the BC network.

Oracle offers both a managed cloud version (Oracle Blockchain Platform) of OBP (blockchain-as-a-service) and a customer-managed OBP Enterprise Edition for on-premise (or third-party cloud) deployment, and nodes can be deployed using both for a hybrid network deployment (see Fig. 10). The cloud SaaS version was used for this project. To access this platform, users must log in with authenticated credentials in Oracle cloud infrastructure. Once logged in, users can provision an instance, which comes with a default channel and participant nodes, along with "orderers" that are responsible for maintaining the order of the ledger. An operations console is provided, and users are not required to download any external software to work with the platform, other than an integrated development environment (i.e., Visual Studio Code) to develop the chaincode and the REST API testing tool, such as Postman and/or HLF software development kit, which is downloadable from the OBP console under the developer tools tab.

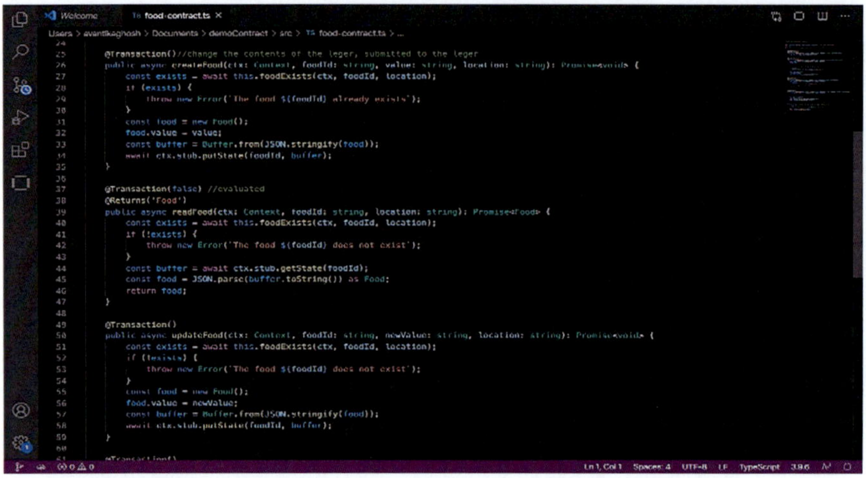

Fig. 9 Sample smart contract for tracking food shipments (language, Typescript)

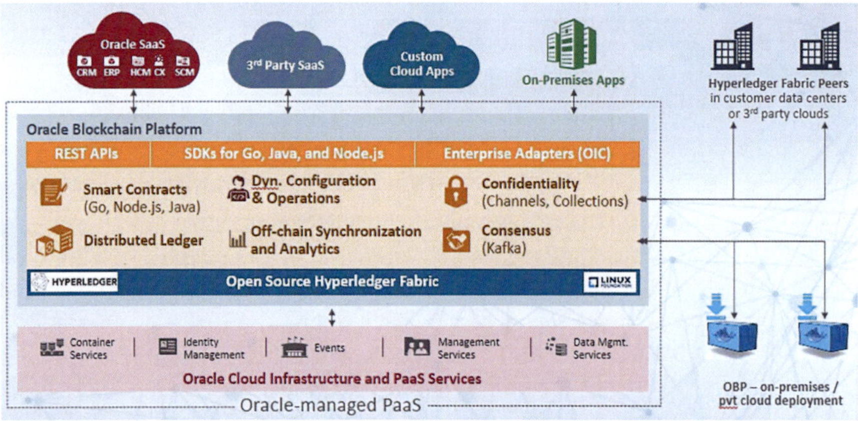

Fig. 10 Oracle blockchain platform cloud service architecture

The Oracle Blockchain Platform (see Fig. 10) comes with an API gateway that supports REST API so that developers can invoke a transaction, invoke a query, subscribe events with a registered callback, and view the status of a transaction within the ledger as well as a set of DevOps REST APIs for administration, configuration, and monitoring tasks.[1]

[1] The team was given access to the Oracle Cloud Platform thanks to the NPS liaison relationship with the Oracle Blockchain team.

Current Use Cases Using Linux Version of HLF for System Safety

While both IBM and Oracle are HLF based, their complete solutions that use their respective cloud services are enhanced by their specific products. For our new set of system safety use cases (a work in progress), we installed HLF on a Naval Postgraduate School virtual Red Hat server and installed HLF from the Linux open-source foundation, which provides all needed images and tools to set up a BC. Unix tools include the Git client, CURL, and Docker with Docker Compose without Kubernetes, which are key components to build the BC network rapidly. This model suits the researcher who wants to study and test out the concepts before moving to production, at which point a vendor-supported option can better address the challenges. Typical enterprise BC platforms provide dashboards for BC management such as the status and health of the HLF network. In the case of the open-source version, no such tools are provided; instead, everything is done via command; thus, one has to have a good idea of Unix command line tools and scripting languages like bash. Both IBM and Oracle allow you to use an IDE to build the applications. All three platforms offer interfaces via APIs to programming languages like JavaScript, Java, Microsoft Visual Studio, and others. For most production instances, we think a cloud-based BC is usually the right way to go for maintainability, support, ease of use, and security.

For the Linux Foundation version of HLF, the complete install includes commands to set up an HLF network, issue certificates, set up the ledger, create channels, install chaincode, and more. A sample BASH script is provided that goes over all these steps and can be customized for new projects such as for our three system safety use cases. The Docker container-based platform allows one to have several HLF projects to coexist. The test network is shown in Fig. 11 with two organizations, R1 and R2. Organization R0 owns the ordering service (O) of channel C1. A copy of the ledger L1 is on all nodes. The root CA issues the certificates CA0, CA1, and CA2 for the three organizations.

Use Case 1 First, a channel is created, and member organizations are added to the channel. The ledger contains the needed URLs to access different binaries. Using

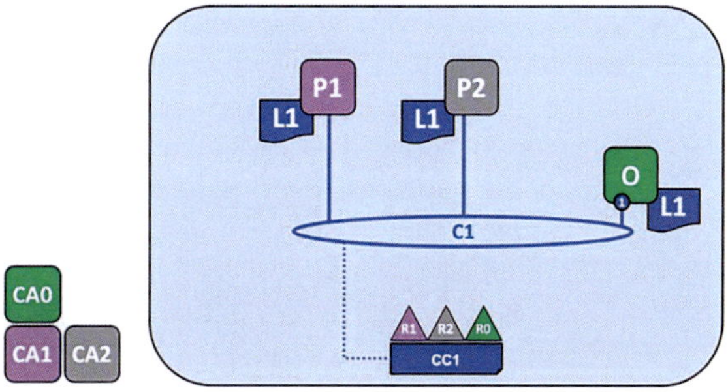

Fig. 11 Initial blockchain network on open-source HLF

APIs, depending on the requester, the chaincode will craft a unique response to be sent back. When the response is received, the URL and text are preprocessed; this happens in a middleware server outside the HLF; thus, a custom Web page is created and served to the end user. This Web page has links to authenticated repositories (database back end). Once authenticated access is granted to the data scientist (data and training sets used in research and stored in a database), an encrypted anchor or URL may be sent to the data scientist to download the dataset.

Use Case 2 When a member organization needs to see a part of the ledger, then channels have to be created. Membership to a channel is restricted to a subgroup of the organizations. In using chaincode, the metrics part of the ledger is provided to members of the metrics channel. There exists another channel where members can see the raw data using queries—again dictated by the chaincode. Membership to the two channels is a different set of organizations. Similar development methods used in Use Case 1 apply to Use Case 2.

Use Case 3 In this scenario again, a channel is created for a certain member organization (not all). These members will be able to access the data in CouchDB via API queries. Chaincode will decide which source code (stored in the CouchDB) is provided as a returned result of the query. The database is replicated on every node, which might be an advantage on the edge and an I/O to IoT devices on the edge. Specific use cases for this capability haven't been developed.

6 Findings

We demonstrated through IBM and Oracle examples that HLF could meet logistics/audit and security requirements through smart contracts and the inherent trust systems with embedded certificates. Data entry errors could be reduced through smart contracts, which is an inherent feature of HLF. We believe a consortium BC through HLF would be a way to go to be able to share information (through the ledger) with suppliers and other third parties but also have the capability not to share when appropriate. BC could add the capability for secure transactions through certificates and the immutability of the transactions on the BC. The additional capability of BC on Navy logistics and supply would be able to catch some data entry errors, to trace back to the source, and basically to better know the what, the who (verified), and the where of various transactions generated by the supply chain.

We found that both IBM and Oracle BC platforms may be used to create a secure network of peer nodes or naval hotspots that can generate a consensus for the legitimacy of the shipment ledger, which can only be modified using smart contracts. Since a key component of both platforms is maintaining the accuracy and security of the ledger, all users must consistently export and import the smart contracts and ledgers onto their respective peer nodes every time an update is made on the ledger or if the transaction protocol on the smart contract is changed. A special concern with Navy logistics is the possibility of unreliable networks, especially from shore to ship. The BC protocol creates a multitude of copies of the blocks (the public ledger), and if connectivity is lost, the blocks will be updated once the network node communications are reestablished. Both IBM and Oracle BC platforms were accessed through the cloud, but the option is for the Navy to put either platform on its implementation of the cloud or on servers.

There were differences between IBM and Oracle implementation of HLF—such as how the whole network infrastructure was implemented, user interfaces, developer tools and application programming interfaces provided, and how the implementation would connect to the Navy's legacy systems to reach the last mile, such as on the ship. These were real value-added capabilities since HLF alone cannot make an enterprise BC system that supports the existing logistics information system.

We found a "consortium BC" with a BC consensus network to be the best fit for the use cases. A consortium allows both private and public users to use the BC, while control is maintained by the private users (the Navy) through a consensus network, which means by the consensus of trusted Navy entities. This is contrasted by PoW BC networks used in cyber currency, which are inefficient and not appropriate for a government entity. BC technology has the potential for revolutionizing the logistics process by ensuring the quality and trustworthiness of logistical generated data as well as providing provenance of parts and food, but it is new and risky.

The team also compared the IBM and Oracle BC platforms on efficiency and maintainability of a ledger of shipments and discovered that it was easier to use the IBM platform to create and export smart contracts and ledger; however, in September 2021, Oracle will provide similar capabilities for developing and deploying smart contracts. The IBM platform required users to develop their smart contract on the Visual Studio Code environment, export the contract as a .JSON file, log in to the online BC network, and import the contract and ledger as a .CSV file using a converter.

The Oracle Blockchain Platform, on the other hand, allowed users greater flexibility to join ledgers more cohesively. The Oracle platform allowed users to log in to the Oracle cloud after they were approved by an administrator and used simple software like integrated development environment (IDE) and the software development kit. Furthermore, the Oracle Blockchain Platform employed chaincode as a smart contract for transactional protocols in the network. A chaincode is written in either Java, Node.js, or Go and packaged into a ZIP file, which can be installed on the network. This is similar to how smart contracts are exported as .JSON files and uploaded on the IBM network as .CSV files. More specifically, chaincodes outline the structure of the ledger, initialize it, create updates (such as reading or updating entries), and respond to queries.

Should HLF be used for software safety for ML and AI development? BC is general-purpose technology (GPT) like the Internet, so BC isn't a solution in and of itself, but it acts as an enabler that provides a trusted, distributed ledger that could be used for smart repositories and software safety. If other technologies are better, then why aren't they commonplace? BC isn't *the* solution but, along with off-chain technology, may be a technology that enhances existing business processes.

7 Summary and Conclusions

From our work, the following lessons learned can be applied to protecting datasets such as training sets for AI:

1. Various versions of HLF will work adequately, but due to complexity, we recommend not using open-source but software vendors such as Oracle, IBM, Microsoft Azure BC, and others. BC is not a DoD core competency; therefore, contractor support is needed.
2. HLF or other BCs alone are not the entire solutions, since BC is an enabling or general-purpose technology (GPT)—so in itself, it is not a solution. You must use a BC protocol within an integrated network infrastructure that also provides for the last mile to bring the data to the user, and this is through APIs. We recommend, ceteris paribus, that you consider using the same company that runs your relational databases or ERP, as your team will be familiar with that architecture.

We used a qualitative methodology that included three general logistic use cases: (1) financial and inventory transaction audit trails, (2) serial number tracking, and (3) maintenance log integrity. These were used in consultation with the topic sponsor. We created simple scenarios where items were tracked through a BC network, and smart contracts would check for certain conditions that would simulate quality control and tracking. We selected two enterprise HLF platforms, Oracle and IBM, and evaluated them in terms of functionality, development ease, and security.

We found that both IBM and Oracle BC platforms may be used to create a secure network of peer nodes and a consensus for the legitimacy of the shipment ledger, which can only be modified using smart contracts. A special concern with Navy logistics is the possibility of unreliable networks, especially from shore to ship. The BC protocol creates a multitude of copies of the blocks (the public ledger), and if connectivity is lost, the blocks will be updated once the network node communications are reestablished. Both IBM and Oracle BC platforms were accessed through the cloud, but the option is for the Navy to put either platform on its implementation of the cloud or servers.

There were differences between IBM and Oracle implementation of HLF, such as how the whole network infrastructure was implemented, user interfaces, developer tools and application programming interfaces provided, and how the implementation would connect to the Navy's legacy systems to reach the last mile—such as on the ship. These were real value-added capabilities since HLF alone cannot make an enterprise BC system that supports the existing logistics information system.

BC technologies offer the potential to reduce costs and logistical friction by providing a trusted ledger in support of logistic transactions and processes. Errors can be reduced through smart contracts, as demonstrated in both IBM and Oracle BC platforms. BC tracks assets, and therefore, BC can track data assets just as well as a partial solution to software safety.

Intermittent Communications

The Navy primarily operates at sea, which means the communications infrastructure supporting the BC network may not always be available or reliable or provide bandwidth. A significant concern when implementing BC technology in cargo shipments is its dependence on a continuous connection to the fabric environment. However, HLF is a robust distributed database (ledger) that has many copies of itself.

The BC platform does require you to be connected to the fabric environment at all times or to consistently re-upload the ledger to peer nodes to have a constant accurate ledger. BC provides an update method that if a node is off-line, it will have an update of its BC once reliable network is reestablished.

The Issue of Governance

Figure 4 showed a simple notational circle labeled "Governance," but this issue is far from simple and is the key to any implementation of BC in support of data. While a detailed discussion of governance is beyond the scope of this paper, Gaur and Gaur [17] presented a variety of frameworks, some of which would apply to permissioned BC networks. Previous discussions of BC governance tended to be about public BCs supporting cyber currencies. They noted that while BC is about decentralization, there will have to be some aspects of centralized governance—especially ones involving policy and legal aspects in the storage and use of data. For example, governance could include safeguards through smart contracts that could flag possible AI bias, especially ones used for human resources. Governance can consist of different layers, and one classification recognizes different levels the data serves and is classified as strategic, operational, and tactical governance. Since BC is decentralized by nature, the governance should be at the lowest level if diversity and flexibility are important. Ziolkowski et al. [18] looked at governance that includes demand and data management, system architecture design and development, membership, and data ownership. Each one represents a possible off- or on-chain solution that involves technical and policy considerations—both of which may include smart contracts as solutions and resources [19]. System architecture design and development are not trivial tasks and are based to a great extent on governance and policies. To resolve this, IT network engineers must work as part of a consortium to determine the appropriate way to expose their peers to other organizations to receive transaction endorsement proposal/simulation requests while minimizing an attacker's ability to gain access to sensitive information stored in the simulating peer's database [19]. The level of rigor is ultimately determined by the policies derived by governance. Data accessibility is also a key, so governance should have policies that allow scientists working for the DoD to find data not through randomness but structure, without undue delay, and data that complies with software safety. BC supporting "smart repositories" may facilitate this goal. The default should be to allow our data scientists and analysts timely access to data unless there is a good reason not to. Our adversaries work for AI superiority, and withholding data from their researchers is something they avoid. We refer to unclassified and non-PII/medical data.

References

1. K. Abbas, L.A. Tawalbeh, A. Rafiq, A. Muthanna, I.A. Elgendy, A. El-Latif, A. Ahmed, Convergence of blockchain and IoT for secure transportation systems in smart cities. Secur. Commun. Netw. **2021**, 1–13 (2021)
2. A.A. Abd El-Latif, B. Abd-El-Atty, I. Mehmood, K. Muhammad, S.E. Venegas-Andraca, J. Peng, Quantum-inspired blockchain-based cybersecurity: Securing smart edge utilities in IoT-based smart cities. Inf. Process. Manag. **58**(4), 102549 (2021)
3. E.M. Abou-Nassar, A.M. Iliyasu, P.M. El-Kafrawy, O.-Y. Song, A.K. Bashir, A.A. Abd, El-Latif., DITrust chain: Towards blockchain-based trust models for sustainable healthcare IoT systems. IEEE Access **8**, 111223–111238 (2020)
4. G.N. Nguyen, N.H. Le Viet, M. Elhoseny, K. Shankar, B.B. Gupta, A.A. Abd El-Latif, Secure blockchain-enabled cyber–physical systems in healthcare using deep belief network with ResNet model. J. Parallel Distrib. Comput. **153**, 150–160 (2021)
5. A. Athalye, L. Engstrom, A. Ilyas, K. Kwok, Synthesizing robust adversarial examples (2017)
6. X. Jia, N. Hu, S. Yin, Y. Zhao, C. Zhang, X. Cheng, A2 chain: A blockchain-based decentralized authentication scheme for 5G-enabled IoT. Mob. Inf. Syst. **2020** (2020). https://doi.org/10.1155/2020/8889192
7. A.H. Eden, J.H. Moor, J.H. Soraker, E. Steinhart, *Singularity Hypotheses: A Scientific and Philosophical Assessment* (Springer, Berlin, 2013)
8. T. Everitt, Towards safe artificial general intelligence, Doctoral thesis, Australian National University, 2018. https://www.tomeveritt.se/papers/2018-thesis.pdf
9. H. Roland, B. Moriarty, *System Safety Engineering and Management*, 2nd edn. (Wiley, New York, 1990)
10. G. Chauhan, AI safety. Towards data science, 14 September 2018. https://towardsdatascience.com/ai-safety-9aeb9ca42907#:~:text=AI%20Safety%20is%20collective%20termed,of%20real%2Dworl d%20AI%20systems
11. Defense Standardization Program Office, *System safety (MIL-STD 882E)* (Pentagon, 2012)
12. M.T. Simerly, D.J. Keenaghan, Blockchain for military logistics. Army Sustain. **51**(4), 48–49 (2019)
13. K. Sarpatwar, R. Vaculin, H. Min, G. Su, T. Heath, G. Ganapavarapu, D. Dillenberger, Towards enabling trusted artificial intelligence via blockchain, in *Policy-Based Autonomic Data Governance*, ed. by S. Calo, B. Seraphin, D. Verma, 1st edn., (Springer International Publishing, 2019), pp. 137–153. https://doi.org/10.1007/978-3-030-17277-0_8
14. M. ur Rehman, K. Salah, E. Damiani, D. Svetinovic, Towards blockchain-based reputation-aware federated learning, in *IEEE Conference on Computer Communications Workshops*, (2020), pp. 183–188. https://doi.org/10.1109/INFOCOMWKSHPS50562.2020.9163027
15. B. Johnson, Intelligent and adaptive systems of systems for engineering desired self-organization and emergent behavior, in *Proceedings of the Future Technologies Conference* (Springer, Cham, November 2020), pp. 126–146
16. K. Lampropoulos, G. Georgakakos, S. Ioannidis, Using blockchains to enable big data analysis of private information, in *IEEE 24th International Workshop on Computer-Aided Modeling and Design of Communication Links and Networks, 2019*, (2019), pp. 1–6. https://doi.org/10.1109/CAMAD.2019.8858468

17. N. Gaur, N. Gaur, *Hands-on Blockchain with Hyperledger: Building Decentralized Applications with Hyperledger Fabric and Composer*, 1st edn. (Packt Publishing, Birmingham, 2018)
18. R. Ziolkowski, G. Miscione, G. Schwabe, Decision problems in blockchain governance: Old wine in new bottles or walking in someone else's shoes? J. Manag. Inf. Syst. **37**(2), 316–348 (2020). https://doi.org/10.1080/07421222.2020.1759974
19. L. Feagan, Hyperledger fabric myths and reality. Object Computing, April 2020. https://objectcomputing.com/resources/publications/sett/april-2020-hyperledger-fabric-myths-and-reality

Developing Instrument for Investigation of Blockchain Technology

Dmitry Kushnir, Maxim Kovtsur, Ammar Muthanna, Anastasiia Kistruga, Mark Akilov, and Anton Batalov

1 Introduction

Blockchain originally appeared as a distributed registry of the Bitcoin system [1]. Such a register allowed to solve the problem of double spending of cryptocurrency. One of the key features of the blockchain is the immutability of the distributed registry. This property allows for the exchange of data between interested parties, ensuring trust among the initially distrustful parties of information interaction. The rapid success of Bitcoin has attracted a lot of attention of researchers to the new technology underlying it, the blockchain. Despite the fact that the blockchain originated as an infrastructure for cryptocurrency, it has become a technology of distributed systems, which has led to a shift in emphasis from centralized systems to decentralized ones. A decentralized and open architecture is formed on the basis of the blockchain, since it is implemented on a large number of distributed nodes, each of which contains a copy of cryptographically linked records. Such records are organized into blocks agreed by some consensus protocols among blockchain nodes. A cryptographically linked block chain, together with a distributed consensus protocol, ensures the immutability of the blockchain. The openness and immutability of the blockchain allow anyone to check the history of records in the blockchain, which prevents any attempts to interfere with the stored data and protects the information from being changed after being added to the blocks. Thus, trust among untrusted parties of information interaction can also be created on the basis of a decentralized architecture. The decentralized, open, and unmodifiable nature of the blockchain makes it a transparent, publicly verifiable system. In addition, since records are replicated to many distributed nodes, the

---

D. Kushnir (✉) · M. Kovtsur · A. Muthanna · A. Kistruga · M. Akilov · A. Batalov
M. A. Bonch-Bruevich Saint Petersburg State University of Telecommunications, Saint Petersburg, Russia

Y. Maleh et al. (eds.), *Advances in Blockchain Technology for Cyber Physical Systems*, Internet of Things, https://doi.org/10.1007/978-3-030-93646-4_12

blockchain architecture allows you to get rid of the problem of a single point of failure. The combination of these properties allows us to consider the blockchain technology as the basis for a wide range of applications. Such applications can be solutions for the Internet of things (IoT) and cyber-physical systems (CPS) in which the possibility of interaction between a huge number of heterogeneous devices is required. Traditional centralized solutions may face such difficulties as distrust of information exchange nodes to each other, a huge number of interactions, the presence of a single point of failure, and a number of others. If we pay attention to CPS, it is also important to note that many existing industrial networks have only now begun to depart from the standards created in the 1970s of the last century, and the search for solutions based on blockchain can potentially effectively solve many problems in this area.

The widespread use of various systems that use or claim the use of blockchain for the implementation of certain tasks creates a new reality in the modern digital world. If the first of the well-known blockchain implementations was associated with the creation of a cryptocurrency [1], then today's projects have long gone beyond this narrow framework. A huge number of projects [2–4] created on the basis of long-existing platforms or their own implementations create significant difficulties in classifying certain systems and evaluating potential opportunities. In addition, it makes sense to note that from time to time there are projects that only claim to be based on a functioning blockchain or any other mechanisms with similar functionality, but do not have the appropriate technologies. Such projects try to exploit this problem and attract potential customers to themselves, diverting significant resources from real tasks. One of the most famous examples of this kind is a pyramid scheme, which was covered by the pseudo-cryptocurrency OneCoin [5].

In addition, it is important to note that even the presence of a successfully working blockchain does not guarantee the success of a project implemented on its basis. Depending on the application, different parameters may be required from the blockchain. Such parameters can be the amount of data that can be written to blocks and speed and delay during recording. High speed of reaction to events, in particular, is necessary for many solutions in IoT and in CPS, for example, in the Internet of vehicles (IoV). It is also necessary to keep in mind that in some cases, the transition to the blockchain or an unsuccessful choice of its specific implementation can only increase overhead costs and efficiency of interaction within the system. However, small improvements to traditional solutions could more effectively overcome existing problems.

In this regard, one of the tasks facing researchers is the creation and active use of tools that implement and analyze the main mechanisms underlying the construction of the blockchain. In the future, this will allow us to build the formation of approaches that make it possible to assess the availability and effectiveness of certain blockchain construction technologies in software to solve specific problems when building IoT/CPS solutions.

2 Problematics

Currently, the construction of various systems operating on the basis of the blockchain has become a noticeable phenomenon in digital data storage systems. In this regard, an important aspect is the analysis of existing and future solutions for the correct implementation of the declared functionality and, in addition, the very fact of the presence of such functionality.

A number of studies [6, 7] concentrate on the possibility of building various blockchain-based application solutions in the field of IoT and CPS. A feature of such developments is the focus on the already-made selection in advance of one or two or three specific distributed ledgers for IoT/CPS applications, such as Ethereum, Hyperledger Fabric, and IOTA. On the one hand, this approach is justified, since it is impossible to build a solution without taking into account the peculiarities of specific components, but on the other hand, such an approach can potentially limit the functionality of the final solution by excluding certain mechanisms of blockchain formation in the solutions being developed.

However, there are studies in the field of building solutions with a choice of blockchain for IoT in smart city, in which the development of their own blockchain architecture is carried out [8]. The development is justified by the redundancy of the traditional blockchain and the high computational load on individual nodes of the system, which a significant number of smart city elements cannot afford. This approach may be quite appropriate in some applications, but it can significantly complicate the development of the final solution, introduce additional errors into it, and delay implementation.

In this regard, the question of the possibility of a deliberate selection or development of a blockchain with the necessary characteristics in the framework of solving a specific problem remains relevant. Also, understanding the essence of the functioning of the blockchain is important for all participants in building a complete solution from developers at all levels to service personnel and even users, and only then can the maximum effect be achieved from the implementation of distributed ledgers in various IoT/CPS solutions.

One of the approaches that allows us to solve some of these problems is a demonstration implementation of the main functionality associated with the formation of a block chain. The projects existing in the research area in most cases are ready-made solutions that are difficult to analyze. This applies both to cryptocurrencies and related blockchains [1, 9, 10] and to other decentralized systems [11].

There are a number of solutions, both educational and demonstration, and researches, which allow analyzing the main stages of block formation, node interaction, and consensus building [12–14]. However, all these solutions either simplify the essence of block formation too much or work for the user in the form of a black box, which, in fact, without changing the source code, does not allow them to be used effectively for research purposes. Table 1 summarizes the main characteristics of the designated solutions in the field of research modeling of

Table 1 Characteristics of solutions in the field of research modeling of blockchain construction

Solutions	Config/install required	Ability to add nodes arbitrarily	Implementation of the consensus-building mechanism	Viewing block parameters
Blockchain demo 2.0	X	√	X	~
Visual demo of blockchain technology	X	X	~	√
Building a blockchain by Daniel van Flyman	√	√	~	~

√ yes, X no, ~ partially implemented or requires code editing for analysis

blockchain construction. In this paper, we analyze a software model of blockchain formation, designed to partially fill these gaps.

Developed tools should have the following functionality:

- Do not require complex configuration and installation
- Simulate a decentralized distributed system
- Be able to add nodes arbitrarily
- Conduct transactions
- Send transactions to all network participants
- Check transactions received from other network participants
- Support a consensus-building mechanism
- Create blocks
- Send blocks for verification
- Check blocks, including displaying their parameters
- Add blocks to the database, if the check is successful
- Have a graphical representation for clarity of work

3 Materials and Methods

Various development/programming environments can be chosen to study the methods of building a blockchain, but from the point of view of development efficiency and a number of requirements for speed indicators due to the implementation of different nodes of a distributed network on one researcher's computing device, the choice of C++ with the cross-platform Qt framework looks more preferable. For this study, the task of ensuring cross-platform compatibility is important from the point of view of expanding the use of the proposed methods to various platforms.

Since the study of the main stages of building a blockchain requires the implementation of cryptographic functions, a library is needed in which there

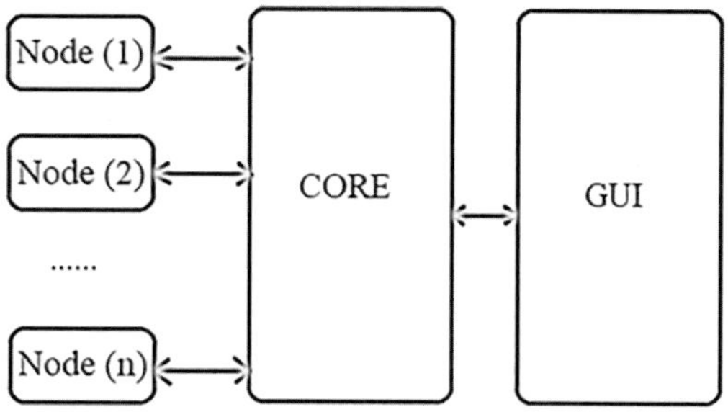

Fig. 1 Implementation of all network nodes within the framework of the developed toolkit

would be functionality for working with cryptography. OpenSSL is a universal cryptographic library. It supports almost all low-level hashing, encryption, and electronic signature algorithms and also implements most popular cryptographic standards. OpenSSL is written in the C programming language, which allows it to be used in C++ projects and guarantees high performance and speed of work.

The possibility of researching the constructed block chain assumes that each node stores blocks, both created independently and received from other participants (in this analysis, each node is assumed to be complete, i.e., it stores the entire block chain). In this case, the research is supposed to be carried out within the framework of a single computing device, and all nodes are formed on the same computer (see Fig. 1). This condition allows you to approach the choice of a data storage system for the block chain on each node using an embedded cross-platform database management system (DBMS), such as SQLite. Choosing this solution for data storage allows you to potentially increase the performance of the final solution, in particular, due to the absence of a client-server architecture, which is not required in this case.

4 Methodology for Analyzing Blockchain Model

As part of the study, it is necessary to determine the functions that will be analyzed [15, 16]. The corresponding functions are shown in Fig. 2.

The necessary functions within the framework of the developed solution include the following:

- Forming nodes:
  - Generating a secret key.
  - Calculating the public key.
  - Formation of the node address.
- Generating data for writing to blocks:
  - Preparation of the data itself.
  - Checking the correctness/balance.
  - Creating a digital signature.

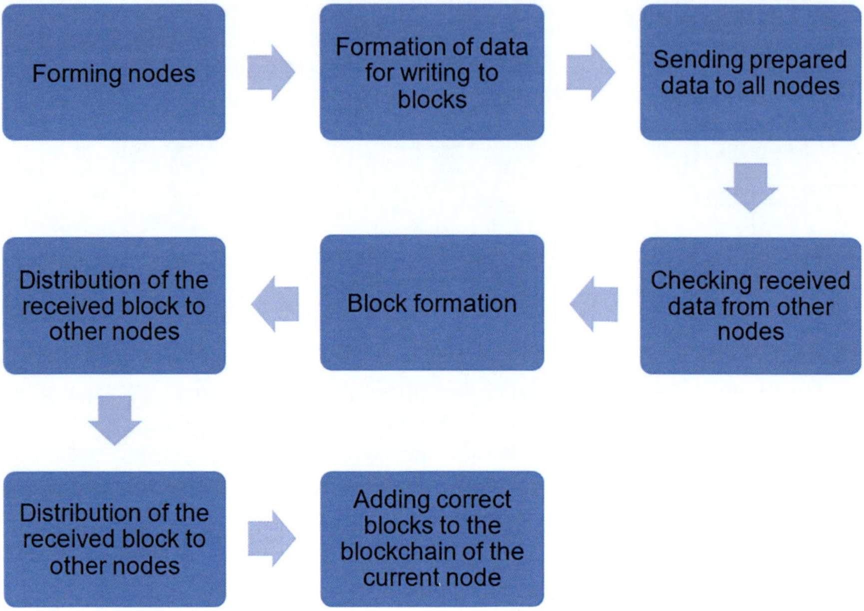

Fig. 2 The main stages of modeling a decentralized distributed system

- Distribution of the prepared data to all nodes.
- Checking the received data from other nodes.
- Forming a block:
  - Collecting data from other nodes in the preliminary version of the block
  - Building a Merkle tree for block data
  - Adding service information to a block, including the hash of the previous block
  - Selecting the nonce value to fulfill the condition for the hash value at the current complexity parameter
- Distribution of the received block to other nodes
- Checking the blocks received from other nodes
- Adding correct blocks to the blockchain of the current node

The nodes will be responsible for the main functionality, i.e., work with the formation of data for writing to blocks (transactions), blocks, and the database. Each node will have its own database to demonstrate that the network is coming to a consensus. In checking, the databases must be the same.

The nodes and the program core will model a decentralized distributed system [17]; the nodes will communicate with each other using the core. In this case, the kernel can be represented as a data transfer medium between network nodes.

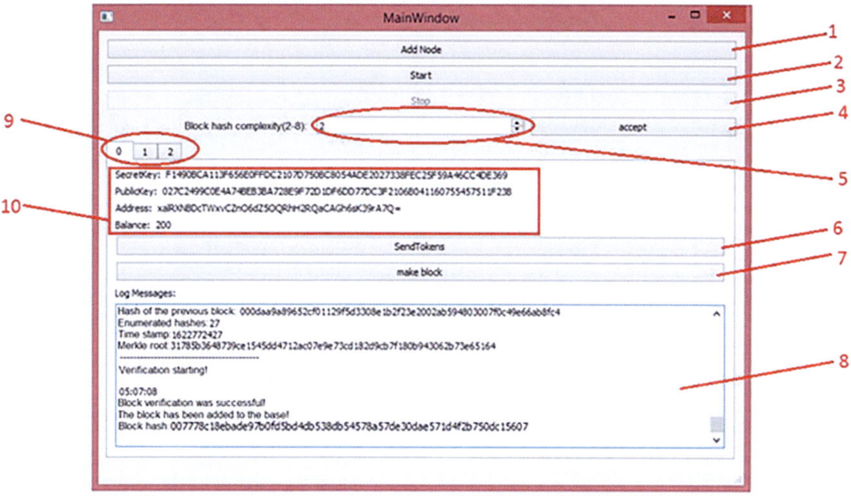

Fig. 3 The main window of the program

The control will be carried out using the graphical interface. Control signals will be sent to the program core, and the core will either redirect the signal to the nodes or perform other necessary actions.

The graphical interface has the form shown in Fig. 3.

The interface has the following elements:

1. "Add Node" button – this adds a new node.
2. The "Start" button – when pressed, all added nodes begin to continuously calculate blocks.
3. The "Stop" button – this stops the calculation of blocks.
4. The "Accept" button – this applies the hash complexity specified in (5) for the block.
5. The widget for entering the complexity of the calculated block hash is set by a number from 2 to 8. This number determines the number of zeros at the beginning for a 16-bit hash entry.
6. "Send Tokens" – this opens a window for making a transaction.
7. "Make block" – when clicked, all nodes start calculating the block hash, but only once. This function is used for step-by-step operation.
8. The log field of the selected node – it displays information for tracking the current operation of this node.
9. Tabs for switching between network nodes.
10. Information about the currently selected node (Fig. 4).

1. Information about the sender
2. Field for entering the recipient's address
3. Input field for the number of tokens to be transferred

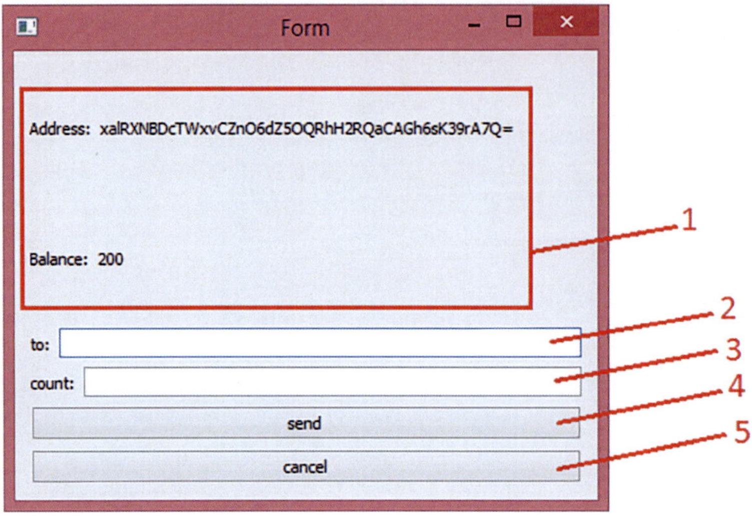

Fig. 4 Window for preparing data for entering into the block (performing a transaction)

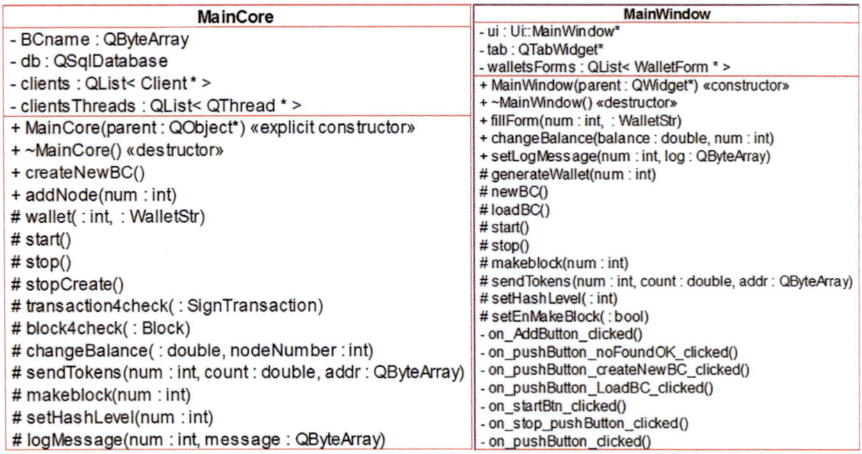

Fig. 5 Diagram of the mainCore class and the MainWindows class

4. Send button
5. Cancel button

To implement the program, six classes were created in total, of which three main classes can be distinguished, mainCore, acting as the application core (see Fig. 5); MainWindow, representing the main window (see Fig. 5); and Client, implementing node functions (see Fig. 6).

Let us consider some features of the developed tools.

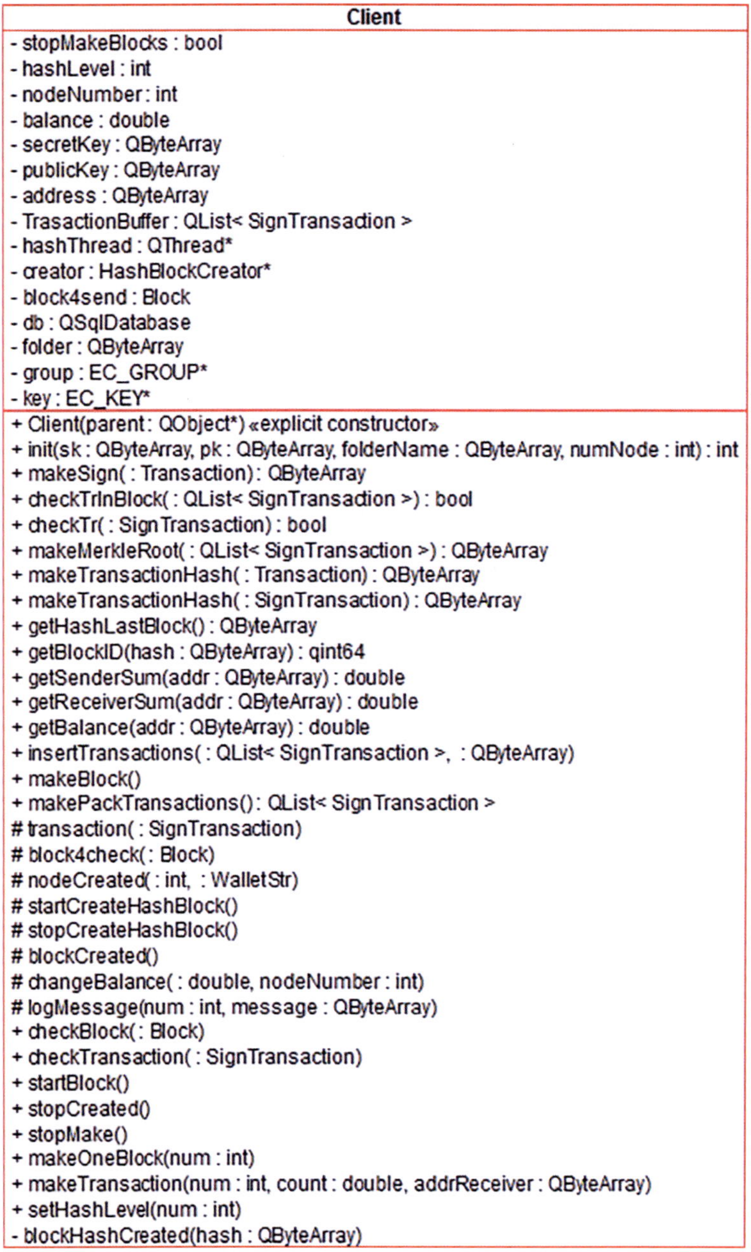

Fig. 6 Diagram of the Client class

In Qt, the signal and slot technique is used for communication between objects. A signal is emitted when a certain event occurs. A slot is a function called in response to a certain signal. The signal and slot mechanism is a central feature of Qt and probably the part that differs the most from the functions provided by other frameworks.

Signals are emitted by an object when its internal state has changed in any way. Signals are public functions and can be emitted from anywhere, but it is recommended to emit them only from the class that defines the signal and its subclasses.

The slot is called when the associated signal is issued. Slots are ordinary C++ functions and can be called in the usual way; their only feature is that signals can be connected to them.

mainCore and MainWindow interact with each other with the following signals: news, generateWallet, wallet, start, changeBalance, sendTokens, stop, makeBlock, setHashLevel, and logMessage.

The generateWallet(int num) signal is emitted when the "add node" button is pressed. This signal is associated with the addNode slot(int num) and passes the "int num" parameter – this parameter means the number of the nodes to be added, and it is needed for interaction between the nodes and the graphical interface. When the generateWallet signal is emitted, the "addNode" slot is executed. Let us look at it in more detail.

"addNode" creates new nodes. Keys are generated for new nodes, and this is where the OpenSSL library is used. The EC_KEY_new_by_curve_name () function allocates memory and binds the EC_KEY object to the specified curve. In our case, the secp256k1 curve is selected. EC_KEY_generate_key () generates a new private and public key. We get the address by taking the SHA-256 hash from the public key.

The QSqlDatabase db object is used to create a database for a new node. The database is created under the name corresponding to the public key of the node.

Next, a new node is created, and its initialization is performed using the generated keys. For each new node, a separate thread is created in which it will work.

The client type object shown in Fig. 7 simulates client actions and also uses the startBlock slot and the makeBlock function to generate new blocks.

First, a signal is emitted with a log message that the calculation of blocks begins, and then the makeBlock () function is called, shown in Fig. 8.

In this function, the block header is formed, and creator is initialized by the block header structure, and the startCreateHashBlock () signal is emitted.

Creator is an object of the HashBlockCreator type shown in Fig. 9. HashBlock-Creator is a class created for selecting a hash of a given complexity. The init()

```
void Client::startBlock()
{
    emit logMessage(nodeNumber,"Start calculate blocks\n");
    stopMakeBlocks = false;
    makeBlock();
}
```

Fig. 7 Client type object

```
void Client::makeBlock()
{
    block4send.prevBlockHash = getHashLastBlock();
    block4send.transactions = makePackTransactions();//add transactions
    block4send.merkleRoot = makeMerkleRoot(block4send.transactions);
    block4send.nonce = -1;
    headBlockForHash str;
    str.prevBlockHash = block4send.prevBlockHash;
    str.merkleRoot = block4send.merkleRoot;
    str.timeStamp = QString::number(QDateTime::currentSecsSinceEpoch())
                                                              .toUtf8();
    str.nonce = -1;
    creator->init(str);
    emit startCreateHashBlock();
}
```

Fig. 8 makeBlock function

```
void HashBlockCreator::start()
{
    stopCreate = false;
    do{
        QByteArray headArr;
        QDataStream s(&headArr,QIODevice::WriteOnly);
        str.nonce++;
        str.timeStamp = QByteArray::number(QDateTime::
                                        currentSecsSinceEpoch());
        s<<str;
        hash = QCryptographicHash::hash(headArr,
                               QCryptographicHash::Algorithm::Sha256);
        QByteArray checkBytes = hash.toHex();
        checkBytes.remove(8,checkBytes.length());
        levelFlag = true;

        for(int i = 0; i < hashLevel;i++){
            if(checkBytes.at(i)!= '0'){
                levelFlag = false;
            }
        }
    }while((!levelFlag) && (!stopCreate) );
    if(!stopCreate){
        emit createdCorrectHash(hash);
    }
}
```

Fig. 9 HashBlockCreator source code

method initializes a new block header. The hash is selected in the start slot, and this slot is called when the start Create Hash Block () signal is triggered.

In this slot, the block header is hashed, the block structure is written to the QByteArray header object, and hashing is performed using the QCryptographicHash::hash function, according to the Sha-256 algorithm. The resulting hash is converted to a 16-bit form and checked. If the received hash does not satisfy the specified complexity and there is no signal that it is necessary to stop the hash selection, the actions are repeated, but the fields of the block header structure such as nonce and timestamp are changed. The nonce field is incremented by 1, and timestamp gets a new timestamp. Otherwise, it is checked whether a stop signal has been received. If not, a createdCorrectHash signal is emitted that a hash of the desired complexity has been selected.

The start slot is executed in a separate thread, so that at the time of hash selection, the node can process transactions and receive signals from other nodes.

5 Developed Application of the Analysis Blockchain Technique

The tool for analyzing the main stages of building a blockchain at the first stage forms the necessary number of nodes involved in further research (Fig. 10).

The next step is to create new blocks. Blocks can be created both automatically and step-by-step. Blocks are created, validated, and added to the node databases. The network comes to a consensus. This can be seen from the same data in the databases of the nodes.

For the possibility of a detailed study of the performed actions at each step of the simulation, all operations are saved in a log file for further analysis.

The introduction of arbitrary records into the blockchain in this simulation is implemented in the form of creating transactions. Transactions are formed, signed, added to blocks, and written to node databases. The balance of nodes changes, and tokens come to the destination address (Figs. 11 and 12).

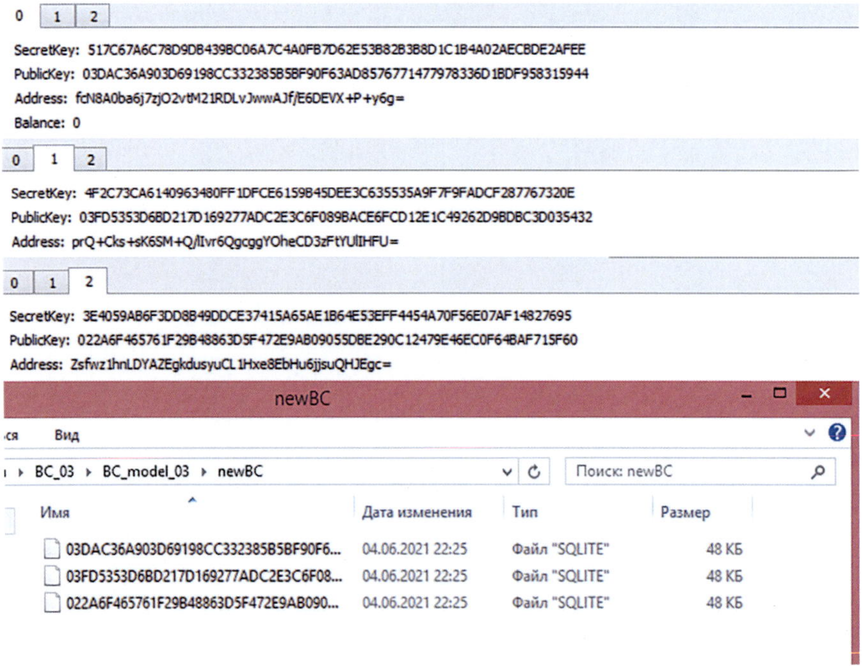

Fig. 10 Adding new nodes. Checking the creation of the corresponding databases

SecretKey: 3E4059AB6F3DD8B49DDCE37415A65AE1B64E53EFF4454A70F56E07AF14827695
PublicKey: 022A6F465761F29B48863D5F472E9AB09055DBE290C12479E46EC0F64BAF715F60
Address: Zsfwz1hnLDYAZEgkdusyuCL1Hxe8EbHu6jjsuQHJEgc=
Balance: 50

0 | 1 | 2

SecretKey: 3E4059AB6F3DD8B49DDCE37415A65AE1B64E53EFF4454A70F56E07AF14827695
PublicKey: 022A6F465761F29B48863D5F472E9AB09055DBE290C12479E46EC0F64BAF715F60
Address: Zsfwz1hnLDYAZEgkdusyuCL1Hxe8EbHu6jjsuQHJEgc=
Balance: 150

Fig. 11 The balance of node number 2 before and after the formation of a new block

Block_id	Transaction_Hash	PublicKey	Signature	Summ
Фильтр	Фильтр	Фильтр	Фильтр	Фильтр
12	58d43fed107c54776072d87e34d0d48e1e18d256d524a615de62ef286be90447	03FD5353D68D217D169277ADC2E3C6F0898ACE6FCD12E1C49262D98DBC3D035432	3046022100dcfaf266...	100.0

	Transaction_Hash	sender_address	tokens_count
	Фильтр	Фильтр	Фильтр
16	58d43fed107c54776072d87e34d0d48e1e18d256d524a615de62ef286be90447	prQ+Cks+sK6SM+Q/lIvr6QgcggYOheCD3zFtYUlIHFU=	100.0

	Transaction_Hash	Receiver_address	tokens_count
	Фильтр	Фильтр	Фильтр
16	58d43fed107c54776072d87e34d0d48e1e18d256d524a615de62ef286be90447	Zsfwz1hnLDYAZEgkdusyuCL1Hxe8EbHu6jjsuQHJEgc=	100.0

Fig. 12 Transaction records in the database

An important parameter for each blockchain system is the average block creation time and possible deviations from the expected value. These parameters can affect the time to reach a network consensus, the necessary time delays for recognition by system participants, and the possibility of certain types of attacks. In some technologies of blockchain formation, the main parameter that affects the formation time of the block creation is the complexity parameter. In the model under study, the complexity is set as the upper bound of the calculated hash value, given as the number of leading zeros in the hexadecimal representation of the boundary, and an additional parameter will be the number of nodes involved. The effect of complexity on the time of making entries in the distributed registry for a network of four nodes and complexity 6 is shown in Fig. 13, and for complexity 7 in Fig. 14. The dots indicate the time spent on creating the next block in the system.

The graphs clearly show the dependence of the average time spent on creating the next block on the complexity parameter with characteristic outliers corresponding to the Poisson distribution law.

6 Conclusion

The analyzed tool for performing research on blockchain technologies combines the ease of use and the ability to track each step of building a blockchain with

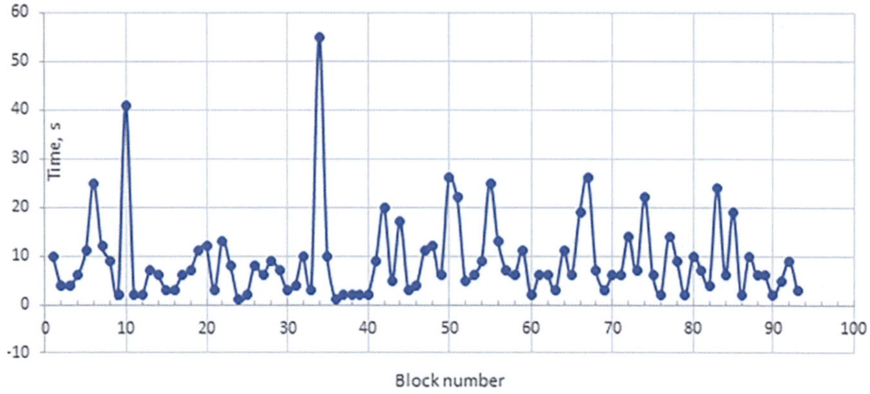

Fig. 13 The time spread during the formation of blocks on difficulty 6

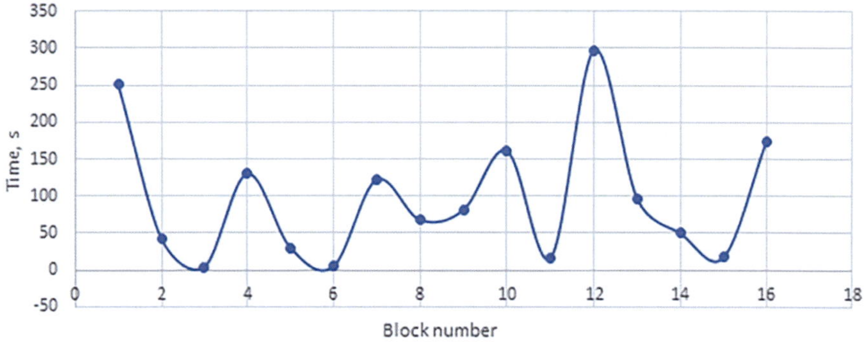

Fig. 14 The time spread during the formation of blocks on difficulty 7

checking current parameters and obtaining aggregating data, which gives certain advantages over similar solutions considered earlier. The presented stages allow us to evaluate the possibilities of traditional approaches and the potential advantages of the latest developments in this area. The paper presents and studies, in particular, such components of this technology as follows:

- Creating keys to confirm the authenticity of entries in the distributed registry
- Formation of nodes, i.e., participants of a decentralized network
- Managing the network complexity parameter, to influence the average speed of generating records in a distributed registry
- Construction of the Merkle tree, formation of the source block, and selection of the nonce parameter to obtain the final block of the system
- Checking the created blocks for compliance with the current system rules by the block creator
- Independent view of the status of the block chain at each node of the network
- Checking the correctness of the created blocks by other participants of the distributed network

An important feature of the analysis is the ability to switch the mode from analyzing the general parameters of the system to the step-by-step execution mode and monitoring all parameters of operations at each completed step. Thus, the results presented in the paper allow, on the one hand, to advance in the field of evaluating the functionality of various implementations of blockchain construction methods and, on the other hand, to solve the problem of detailed analysis and demonstration of the capabilities of the fundamental components of blockchain technology. The result of the analysis of distributed registry methods and technologies underlying the implementation of the blockchain allows you to choose a solution for specific tasks in IoT/CPS or smart city.

References

1. S. Nakamoto, Bitcoin: A peer-to-peer electronic cash system (2008), 9 p [Online]. Available: https://bitcoin.org/bitcoin.pdf. Accessed on 20.07.2021
2. A. Tabernakulov, J. Koifmann, *Blockchain in Practice* (Alpina Publisher, Moscow, 2019)., 260 p
3. A.A. Litvin, S.V. Korenev, E.G. Knyazeva, V. Litvin, Possibilities of blockchain technology in medicine (review). Mod. Technol. Med. **11**(4), 191 (2019)
4. Instruction of the Prime Minister Dmitry Medvedev on the possibility of using blockchain technology in the system of public administration and the economy of the Russian Federation on March 6, 2017 [Online]. Available: http://government.ru/orders/selection/401/26653
5. Cryptoqueen: How this woman scammed the world, then vanished. 2019.11.24 [Online]. Available: www.bbc.com/news/stories-50435014
6. J. Lee, M. Azamfar, J. Singh, A blockchain enabled Cyber-Physical System architecture for Industry 4.0 manufacturing systems. Manuf. Lett. **20**, 34–39 (2019)
7. M. Pustišek, A. Kos, Approaches to front-end IoT application development for the ethereum blockchain. Procedia Comput. Sci. **129**, 410–419 (2018)
8. Z. Dlimi, A. Ezzati, S.B. Alla, A lightweight blockchain for IoT in smart city (IoT-SmartChain). CMC-Comput. Mater. Continua **69**(2) Tech Science Press, 2687–2703 (2021)
9. E.A. Pekhtereva, Innovations in the financial sphere and the practice of their application: Blockchain technology and cryptocurrency in Russia. ESPR (1), 51 (2019)
10. L.S. Zvyagin, Digital economy of cryptocurrency: challenge or threat to traditional society. E-Management **1**(2), 80 (2018)
11. A. Tapscott, D. Tapscott, *Blockchain Revolution: How the Technology Behind Bitcoin Is Changing Money, Business, and the World*, 1st edn. (Penguin, London, 2016)., 324 p
12. Blockchain Demo 2.0 [Online]. Available: https://blockchaindemo.io. Accessed on 20.07.2021
13. Anders Brownworth Visual demo of blockchain technology [Online] Available: http://anders.com/blockchain. Accessed on 20.07.2021
14. D. van Flyman (ed.), *Learn Blockchain by Building One: A Concise Path to Understanding Cryptocurrencies* (Apress, New York, 2020)., 185 p
15. B. Singhal, G. Damedja, P.S. Panda, *Beginning Blockchain: A Beginner's Guide to Building Blockchain Solutions*, 1st edn. (Apress, New York, 2018)., 401 p
16. A.M. Antonopoulos, *Mastering Bitcoin: Programming the Open Blockchain*, 2nd edn. (O'Reilly Media, Beijing, 2017)., 416 p
17. A.A. Minyaev, A.V. Krasov, D.V. Saharov, The method and methodology of efficiency assessment of protection system of distributed information systems, in *The Collection 2020 12th International Congress on Ultra Modern Telecommunications and Control Systems and Workshops (ICUMT)*, (2020), pp. 291–295

Trust Models for Blockchain-Based Self-Sovereign Identity Management: A Survey and Research Directions

Shu Yun Lim, Omar Bin Musa, Bander Ali Saleh Al-Rimy, and Abdullah Almasri

1 Introduction

Identity management (IDM) refers to the mechanism and standards for creation, maintenance, and de-provisioning of user accounts. It covers the administrative area that identifies and authenticates users and controlling the user's access to resources such as applications, systems, or online services. Identity management has evolved from centralized identity, where user credentials are owned and managed by a single entity, to federated identity that provides authentication and authorization capabilities across organizational and system boundaries.

In centralized identity system, users struggle to maintain different sets of credentials for different services. They lose control of their personal data when the information is duplicated across different providers. When federated identity is adopted, a privacy invasion issue arises because users are subject to profiling and analytics when their data resides with providers [1]. Identity providers, on the other hand, are facing constant security attacks on their centralized databases; therefore, high costs are incurred to build multifactor authentication and secure their perimeter network. Identity providers can also be held liable for data breaches under existing data protection acts [2]. The security, privacy, and usability challenges faced by

S. Y. Lim (✉) · O. B. Musa
Faculty of Business and Technology, UNITAR International University, Petaling Jaya, Malaysia
e-mail: lim_sy@unitar.my; omarm@unitar.my

B. A. S. Al-Rimy
School of Computing, Faculty of Engineering, Universiti Teknologi Malaysia, Johor Bahru, Malaysia
e-mail: bander@utm.my

A. Almasri
School of Mathematical & Computer Sciences, Heriot-Watt University, Putrajaya, Malaysia
e-mail: a.almasri@hw.ac.uk

Y. Maleh et al. (eds.), *Advances in Blockchain Technology for Cyber Physical Systems*, Internet of Things, https://doi.org/10.1007/978-3-030-93646-4_13

both users and providers are expected to be resolved with the introduction of self-sovereign identity management (SSIDM) [3].

Self-sovereign identity is the concept that users should be able to control their own digital identity [4, 5]. Individual users or organizations can store their own identity data on their own devices and provide their identity to a verifier without relying on a central repository of identity data. Since this is independent from any individual silo, it gives the user full control, security, and full portability of their data [6]. Blockchain technology can be used to deliver this secure solution without the need for a trusted, central authority. It can be used for creating an identity on the blockchain, giving them greater control over who has their personal information and the way the information is being accessed [7].

One of the pioneers in blockchain-based SSIDM, Sovrin Foundation, describes self-sovereign identity as an Internet for identity where no one owns it, everyone can use it, and anyone can improve it [8]. By removing the need for a trusted third party, blockchain enables the creation of decentralized identity management without a central identity provider. In the light of this, decentralized IDM based on blockchain has different trust requirements compared to traditional IDM. There are various roles and objects that replace the centralized trusted third party; hence, trust must be managed in a dynamic and granular manner [7].

A SSIDM trust model should be able to assign a trust rating or trust score to every stakeholder based on observations from past transactions. The National Institute of Standards and Technology (NIST) presented in its cybersecurity white paper [9] a comprehensive list of entities and their roles in identity management. The stakeholders defined are requester, issuer, subjects and holders, verifier, and relying party. Every role in this ecosystem is involved in requesting credential, issuing credential, disclosing presentation, verifying presentation, and credential revocation (Fig. 1).

When a trust model is implemented on blockchain, a smart contract can be used for transparent, efficient, and secure calculation of trust rating. Automation using a smart contract should incur minimal overhead in terms of latency and throughput.

Trust is a pervasive and significant phenomenon in social societies with a diverse and manifold range of meanings and definitions [10]. Trust is also a fundamental for cooperation, conversation, and mutual interaction between entities. In recent decades, trust has been studied in many different disciplines and used as the basis for decision-making in different contexts [11].

Trust modeling uses the methodology of mathematics to obtain peers' trust intention and reliability information based on the definition of trust [12]. The trust engine, on the other hand, leverages multiple data sources to compute a risk score or credit score [13]. In mobile gaming, trust modeling is used to determine the authenticity of players' geo-position [14]. In wireless communication, trust refers to the relationship value computed based on the rate of successful transactions between network nodes [15]. There is much research on trust modeling, and most of them are in areas such as Internet of things, cybersecurity, social network, online services, and cloud computing, to name but a few [16–21].

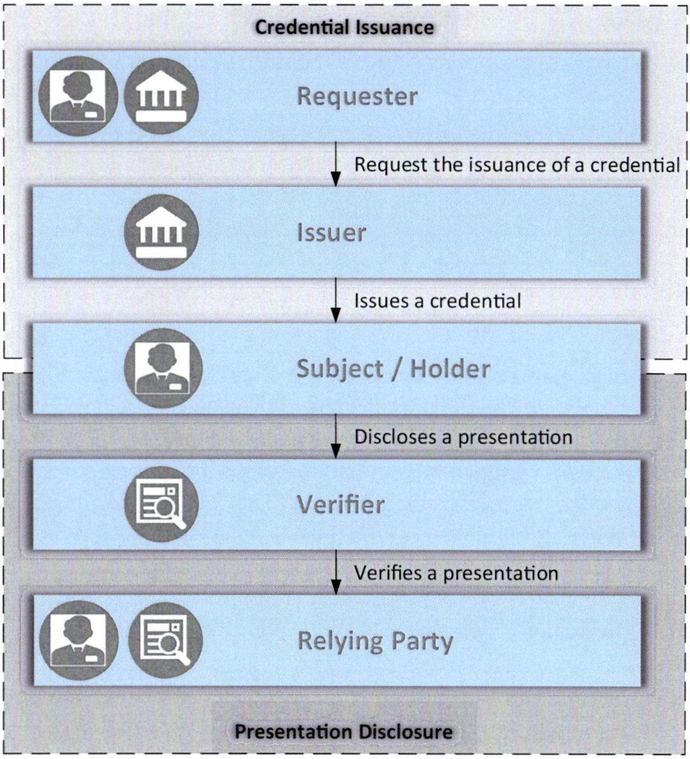

Fig. 1 Identity management roles defined by the NIST [9]

However, these solutions are not suitable when applied to Self-Sovereign IDM. With the introduction of blockchain-based Self-sovereign IDM, a different approach is needed for the computation of trust in digital identities. Self-sovereign IDM calls for specific requirements of trust not just for digital identities but also for claims and attestations made by entities. Existing blockchain-based IDM solutions can be further improved to determine the trustworthiness of claims and digital identities [22].

Many existing trust models use a static, preconfigured trust relationship to interact, such as the web of trust approach with pre-defined trust anchors in the Sovrin project [8]. The trust anchors' trustworthiness is assumed, rather than derived. However, trust can change dynamically according to actions and behaviors of entities [23].

Therefore, a distributed and dynamic approach for managing trust among identity management roles in Blockchain-based Self-sovereign IDM is needed. There are many parameters that could be considered for the trustworthiness of identities, claims, and attestations. A richer set of trust clues or parameters will lead to less fraud in identity transactions and management.

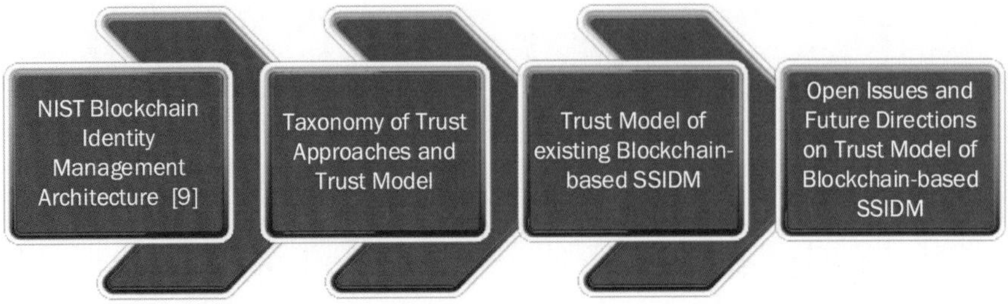

Fig. 2 Research methodology

We provide a survey of trust model in blockchain-based SSIDM with the methodology shown in Fig. 2. First of all, the NIST blockchain identity management system architecture is adopted and referenced for the baseline terms and definition. Next, a taxonomy of trust approaches and trust models were examined. In the third phase, related works were investigated with a focus on their trust model. Finally, we use input from the previous phases to derive the open issue and the future directions of trust modeling for blockchain-based SSIDM.

The paper starts by introducing the evolution of online identities and the concept of blockchain-based self-sovereign identity. In Sect. 2, the components of blockchain-based SSIDM are described. The components include verifiable credentials, verifiable presentations, digital wallet, decentralized identifiers, the underlying Blockchain network, and the Trust infrastructure of SSIDM. Section 3 presents a taxonomy of trust approaches and trust models. A summary of related SSIDM projects and respective trust models are presented in Sect. 4. The paper closes with research directions in Sect. 5 and a conclusion.

2 Architecture of Blockchain-Based SSIDM

Blockchain-based SSIDM consists of several components at different layers of the architecture. All identity data such as *claims*, *verifiable credentials*, and *verifiable presentations* are held in a *digital wallet* by an identity holder. The digital wallet is identified by a public key, facilitated by *decentralized identity (DID)* layer. A *smart contract* runs on top of the blockchain to implement the business logic. Beneath all components is the *distributed ledger*, a shared and tamper-proof record of transactions. The ledger on different nodes forms the heart of a *blockchain system* that empowers this self-sovereign identity ecosystem. The building blocks are discussed in detail in the following subsections.

A. Stakeholders

The technical paper presented by the NIST [9] provides an overview of stakeholders that interact in a blockchain-based SSIDM. *Subjects or holders* request for

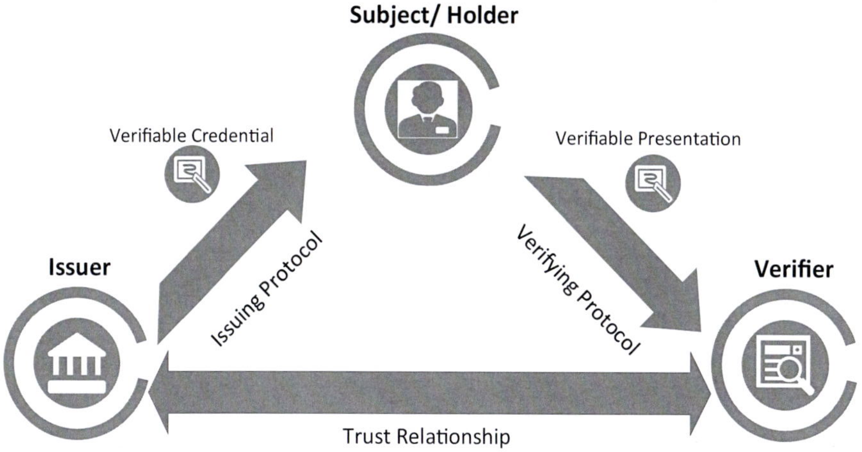

Fig. 3 Stakeholders of blockchain-based SSIDM

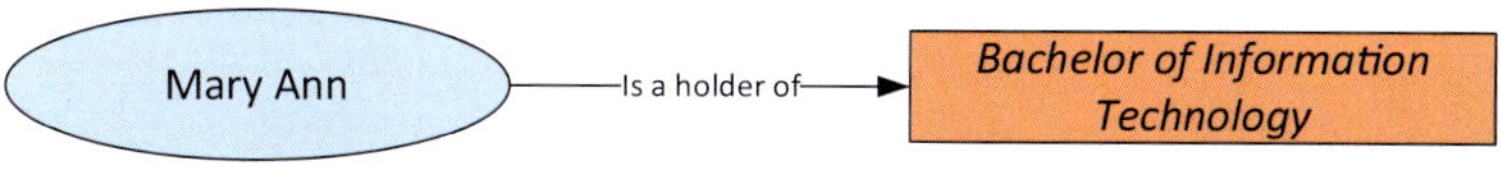

Fig. 4 Subject-property-value relationship [24]

the issuance of a credential. The *issuer* issues a credential to *subjects or holders* based on the request. The credential can later be presented to a *verifier*. The *verifier* will verify the presentation to a *Relying Party*. These roles are not exclusive because both *subject* and *issuer* can play the role of requester and a *subject* and *verifier* can both be a *Relying Party* (Fig. 3).

B. Claim

A claim is an identifier, or a statement made about an entity. An entity in this case can be a distinct person, organization, or device. For example, "Mary Ann is a holder of a bachelor's degree in IT" is described as a subject-property-value relationship in Fig. 4 [24].

The verifiable claims have a specific data model that can be expressed in data representation languages such as JSON (Fig. 5), JSON-LD, WebIDL, and XML [24].

C. Verifiable Credential

On the other hand, credential is more formal than a claim. It can be a set of one or more claims made by an entity. Verifiable credentials (Fig. 6) are digital certifications such as academic degree (Fig. 7), proof of employment (Fig. 8), and proof of income (Fig. 9). Every stakeholder could issue, hold, or verify credentials.

A verifiable credential may contain at least one or a set of claims in the form of metadata that describes the properties of the credential, such as a credential

```
{
    "id": "https://exampleuniversity.edu/credentials/5566",
    "type":["Credential","ProofOfAcademicQualification"],
    "issuer": "https://exampleuniversity.edu",
    "issued": "2005-01-01",
    "claim": {
        "id": "did:29987ab234f4561ffcb23ad5",
        "qualification": "Bachelor of Information Technology"
    }
}
```

Fig. 5 Example of verifiable claim in JSON

Fig. 6 Verifiable credential [25]

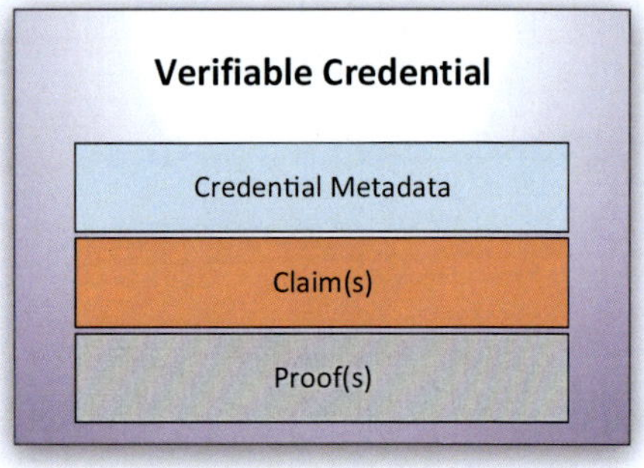

identifier, a public key of the issuer, or a timestamp. These metadata may be signed by the issuer. The issuer will attach a cryptographical signature such as an RSA signature, a nonce, a signature value, a creation timestamp, and an issuer's public key. These parameters are required for a third party to verify a credential.

D. Verifiable Presentation

Verifiable presentations are created out of claims and verifiable credentials. They serve to present personal identity information in a trusted way to third parties, revealing only as much information as required, to preserve the identity owner's privacy. Presentation is based on one or multiple credentials. The relationship between a claim, credential, and presentation is depicted in Fig. 10.

E. Digital Wallet

A subject or holder stores credentials in a personal device and software such as digital wallet, as in the real world where people keep their IDs in their physical wallet [27, 28]. A digital wallet serves as an agent in SSIDM ecosystem [7]. The wallet is used to perform authentication and prove ownership using the public and private key pairs generated. Since credentials are issued off-chain, the wallet contains all the self-attested information and credentials regarding the identity

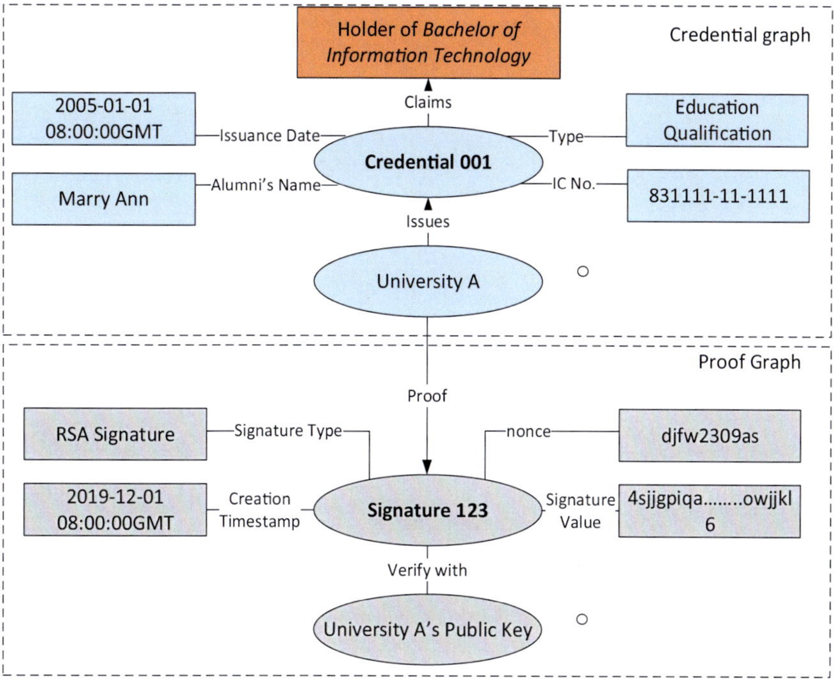

Fig. 7 Credential graph showing credential metadata and proof graph for credential presentation. University A is the issuer of credential as a degree awarding institution. This is a proof of academic qualification

owner. The credential can be presented to a third party for authentication or authorization to use a service. A holder could present entire credentials, parts of them, or combinations of multiple credentials in the form of proofs to verifiers. Thus, the holder has full control over which data is shared and how it is used.

F. Decentralized Identity

The decentralized identity (DID) layer allows an entity to be publicly identified in SSIDM solutions. DID methods allow users to request or issue verifiable credentials by providing the operations to create, read, update, and delete credentials in a decentralized way without the need of a central authority. There are emerging standards for recording credential metadata such as decentralized identifiers (DIDs) from W3C [26], DID Auth from the Rebooting the Web-of-Trust (RWOT) working group [29], Universal Resolver and Identity Hubs from the Decentralized Identity Foundation (DIF) [30], and Open Badges from Mozilla and IMS Global [31]. A DID standard will decide what credential metadata is recorded on the distributed ledger. Instead of storing credential metadata directly into the ledger, an identifier is used because the underlying blockchain is immutable.

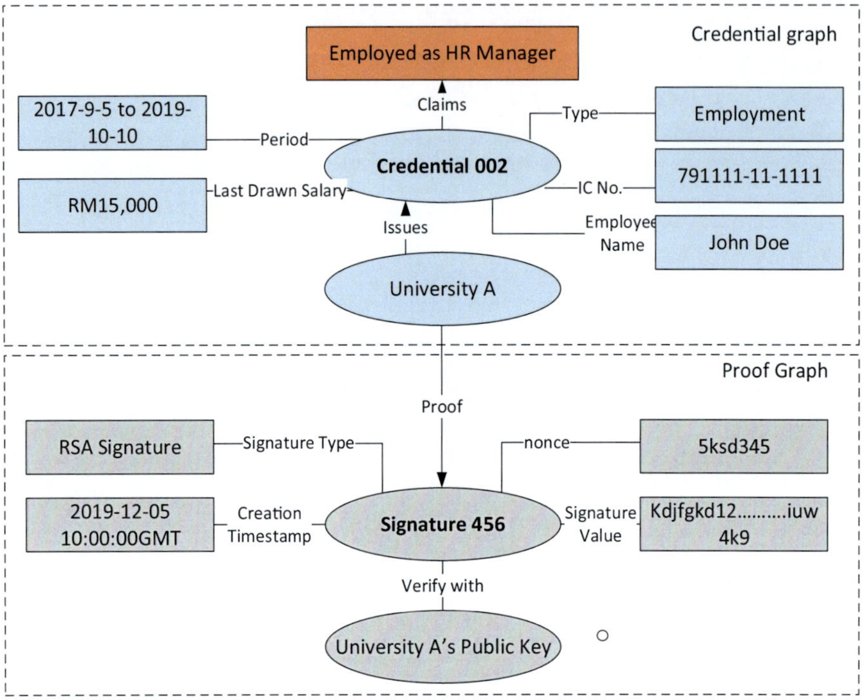

Fig. 8 Credential graph showing credential metadata and proof graph for credential presentation. University A is the issuer of the credential as an employer. This is a proof of employment

G. Smart Contract

A smart contract [32] defines the interactions between transacting parties and implements logic agreed by all nodes in a blockchain network. It is sometimes referred to as chaincode, but a smart contract is in fact defined within a chaincode. Multiple smart contracts for related business processes can be deployed in the same chaincode.

A smart contract comprises trigger conditions and response rules. Input to the smart contract can be time, event, transaction, action, etc. It performs evaluation of contract clauses and auto-executes contract statements once triggered. Upon completion, the output based on conditions and response rules will be written on a new block (Fig. 11).

H. Blockchain and Distributed Ledger Technology (DLT)

Blockchain is one of the main pillars of SSIDM, alongside verifiable credentials, verifiable presentations, decentralized identifiers, and smart contract. Underneath smart contract is the blockchain network and distributed ledger where the immutable and transparent records reside. The characteristics of blockchain make it a good fit

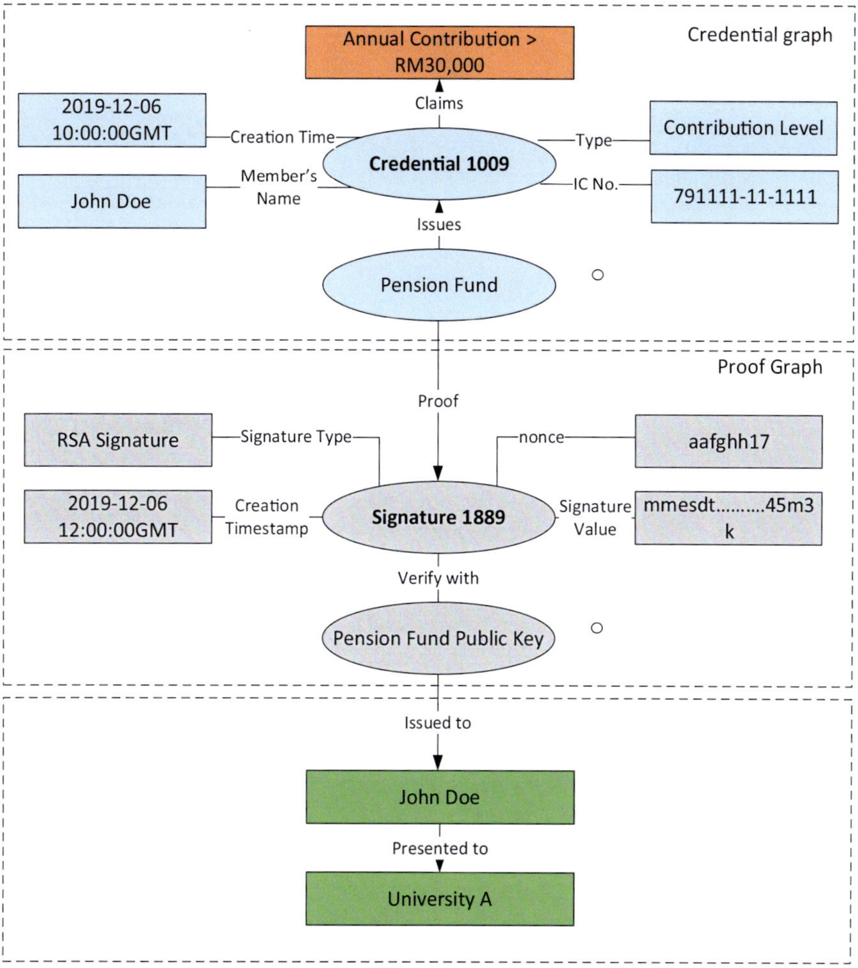

Fig. 9 University A, a potential employer for John Doe, is the verifier of a credential issued by pension fund to determine annual remuneration of John Doe. This is a proof of income level.

for creating an advanced identity management ecosystem in a decentralized manner which satisfies the principles of self-sovereign identity (Fig. 12).

The consensus layer is critical for any blockchain network. Consensus ensures that all nodes in blockchain agree to the truth. For a blockchain with cryptocurrency like Ethereum, consensus also rewards the nodes for validating the transactions and maintaining the blockchain network. Proof of work (PoW), proof of stake (PoS), and Practical Byzantine Fault Tolerance (pBFT) are excellent consensus algorithms for nodes to agree on the records on blocks. Hyperledger Fabric, Indy, and Iroha implemented voting-based consensus. For instance, Hyperledger Fabric uses RAFT algorithm for the log replication process [34]. Once a leader is elected, all messages

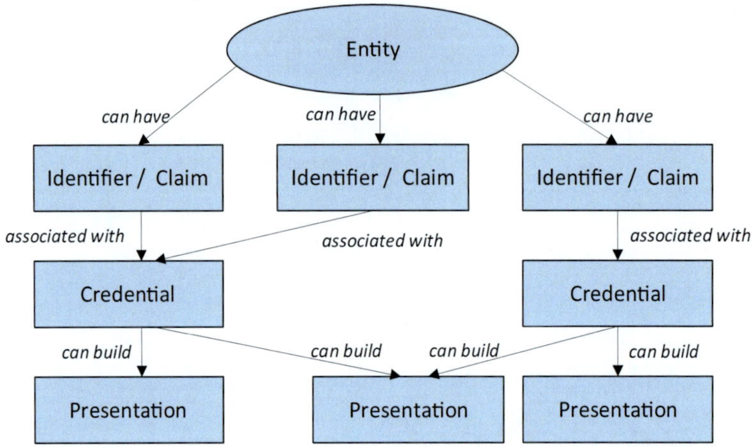

Fig. 10 Relationship between verifiable claim, verifiable credential, and verifiable presentation [26]

are sent via the leader. The leader will propagate the messages to all nodes, and the nodes then validate and write the messages. The nodes will also send a response to inform the leader that the message has been validated and written. Hyperledger Indy uses plenum [35] algorithm, which is an improved version of Redundant Byzantine Fault Tolerance. The consensus algorithm uses three-phase commit on the request to ensure that the ledger contains entries that are ordered and validated.

Network layer is used for information dissemination between participating peers [36]. *Data layer* consists of Merkle tree, a binary tree of hashes to offer integrity and non-repudiation for blockchain. Transactions are digitally signed in data layer using asymmetric cryptography.

Infrastructure layer is where all peer nodes reside. Organization uses certificate authority to assign X.509 digital certificates to all participating nodes recognized by the blockchain network. The nodes with virtualization using virtual machines or containers can support messaging services and storage of data [37].

Many blockchain networks have been developed for identity management. Notable works on identity management have been primarily conducted on the Hyperledger blockchain. Hyperledger Indy [38] is specifically created for self-sovereign identity management. This blockchain provides tools, libraries, and reusable components for providing digital identities rooted on blockchains or other distributed ledgers so that they are interoperable with other blockchains. Indy provides built-in support for zero-knowledge proofs to avoid unwanted disclosure of identity attributes. When a verifiable claim is not considered true, zero-knowledge proofs enable identity owners to authenticate the possession of a credential without displaying the credential itself with the help of anonymous credential scheme [39].

Hyperledger Aries [40] is a spin-off of Hyperledger Indy, to realize interoperable self-sovereign identity which covers more on the client side components such as

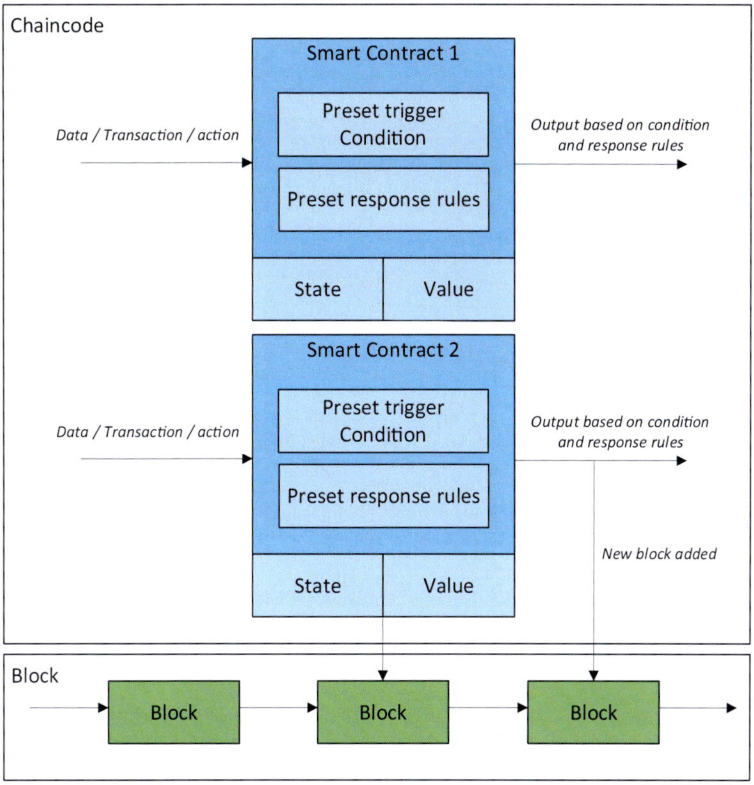

Fig. 11 Structure of smart contract and chaincode [32]

Fig. 12 Architecture of blockchain DLT [33]

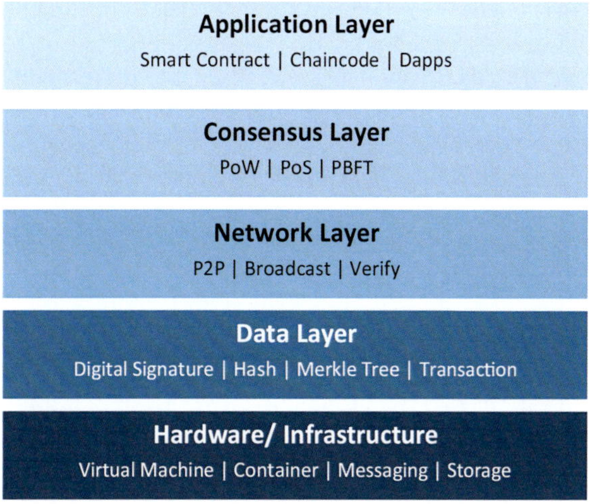

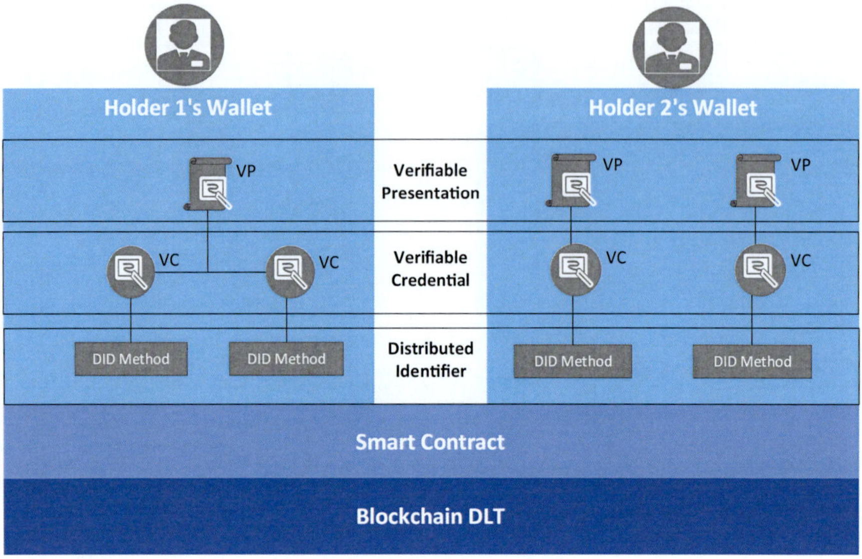

Fig. 13 Building blocks for blockchain-based self-sovereign identity management

wallet services and agent protocols. The blockchain focuses on providing tools and features to create, transmit, and store verifiable credentials in a wallet. This project utilizes cryptographic libraries and key management schemes provided by Hyperledger Ursa [41].

Another project under the Hyperledger project umbrella is Hyperledger Fabric [42, 43] which supports digital assets, distributed logic through chaincode, and the use of custom consensus through endorsement policies. Initially Fabric still lacked a key component for a decentralized identity, but TrustID was later incorporated in Hyperledger Fabric to simplify identity management in blockchain networks.

The trust framework is not shown in the architecture (Fig. 13) because trust can be implemented at all layers. The blockchain DLT serves as a root of trust in the architecture. Trust can also come from the decentralized identity layer and is managed depending on adopted DID standard. Verifiable credentials and verifiable presentations can have their own trust features implemented at a higher layer.

3 Types of Trust Model in Identity Management

One of the most cited definitions of trust is by Mayer et al. [44] "the willingness of a party to be vulnerable to the actions of another party based on the expectation that the other will perform a particular action important to the trustor, irrespective of the ability to monitor or control that other party." Trust is determined by the trustor's

Table 1 Type of trust models based on flow of control

Type of trust model	Description
Centralized trust model	Top-down. Entities report trust rating to a trusted party
Decentralized trust model	Bottom-up. A peer-to-peer system for entities to determine trust rating
Distributed trust model	Bottom-up. Trust rating is shared among entities

Table 2 Types of trust models based on control of transaction

Type of trust model	Description
Static trust model	Rules pertaining to trust are defined by trust administration system
Dynamic trust model	Defines trust rating based on changing parameters

propensity to trust, ability, benevolence, and integrity of the trustee in their proposed model of trust.

Propensity to trust is the willingness to trust others across a broad spectrum of situations and trust targets. This suggests that every individual has some baseline level of trust that will influence that willingness to trust. *Ability* is also referred to as the competence of the trustee to do a given task. *Benevolence* is the disposition of goodwill toward the trusting party. And lastly, *integrity* is the trustor's perception that the trustee adheres to a set of principles that are acceptable to the trustors.

The goal of trust is to determine what course of action, if any, the trusting party is willing to take in relation to the trusted party. Based on the level of trust and the perceived risk, the trusting party may decide to take some action that involves some degree of risk taking. Trust level has a corresponding risk rating; a lower risk translates to higher level of trust.

Trust models are classified based on how they are controlled [20] as shown in Table 1. The NIST [25] defined the two main approaches as top-down and bottom-up, with the latter frequently associated with SSIDM principles. Top-down approaches to trust lead to centralization of information, control, and loss of individual privacy. The bottom-up approach to trust is taken to avoid these pitfalls.

These two approaches form a spectrum of trust models, i.e., centralized, decentralized, and distributed models which can support different types of governance structures and power delegation mechanisms. In a centralized system, trust level is exerted by just one entity (i.e., trust anchor, CA, board of trustees). In a decentralized system, there is no single controlling entity, and every entity makes their own decision on trust level. In distributed approach, the trust level is shared among entities, and trust computation is distributed across nodes. Nodes interact with each other to determine trust level.

A trust model can also be categorized based on control of transactions (Table 2). The static model follows pre-defined rules, but the dynamic model adjusts with different parameters and progress based on the previous cached data stored in a data store.

Table 3 Types of trust approach

Type of trust approach	Description	Advantages	Disadvantages
Reputation based	Reputation of an entity is the collected estimation of public's trust	Public trust is ingrained in all communities	Reputation of an entity is assumed, not earned
Policy based	Formal trust methodologies which play a main role in PKI	Highly scalable and manageable	Rogue certificates issued by CA
Evidence based	Performance of entities from previous transactions determines the trust level	Higher accuracy since trustworthiness is dynamically deduced from past behaviors	Higher computation cost and performance issues

The types of trust approach are categorized as reputation-based trust, policy-based trust, and evidence-based trust [20]. In reputation-based trust, the reputation of an entity is the collected estimation of the public's trust toward that entity. Generally, many entities in a community trust an entity that has a high reputation; an entity, which is required to build trust decision on a trustee, uses the reputation to compute or approximate the trust level of the trustee [45]. However, Forrester Research [46] introduced the concept of a zero-trust model which states that no trust should be assumed but instead trust should be continually validated. This concept has been adopted widely in the design and implementation of IT systems.

In policy-based trust, formal trust methodologies are used to support key certification, digital signature, and validation. For instance, in PKI model, a certificate authority (CA) supports data attribute certification and validation. Certificate policies play a main role in PKI trust which has been introduced since the introduction of PGP [47]. In evidence-based trust, performance of entities from previous transactions determines the trust level. Trust level is deduced from past behavior in terms of accuracy and honesty [48].

A summary of the trust approaches is presented in Table 3 with respective advantages and disadvantages.

4 Related Works on Blockchain-Based SSIDM Trust Models

This new approach to manage identity has many opportunities going forward. Initiatives to research and explore the possibilities of this technology come from individuals and companies as well as governments. There are several trust models that have been introduced by various researchers and organizations incorporating their best parameters and efficiency. Limitation and summary of related works are exhibited in Tables 4 and 5.

Table 4 Limitation of existing trust models

Solution	Trust model	Limitations
Sovrin [8]	Trustees and trust anchors play a role in building the Sovrin web of trust. This framework uses delegation of trust from pre-defined trust anchor	Not efficient because every new node in the network will add to the existing long chain. It is costly to maintain the trust chain, and a mesh of cross-certifying nodes does not scale well
uPort [51]	Trust management platform where enterprises can assign trust rating for digital identities using the tools that come with the product suite	Static, top-down approach of trust assignment, which is assumed, not earned
Evernym [53]	Operating as a trust management platform with a verifiable credential trust triangle between issuer, holder, and verifier	The trust control is static, but trust level can change dynamically according to actions and behaviors of entities. Entities should not be trusted by default
Jolocom [54]	Static trust management platform	No mechanism to compute trust in a decentralized manner
Quantifiable trust model [56]	Aggregated trust into attestation issuers. Uses calculated numerical trust metric instead of dedicated evaluation of a trusted third party	Security assumption of the trust model is based on preconfigured trust of identities
WiP [57]	Dynamic trust control which does not require entities' preconfigured trust relationships. Trustworthiness is computed based on their behavior over time	The proposed credibility value is a preset range which lacks tests and experiments to ensure its accuracy and usability in the environment
SCPKI [58]	Gradually builds a web of trust where users vouch for each other's identity attributes	It does not provide the trustworthiness of verifiable claims. Cost incurred to process transactions on Ethereum blockchain. Actions may be delayed by transaction processing time
Centralized trust registry [59]	Decentralized exchange of data but a centralized issuance of trustworthiness by having a trust registry	Trust management is centralized, hence inheriting all problems of a centralized trust model

Table 5 Summary of blockchain-based SSIDM and trust models

Solution	Description	Project type	Blockchain	Network	Trust flow	Trust control	Trust approach
Sovrin [8]	Decentralized global public utility for self-sovereign identity	Nonprofitfoundation	Hyperledger Indy	Public permissioned	Decentralized	Static	Policy based, reputation based
uPort [51]	Ethereum-based identity management platform	Company	Ethereum	Public/private	Decentralized	Static	Policy based, reputation based
Evernym [53]	Identity and trust management platform	Company	Hyperledger Indy	Public/private	Decentralized	Static	Policy based, reputation based
Jolocom [54]	Open-source protocols for decentralized identity management	Open source	Generic	Public/private	Decentralized	Static	Policy based, reputation based
Quantifiable trust model [56]	Qualitative assurance levels based on preset intervals	Academic	Generic	Not specified	Distributed	Dynamic	Evidence based
WiP [57]	Trustworthiness of entities and verifiable claims	Academic	Generic	Not specified	Distributed	Dynamic	Evidence based

SCPKI [58]	A smart contract-based PKI and identity system	Academic	Ethereum	Public/private	Decentralized	Static	Policy based
Centralized trust registry [59]	Blockchain-based SSIDM with decentralized exchange of data and centralized trust registry	Academic	Ethereum	Not specified	Centralized	Dynamic	Policy based

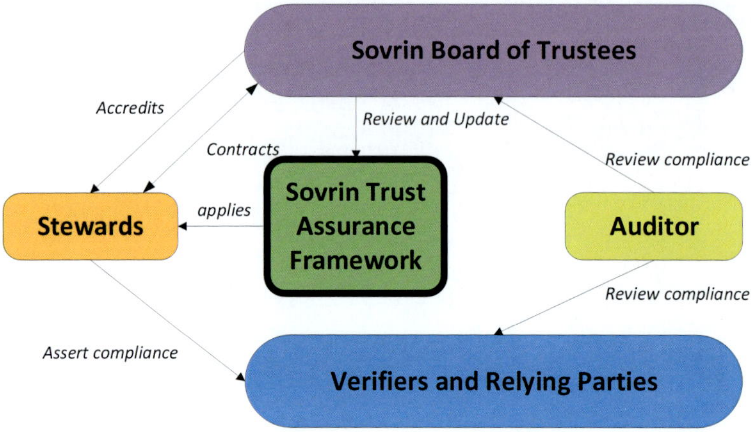

Fig. 14 Sovrin trust assurance framework

A. Sovrin

Sovrin [8] is a private, global, nonprofit foundation to govern self-sovereign identity network. It is the first of its kind trust framework for advocating self-sovereign identity. The foundation believes in portable identity which allows general users to perform verification and authentication of identity, while preserving personal information. The foundation proposed the idea of having identity claim, credentials to replace the use of physical documents. Identity data includes social security number, name, address, education, employment data, etc.

The Sovrin protocol is built on public permissioned blockchain using open standards and the open-source Hyperledger Indy project. All Sovrin identifiers and public keys are pseudonymous by default. Sovrin uses pairwise-pseudonymous identifiers, a separate decentralized identity (DID) for every relationship.

Sovrin network comprises an identity network and a trust network. The trust network executes a proprietary trust framework (Fig. 14). Identity owners can use their Sovrin identities to establish a basic level of trust [49]. Trustees and trust anchors play a role in building the Sovrin web of trust. The web of trust mechanism is not the most efficient because every new node in the network will add to the existing long chain. This makes it very costly to maintain the trust chain, and a mesh of cross-certifying nodes does not scale well [50].

Sovrin's board of trustees are also required to accredit stewards which later apply the trust assurance framework. Stewards also assert compliance to other verifiers and relying parties. All transactions will be reviewed for compliance by the auditor.

Another problem with the Sovrin web of trust is the use of delegated trust similar to PGP 5.0. This concept involves the delegation of trust from a pre-defined trust anchor. Delegated trust is hierarchical and centralized, hence inheriting all problems of a centralized trust model.

B. uPort

uPort [51] aims to be an open self-sovereign identity system that operates on the Ethereum blockchain. uPort enables users to handle their identity and credential in a secure manner like every other SSI project. It provides portability of identity and credential data to other blockchain network such as Bitcoin. uPort utilizes two protocols, namely, the identity and claim protocols. The Identity Protocol is an address on a decentralized network, controlled by a private signing key, and makes use of a decentralized public key infrastructure (PKI) that enables signature validation. On the other hand, the claim protocol refers to a standard message format that enables source attribution and facilitates interoperability between various blockchain and identity networks. The claim protocol supports the JSON Web Token (JWT) and Ethereum transactions. Among products and tools offered by uPort is the self-sovereign wallet. Being unmanaged and fully self-sovereign, there is no entity identity proofing of user accounts in uPort [52].

uPort also offers uPort Serto, a product suite for organizations to set up identity ecosystems. The Serto product suite includes a mobile wallet, a credential management platform, a privacy preserving graph data, and a credential discovery platform. uPort Serto took the approach of mapping verifiable credentials and decentralized identifiers (DIDs) into existing ecosystems based on local law, international agreements, and even internal business rules which gives them the advantage of fulfilling data compliance such as General Data Protection Regulation (GDPR).

uPort itself is a trust management platform; therefore, the trust control is static and archaic. Enterprises can assign trust rating for digital identities using the tools that come with the product suite.

C. Evernym

Evernym [53] is another blockchain-based SSIDM built on Hyperledger Indy [38]. This project introduces a concept called "Trust over IP" (ToIP). This is an architecture that can establish trust between peers over the network. This solution ensures interoperability with Hyperledger Aries [40] and open standards such as W3C DIDs [26] and W3C verifiable credentials [24]. Like uPort, Evernym operates as a trust management platform with a verifiable credential trust triangle between issuer, holder, and verifier. The trust control is also static, using policy- and reputation-based approaches.

D. Jolocom

Jolocom [54] is an open-source project to provide sets of protocols for building a dynamic self-sovereign identity ecosystem. The entire stacks are based on open standards such as W3C DIDs and verifiable credentials. Jolocom also provides a smart wallet for users to create and manage identities in a visual and user-friendly manner. These sets of protocols are compatible with any public permissioned, public permissionless, or private blockchain network. The project aims to realize a truly decentralized and modern digital identity management. Jolocom is playing the role of trust management platform; therefore, trust control is static. There is no mechanism to compute trust in a decentralized manner based on the exchange of verifiable claims for associated identities in the ecosystem.

E. Quantifiable Trust Model

Grüner et al. [55] analyzed decentralized IDM trust requirements based on blockchain. Their paper presented a comparison study of trust requirements for traditional IDM and decentralized IDM through defining topology patterns. The topology pattern reflects the relevant entities and their interaction paths. Trust requirements for isolated, centralized, federated, and decentralized IDM were formally defined, compared, and presented. The authors concluded that the benefit of decentralized trust model is reduced reliance of trust toward the identity and attribute providers.

The authors also proposed the concept that replaces trust with a central identity provider by aggregated trust into attestation issuers [56]. The calculated numerical trust metric serves as an independent basis for the definition of assurance level to simplify and automate reasoning about trust by service providers without requiring a dedicated evaluation of a trusted third party. However, the security assumption of the trust model is based on preconfigured trust of identities.

F. WiP

Bendiab et al. presented a blockchain-based decentralized model [57] to provide authentication and trust computation. This trust model does not require entities' preconfigured trust relationships, but trustworthiness is computed based on their behavior over time. The behavior data can be captured from the transactions stored in the blockchain. The authors proposed a much-desired dynamic trust control. Nevertheless, the proposed credibility value is a preset range which lacks test and experiments to ensure its accuracy and usability in the environment.

G. SCPKI

Al-Bassam et al. proposed a smart contract-based PKI (SCPKI) [58], an alternative PKI approach that uses smart contracts to build a decentralized web of trust adopted from the Pretty Good Privacy (PGP) [47] system. It addresses the issue of rogue certificates issued by certificate authorities in traditional public key infrastructures. The smart contract allows users to add, sign, and revoke attributes. This gradually builds a web of trust where users vouch for each other's identity attributes, but it does not provide the trustworthiness of verifiable claims. Due to the implementation of smart contracts on the Ethereum platform, charges are incurred for identity transactions as a result of the cost of paying the blockchain miners to process a transaction. Lastly, actions may be delayed by transaction processing time.

H. Centralized Trust Registry

Baars et al. [59] claim that reliability of an identity is only as good as the authority issuing that identity so a system should not be dependent on a trusted third party. Although there are many cases where community-based reputation systems (distributed reputation-based approach) can be useful, most business transactions are required to trace back a chain of responsibility in case things go wrong. The system should also allow acquirers to determine the validity of a claim. The project proposed a decentralized exchange of data but a centralized issuance of trustworthiness by having a trust registry. This way the SSIDM is independent from the systems of the issuer and allows availability of claims even when the issuer itself stops its services.

5 Research Directions

Existing trust models in SSIDM still very much rely on the web of trust, as well as governance and trust frameworks in a centralized manner. There is a need for research in this area to improve trust models of a decentralized nature. More use cases and prototypes are also needed to evaluate their accuracy and usability.

A trust engine automates the computation of trustworthiness of digital identities and verifiable credentials. In recent years, machine learning and deep learning have proven to be remarkably good at solving complex problems such as computer vision, big data, and natural language processing. Machine learning also plays an important role in establishing and measuring trustworthiness [60]. By investigating useful features that are capable of distinguishing successful transactions from unsuccessful ones, sophisticated machine learning algorithms can be applied to analyze past transactions. If these algorithms manage to model efficiently what a successful or unsuccessful transaction is, they can be used to predict the trustworthiness of a potential transaction [61].

Trustworthiness of SSIDM stakeholders can be facilitated by computational trust models, and the accuracy of trust rating can be effectively improved. There are a variety of attributes and multitudes of characteristics to support the computation of trustworthiness in SSIDM, for instance, the transaction history in account provisioning, revocation, and recovery; the number of verifiable claim exchange, claims, or counterclaims issued; and the number of correct or incorrect attested claims.

These are data that are globally readable on the ledger. The immutable data on the blockchain can be trusted by all stakeholders. Therefore, instead of having a centralized certificate authority, the data on ledger can provide a richer set of parameters that could be explored to determine trust rating in a dynamic manner.

Additionally, trust and reputation from other layers such as DIDs and digital wallet in the ecosystem can be considered. Trust rating from blockchain and DLT consensus and peer-to-peer communication layer can also be incorporated to achieve a comprehensive trust framework (Fig. 15).

The SSIDM architecture presented in Sect. 2 is still constantly evolving; therefore, it is difficult to ensure the interoperability of trust model with different

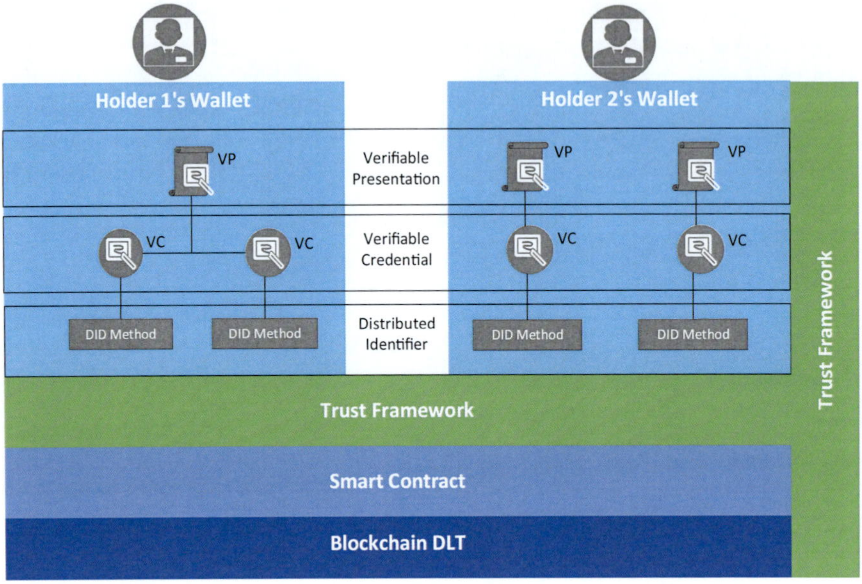

Fig. 15 Trust framework for blockchain-based SSIDM

ledgers. There is a challenge to ensure that the trust framework comprising of trust model and trust engine is working as desired in a variety of SSIDM platforms.

6 Conclusion

In this paper, we presented a comprehensive review of the architecture, components, trust management, and approaches for blockchain-based SSIDM. Despite blockchain being an effective technology for self-sovereign identity management, it does not comprise an effective trust framework. As with any other IDM solutions, blockchain-based SSIDM requires a unique model to ensure trustworthiness of entities in the ecosystem.

Every trust management approach in the literature has its own strengths and weaknesses. Existing solutions are lacking in certain ways especially the trust computation in digital identities and verifiable claims. We believe, with the introduction of a dynamic computation of trustworthiness, this open issue can be addressed and subsequently can break the adoption barrier of blockchain-based SSIDM.

Acknowledgments This work was supported in part by the Fundamental Research Grant Scheme, Ministry of Higher Education Malaysia (FRGS/1/2018/ICT04/UNITAR/03/1).

References

1. Y. Liu et al., Blockchain-based identity management systems: A review. J. Netw. Comput. Appl. **166**, 102731 (2020)
2. S.Y. Lim, M.L.M.K, T.F. Ang, Security issues and future challenges of cloud service authentication. Acta Polytechnica Hungarica **14**(2), 69–89 (2017)
3. M. Swan, *Blockchain: Blueprint for a New Economy* (O'Reilly Media, Inc, 2015)
4. D.R. Andrew Tobin, The Inevitable Rise of Self-Sovereign Identity. 2017
5. S.Y. Lim et al., Blockchain technology the identity management and authentication service disruptor: A survey. Int. J. Adv. Sci. Eng. Inf. Technol **8**(4–2), 1735–1745 (2018)
6. P.D.F. Aaron Wright, *Decentralized Blockchain Technology and the Rise of Lex Cryptographia* (2015)
7. M. Schäffner, Analysis and evaluation of blockchain-based self-sovereign identity systems. Master's thesis (2019)
8. D. Reed, J. Law, D. Hardman, The technical foundations of Sovrin. The Technical Foundations of Sovrin (2016)
9. L. Lesavre, *A Taxonomic Approach to Understanding Emerging Blockchain Identity Management Systems* (2020)
10. A. Jøsang, R. Ismail, C. Boyd, A survey of trust and reputation systems for online service provision. Decis. Support. Syst. **43**(2), 618–644 (2007)
11. J.-H. Cho, K. Chan, S. Adali, A survey on trust modeling. ACM Computing Surveys (CSUR) **48**(2), 1–40 (2015)
12. W. Dou, *The research on trust-aware P2P topologies and constructing technologies [Ph. D. Thesis]* (National University of Defense Technology, Changsha, 2003)
13. E. Gilman, D. Barth, *Zero Trust Networks* (O'Reilly Media, Incorporated, 2017)
14. J.M. De Valmaseda, G. Ionescu, M. Deriaz, TrustPos model: Trusting in mobile users' location, in *International Conference on Mobile Web and Information Systems*, (Springer, 2013)
15. Z. Yan, P. Zhang, A.V. Vasilakos, A survey on trust management for Internet of Things. J. Netw. Comput. Appl. **42**, 120–134 (2014)
16. H. Yu et al., A survey of trust and reputation management systems in wireless communications. Proc. IEEE **98**(10), 1755–1772 (2010)
17. W. Sherchan, S. Nepal, C. Paris, A survey of trust in social networks. ACM Comput. Surveys (CSUR) **45**(4), 1–33 (2013)
18. J.-H. Cho, A. Swami, R. Chen, A survey on trust management for mobile ad hoc networks. IEEE Commun. Surveys Tutorials **13**(4), 562–583 (2010)
19. T. Grandison, M. Sloman, A survey of trust in internet applications. IEEE Commun Surveys Tutorials **3**(4), 2–16 (2000)
20. A.B. Filho et al., A study on trust models in cloud computing, in *2019 IEEE/ACS 16th International Conference on Computer Systems and Applications (AICCSA)*, (2019)
21. A. Albuali, T. Mengistu, D. Che, ZTIMM: A zero-trust-based identity management model for volunteer cloud computing, in *International Conference on Cloud Computing*, (Springer, 2020)
22. K. Bendiab et al., A novel Blockchain-based trust model for cloud identity management. arXiv preprint arXiv:1903.04767 (2019)
23. A. Mühle et al., A survey on essential components of a self-sovereign identity. Comput. Sci. Rev **30**, 80–86 (2018)
24. D.L. Manu Sporny, D. Chadwick, *Verifiable Credentials Data Model 1.0*. 2019.; Available from: https://www.w3.org/TR/vc-data-model/
25. L. Lesavre et al., A taxonomic approach to understanding emerging blockchain identity management systems. arXiv preprint arXiv:1908.00929 (2020)
26. M.S. Drummond Reed, D. Longley, C. Allen, R. Grant, M. Sabadello, *Decentralized Identifiers (DIDs) v1.0 Core Architecture, Data Model, and Representations* (W3C: W3.org, 2021)

27. N. Naik, P. Jenkins, Self-sovereign identity specifications: Govern your identity through your digital wallet using blockchain technology, in *2020 8th IEEE International Conference on Mobile Cloud Computing, Services, and Engineering (MobileCloud)*, (IEEE, 2020)
28. X. Li et al., A survey on the security of blockchain systems. Futur. Gener. Comput. Syst. **107**, 841–853 (2020)
29. T.W. Brent Zundel, M. Varley, M. Csernai, *Peer DID Method Specifcation Report* (Rebooting the Web of Trust VIII, 2019)
30. K.K. Daniel Buchner, *DID Credential Manifest* (GitHub, 2019)
31. T.F.C. Jeff Bohrer, S. Gance, M. Gylling, V. Haag, A. Hripak, N. Otto, J. Pitcher, A. Reis, J. Schmidt, Open Badges 2.0 Implementation Guide IMS Final Release (2018)
32. H. Fabric, *Smart Contracts and Chaincode* (2020); Available from: https://hyperledger-fabric.readthedocs.io/en/release-2.2/smartcontract/smartcontract.html.
33. V. Acharya, A.E. Yerrapati, N. Prakash, *Oracle Blockchain Quick Start Guide: A Practical Approach to Implementing Blockchain in your Enterprise* (Packt Publishing Ltd, 2019)
34. H. Howard, *ARC: Analysis of Raft Consensus* (University of Cambridge, Computer Laboratory, 2014)
35. HyperledgerIndy. *Indy's Plenum Documentation*. 2018; Available from:https://hyperledger-indy.readthedocs.io/projects/plenum/en/latest/main.html.
36. T. Neudecker, H. Hartenstein, Network layer aspects of permissionless blockchains. IEEE Commun. Surveys Tutorials **21**(1), 838–857 (2018)
37. S.S. Gupta, *Blockchain*. IBM Onlone (http://www.ibm.com), 2017
38. LinuxFoundation. *Hyperledger Indy* 2020; Available from: https://github.com/hyperledger/indy-node#about-indy-node
39. D.K. Michael Lodder, *Anonymous credentials 2.0*. 2019.; Available from: https://wiki.hyperledger.org/download/attachments/6426712/Anoncreds2.1.pdf?version=1&modificationDate=1551851745000&api=v2
40. LinuxFoundation. *Hyperledger Aries* 2020; Available from: https://github.com/hyperledger/aries
41. LinuxFoundation. *Hyperledger Ursa* 2020; Available from: https://github.com/hyperledger/ursa
42. Androulaki, E., et al., *Hyperledger fabric: a distributed operating system for permissioned blockchains*, in *Proceedings of the Thirteenth EuroSys Conference*. 2018, Association for Computing Machinery: Porto, Portugal. p. Article 30
43. LinuxFoundation, *Hyperledger Fabric* 2020; Available from: https://github.com/hyperledger/fabric#releases
44. R.C. Mayer, J.H. Davis, F.D. Schoorman, An integrative model of organizational trust. Acad. Manag. Rev. **20**(3), 709–734 (1995)
45. P.S. Challagidad, V. Reshmi, M.N. Birje, *Reputation Based Trust Model in Cloud Computing* (2017)
46. J. Kindervag, *No More Chewy Centers: The Zero-Trust Model of Information Security* (Forrester Research, Inc., dated Mar, 2016) 23
47. S. Garfinkel, *PGP: Pretty Good Privacy* (O'Reilly Media, Inc, 1995)
48. A. Selvaraj, S. Sundararajan, Evidence-based trust evaluation system for cloud services using fuzzy logic. Int. J. Fuzzy Syst **19**(2), 329–337 (2017)
49. Sovrin, Sovrin Provisional Trust Framework (2017)
50. D. Weller, R. Dijksman, *Blockchain's Relationship with Sovrin for Digital Self-Sovereign Identities* (2019)
51. R.H. Christian Lundkvist, J. Torstensson, Z. Mitton, M. Sena, *UPORT: A Platform for Self-Sovereign Identity* (2016)
52. P. Mell, J. Dray, J. Shook, Smart contract federated identity management without third party authentication services. arXiv preprint arXiv:1906.11057 (2019)
53. C. Grinyer, Designing blockchain based services, in *Tensions, Paradoxes+ Plurality: Proceedings of the ServDes. 2020 Conference*, (Linköping University Electronic Press, 2020)

54. J.L. Charleen Fei, E. Rusu, K. Szawan, K. Wagner, N. Wittenberg, *Jolocom: Decentralization By Design* (2018)
55. A. Grüner et al., *A Comparative Analysis of Trust Requirements in Decentralized Identity Management* (Springer International Publishing, Cham, 2020)
56. A. Grüner et al., A quantifiable trust model for blockchain-based identity management, in *2018 IEEE International Conference on Internet of Things (iThings) and IEEE Green Computing and Communications (GreenCom) and IEEE Cyber, Physical and Social Computing (CPSCom) and IEEE Smart Data (SmartData)*, (IEEE, 2018)
57. K. Bendiab et al., WiP: A novel blockchain-based trust model for cloud identity management, in *2018 IEEE 16th Intl Conf on Dependable, Autonomic and Secure Computing*, (IEEE, 2018)
58. M. Al-Bassam, *SCPKI: A Smart Contract-Based PKI and Identity System* (2017), pp. 35–40
59. D. Baars, Towards self-sovereign identity using blockchain technology (2016)
60. X. Liu, A. Datta, E.-P. Lim, *Computational Trust Models and Machine Learning* (CRC Press, 2014)
61. H. Jiang et al., To trust or not to trust a classifier, in *NeurIPS*, (2018)

Shu Yun Lim is currently a senior lecturer and Ph.D. candidate at the Faculty of Business and Technology, UNITAR International University, Malaysia. She obtained a bachelor's degree in information technology majoring in information system engineering from Multimedia University, Malaysia, in 2005 and a master of engineering in ubiquitous network engineering from Dongseo University, South Korea, in 2007. From 2007 to 2010, she was a researcher in British Telecom Malaysian Research Centre, Malaysia. She has published extensively in the area of information security and network security. Currently, her research interests include forensics analysis of cloud services, authentication, and identity management solutions, in particular self-sovereign identity management on blockchain DLT.

Omar Bin Musa is currently an associate professor at the Faculty of Business and Technology, UNITAR International University. He graduated with a bachelor's degree in electronics engineering from SUNY College at Buffalo, NY. He completed an MBA in operations management at Ohio University, Athens, Ohio, in 1988. He started his academic career in 1990 at the Faculty of Economics and Management Sciences, International Islamic University Malaysia (IIUM). He then worked in the IT industry from 1997 to 2004 and returned to academia in 2005. His current research interests are in strategic IT management, business intelligence and analytics, and Blockchain Technologies and Applications. AP Omar is a recipient of several ongoing research and industry grants including the Fundamental Research Grant Scheme from the Ministry of Higher Education, Malaysia.

Bander Ali Saleh Al-Rimy is a senior lecturer at Universiti Teknologi Malaysia (UTM), Johor, Malaysia. He received the B.Sc. degree in computer engineering from the Faculty of Engineering, Sana'a University, Yemen, in 2003; the M.Sc. degree in information technology from OUM, Malaysia, in 2013; and the Ph.D. degree in computer science (information security) from the Faculty of Engineering, Universiti Teknologi Malaysia (UTM), in 2019. His research interests include but are not limited to malware, IDS, network security, and routing technologies. Dr. Al-Rimy has been a recipient of several academic awards and recognitions including but not limited to the UTM Alumni Award, the UTM Best Postgraduate Student Award, the UTM Merit Award, the UTM Excellence Award, the OUM Distinction Award, and the Best Research Paper Award.

Abdullah Almasri is currently an assistant professor at the School of Mathematical and Computer Sciences, Heriot-Watt University Malaysia Campus. He graduated from Universiti Kebangsaan Malaysia (UKM), in 2015. He got his master of computer applications (MCA) from Jamia Hamdard University, New Delhi, India, in 2006 and his diploma of postgraduate in informatics and B.Sc. (mathematics, informatics section) from Al-Baath University, Syria, in 2003 and 2002, respectively. He worked at the Faculty of Business and Technology (UNITAR International University) from 2015 to 2020. His research is focused on artificial intelligence, machine learning, and deep learning and, most recently, blockchain technology.

Blockchain-Enabled Trust Management for Digital Marketing in the Industry 4.0 Era

Fazla Rabby, Ranga Chimhundu, and Rumman Hassan

1 Introduction

An effective marketing campaign is a key factor in the success of any business. Marketing is the only way to expand businesses and attract, engage, and retain consumers by concentrating on different marketing methods and channels [10]. In recent years, online platforms, social media (such as Facebook, Google, and Twitter), and digital technology have emerged as the primary methods for engaging with consumers [25]. Digital marketing is the practice of offering services and products to people via the Internet and digital channels [25]. In recent years, the marketing sector has seen a change, leading to the complete digitalization of marketing activities [13]. Smart technologies such as blockchain, big data, the Internet of things (IoT), augmented reality, and virtual reality are some of the new technologies driven by the expansion of industry 4.0 [45]. Industry 4.0 will be fully automated, enabling superior customer service speed and efficiency and secure data processing and creating a safe and secure cyber environment [53].

Digital marketing has been called a pillar of the "industry 4.0" revolution. It is becoming increasingly popular because it is a great monitoring tool for tracking real-time references to a business product [25]. However, digital marketing has not been possible to attain its full potential because of the increased number of security and privacy breaches, such as identity theft and financial fraud [27]. Businesses must be mindful of the security and privacy risks for their operations and their consumers [16]. The detection and blocking of risks and fundamental awareness are all essential security measures [27]. In the digital marketing approach and Web

F. Rabby (✉) · R. Chimhundu · R. Hassan
School of Business, Faculty of Business, Education, Law & Arts, University of Southern Queensland, Toowoomba, QLD, Australia
e-mail: Fazla.Rabby@usq.ebu.au; Ranga.Chimhundu@usq.edu.au; Rumman.Hassam@usq.edu.au

Y. Maleh et al. (eds.), *Advances in Blockchain Technology for Cyber Physical Systems*, Internet of Things, https://doi.org/10.1007/978-3-030-93646-4_14

presence, security awareness and an appropriate level of skepticism should be the first instruments that marketers employ [28].

Many clients are afraid to conduct online transactions, citing a lack of confidence or concerns about the security of their personal information as the reason [5]. Clearly, digital marketing transactions demand disclosure to the seller of a considerable amount of sensitive personal information, putting consumer identity and financial security at substantial risk [5]. For digital marketing to thrive into the future, it is critical to consider and, more importantly, properly define consumer trust [7]. To reach millions of consumers and generate reliable data, modern Internet technologies such as the blockchain, Internet of things (IoT), artificial intelligence (AI), and big data analytics have largely replaced traditional digital marketing [29]. Privacy, security, and trust are three of the most pressing issues facing digital marketing today [11]. Blockchain technology will have the capacity and potential to disrupt many parts of the digital marketing industry [4]. The research will examine the effectiveness of digital marketing in the integration of industry 4.0 technologies. Today, big companies have a significant role in setting digital marketing conditions, resulting in privacy concerns [31]. Furthermore, there is room for improvement in the accuracy of current digital advertising and targeting strategies. A potential solution to both of these difficulties is incorporating blockchain technology into digital marketing campaigns [54].

2 Research Methodology and Hypotheses

This study identified how blockchain could strengthen digital marketing security, increasing consumer trust and fostering a safe cyberspace environment. The application and execution of blockchain technology are heavily reliant on the goal for which the technology is being used. In this study, an extensive literature review involves a deep look at the analysis used, the problem, and the paper's objectives. In this systematic review of the literature, we used search terms such as "blockchain technology," "industry 4.0," "digital marketing," "cybersecurity," "privacy concern," and "trust" in digital marketing. An extensive list of scholarly peer-reviewed articles was collected and analyzed during the study using scientific databases, such as Scopus, ScienceDirect, Google Scholar, IEEE Xplore, and the University of Southern Queensland library database. Articles were analyzed to address the following hypotheses:

(a) Blockchain technology has the potential to improve privacy protection while also increasing digital marketing security.
(b) Blockchain technology can support the prevention of fraud and the strengthening of consumer trust.

3 Overview of Digital Marketing and Blockchain Technology

The use of digital marketing has also become increasingly popular as a marketing tool for the establishment of key business processes and for influencing consumers' buying patterns and purchasing decisions [25]. While digital marketing provides an opportunity to spread information about the products and services offered by businesses, it also gives a platform for interactions and consideration of consumer expectations or viewpoints [25]. Companies can shift from dialogue to trialogue in their relationships with consumers, in which consumers engage in important relationships with one another and with the businesses [10]. Blockchain technology is an important component in the new industry 4.0 revolution since it offers capabilities foundation that supports various industry applications [13]. The influence of industry 4.0 and blockchain on digital marketing strategy will bring a range of new opportunities for online presence. In addition to increasing operational efficiencies through smart digital marketing strategies, industry 4.0 will create new potential for growth through innovation and tailored solutions that will boost consumer value [13].

A new phenomenon, industry 4.0, has emerged to transform the characteristics of the industrial revolution [45]. Industry 4.0 is concerned with using new technologies to integrate things, individuals, and machines in the organization to develop a new type of networked value chain [45]. In the context of digital marketing, industry 4.0 involves an improved degree of faith and privacy and minimizes risk [21]. E-commerce security refers to components that impact digital marketing, such as computer security, data security, integrity, availability, and other aspects of the information security framework that are more broadly applicable [31]. According to Attaran and Attaran [9], digital marketing is experiencing consistent growth (approximately 19 % per year), and it is a comfortable method of purchasing for many buyers. Trust is essential for successful commerce, and consumers are reluctant to make purchases unless they have confidence in the seller [39].

Blockchain is a digital ledger technology (DLT) that establishes a blockchain that nobody can access [1]. It contains block value, hash, time-stamping, cryptography, consensus algorithm, and peer-to-peer networks that provide decentralized, transparent, and cybersecured services without a trusted intermediary [1]. The model of cryptography was combined with blockchain and other technologies to create modern cryptocurrency by Nakamoto in 2008 [41]. The launch of cryptocurrency "Bitcoin" in 2009 attracted the world to blockchain technology. Dramatic developments and advancements in blockchain technology enable blockchain-based applications across financial sector, healthcare, e-signature and document management systems, digital bonds, remittance, crowdfunding, smart contracts, IoT, and security services [41].

A distributed ledger is a type of distributed ledger technology (DLT) used across multiple locations and people, eliminating the need for a single authority to control manipulation. Distributed ledgers are intrinsically tougher robust as all of the distributed copies need to be attacked simultaneously for an attack to be successful

[3]. Blockchain is a distributed ledger that stores sets of information in blocks, and it is one of the most widely used today [1]. Blockchain is an emerging digital marketing technology that positively influences marketers and shifts the consumer-company relationship to new heights [25]. Blockchain has brought new ways of communicating with clients, and substantial changes to the marketing mix and marketing programs are on the horizon for businesses [54]. Specifically, according to Hariguna et al. [26], a design that integrates these qualities supports higher levels of security. A system that does not rely on intermediaries fosters confidence while also lowering transaction costs. Due to privacy and security concerns over digital marketing, blockchain technology can aid in the development of trust in the digital marketing industry, where transparency and privacy concerns are high in customers' eyes [46]. Additionally, the decentralized ledger helps mitigate risks of a network failure due to malicious attacks.

Blockchain networks record transactions in a shared ledger within the decentralized community without any external authority or entity [4]. Cryptocurrencies are founded on blockchain technology for simplified online digital payment and verification processes [41]. As shown in Fig. 1, there are many key characteristics of the blockchains that are essential for futuristic document verification and management systems. These are listed as follows:

- Each member records the ongoing digital transitions into a shared ledger [4].
- Blockchains provide full transactional history using an append-only ledger [4].
- The values in the ledger are not overridden [4].
- Blockchain allows users to verify data versions making the data tamperproof. Any data that has been recorded cannot be modified after it has been recorded [23].
- The blockchain network provides transparency to all participants the layer of trust [15].
- The distributed nature of blockchains provides a secure network against cyberattacks and intruders [15].
- Blockchains' peer-to-peer architecture has nodes participating in the network, and each node stores an identical copy of the blockchain and is authorized to validate and certify [51].

Blockchain technology in industry 4.0 has evolved rapidly in the last few years. Blockchain-based cryptocurrencies have been accepted for financial transactions worldwide, and this technology has the potential to solve existing and emerging business and marketing problems [41]. Complex algorithms have determined the rights of each participant, and no one has the right to alter the previous transactions. Blockchains are not only integrated into cryptocurrency or economic services, but they also have applications in smart contracts, public services, Internet of things (IoT), supply chain management, healthcare services, security services, and document management services.

The interaction by marketers with consumers has been dramatically reshaped due to the Internet, information technologies, and social networking platforms [10]. The advancement of technology creates a competitive and demanding environment for

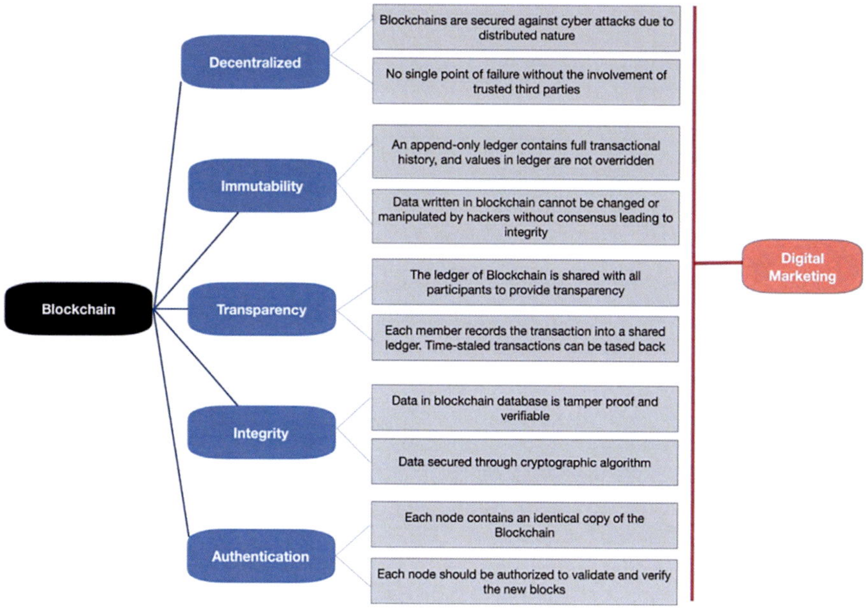

Fig. 1 Characteristics of blockchain network

digital marketers to manage their digital marketing and advertisement campaigns in a digital world [29]. This study aims to examine and understand the constraints of blockchain technology and its influence on digital marketing.

This study reveals that marketers need to consider cautiously how they can be affected when swiftly developing blockchain-enabled digital marketing and how interactions between consumers and marketers are changed. In the next section, the latest trends and transitions in blockchain technology are explained. Section "Blockchain-Enabled Digital Marketing" provides a synopsis of blockchain-enabled digital technology. Blockchain cybersecurity and privacy, trust, and transparency are reviewed in Sections "Cybersecurity and Data Privacy Issues in Digital Marketing" and "Enhancing Trust and Transparency". In Section "Reinforcing Blockchain with Hyperledger Technology", we address how the Hyperledger system in the blockchain contributes to creating a secure environment. Section 4 presents the findings, benefits, and limitations of the study and their implications. Blockchain technology, along with its pros and cons in the context of digital marketing, is presented. Ongoing research and real-world applications of blockchain technology in digital marketing are also explored. The next section investigates how blockchain improves the security and privacy of digital marketing and develops consumers' trust through transparency.

Cybersecurity and Data Privacy Issues in Digital Marketing

Promoting products and services through digital marketing and building client relationships is now possible, and businesses can quickly notify consumers of new products or services. While an Internet presence for businesses is necessary, the marketing and advertising sector has been troubled by scammers [30]. Anyone can learn to use automated scripts, computer programs, or human clickers to impersonate real users and steal advertising budgets [30]. Accessing data or network resources can be difficult for several reasons. Organizational and technical survival requires data protection and high transaction security, and protection is vital for many businesses. Securing digital products and services is critical, because brands need to defend consumer data and prevent data leaks. Data security issues have already hampered digital marketing, and customers are wary about giving up personal information like addresses and credit card numbers [20]. Account takeover, database attacks, and data loss or theft are the most common crimes in digital marketing platforms [30]. False authentication is another digital marketing fraud, and browser cookies and weblog data may endanger privacy [34]. Cybercrime will cost USD 6 trillion globally by 2021 due to security weaknesses [12].

The ultimate goal of a mountain climber is to reach the pinnacle, but safety must always come first. In the same way, digital marketers should prioritize security first to fulfill business objectives [8]. Distributing damaging malware would be devastating if they were to spread [31]—understanding and acknowledging threats are necessary to maintain security awareness [8]. The worst thing that a digital marketing strategy can do is put business websites and related services at risk of being compromised [31]. According to the PricewaterhouseCoopers (PwC 2018) Global Economic Crime and Fraud Survey report, cybercriminals had targeted 45 % of Australian companies in the year between 2017 and 2018. For digital marketers, customer data is the most valuable asset they have [19]. Hackers benefit from this data, in which they exploit by siphoning, selling, and sometimes even compromising user passwords, credit card information, and personal information [19].

Data breaches have compromised millions of consumers' data and resulted in negative publicity, fines, and other legal penalties for targeted companies. Rather than improving the customer experience, many digital initiatives today are designed to cause disruption [47]. Instead of concentrating on what the customer needs, brands concentrate on how to get their messaging safely and securely to as many people as possible [47]. Hackers can take over a digital marketing account and confuse by changing the company profile, adding false and misleading information, and sending spam emails to clients from the company account [16]. Thieves, scammers, and cybercriminals can utilize any digital marketing platform or social media site that a marketer employs to spread their malicious software [27]. Various issues include business information hijacking and theft of customers' personal information [16]. Figure 2 shows how marketing through different channels is affected due to privacy risks. Around 68–82% of consumers don't trust information available on different online platforms.

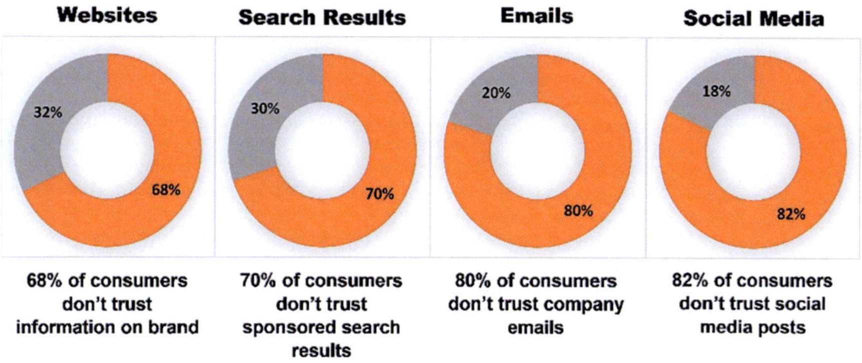

Fig. 2 Consumer trust across various digital marketing platforms

Cybersecurity, transparency, and confidentiality of data are the most important concerns of Internet users [12]. The advancement in Internet technologies and the increasing trend of online shopping put users at the forefront of new challenges like trust, privacy, security, and transparency [49]. Each online transaction on conventional networks leaves behind a digital track of comprehensive information about the consumer, shopping preferences, identity, and shopping habits [49]. A weblog record for tracking consumer online activities and a cookies-based approach can compromise consumers' privacy [49]. Consumers are now aware that website cookies store personal information and track their activity, further increasing consumer concerns around privacy. In the online marketing context, cyberattack on data transactions, unauthorized access of hackers, and illegal use of credit cards are still security threats [34]. Further, the implementation of data mining technologies enables marketing companies to easily identify, track, and collect information from their consumers [49].

These leaks of private information impact consumer trust, leading to the avoidance of online purchasing. Digital marketing heavily relies on big data for advertisement and marketing [32]. The consumer has the right to decide when and how much information related to their data will be shared and disclosed to marketing companies. Only the consumer has the authorization key to share the data with requesting parties [4]. Blockchain enables consumers to protect their identity by pseudonymity, control their data, and protect it from being monetized by third parties [24]. Blockchain technology will help to eliminate the concerns related to data privacy and cybersecurity [2]. In the next section, we will discuss the impacts of blockchain in digital marketing in detail.

Blockchain-Enabled Digital Marketing

Given the potential of blockchains in industry 4.0, a revolution in the digital marketing field may be on its way. Blockchain could allow marketers to use micropayment to encourage consumers to share personal information without involving the intermediary (Facebook or Google) the third party to share information. Advertising fraud is the biggest challenge faced by the digital marketer of current years [52]. Bots can generate fake clicks or fraudulent impressions that simulate the actual clicks. Ad fraud can take many shapes. An estimate of the World Federation of Advertisers suggests that ad fraud could reach more than $50 billion by 2025 if it is not countered by the authorities [52]. A blockchain-based direct link can be established between the advertiser and the publisher using smart contracts to combat these ad fraud activities. In this way, the advertiser knows how the ad is circulated on the ad network and the number and timing of consumers who see the ad. Thus, the landscape of digital marketing can be revolutionized by enabling blockchain technology for digital advertising to attract consumers.

This section will review the impacts of using blockchain technology on digital marketing. Blockchain is a decentralized and distributed ledger that works on cryptographic algorithms and peer-to-peer (P2P)-based networks [51]. The information stored in the blockchain is cybersecure and cannot be tampered with or deleted [3]. Each transaction in the blockchain is time-stamped and shared with each participant/node, where the participant authenticates and validates the transaction. Only one block can be created simultaneously, as the proof of work (PoW) has been calculated by node and verified by other participants. Once a new block has been created in the chain, it cannot be altered or mutable, making the system tamperproof [1]. The nature of blockchain technology makes it much more secure and reliable than centralized platforms. Consumer's data can be tokenized, and buyers can agree to provide their personal information anonymously in a distributed ledger, where firms can "purchase" them without the need for middlemen or centralized systems [55]. Identity applications of blockchain applied to industry 4.0 can assist in validating reviews and increasing their credibility for both businesses and buyers [6]. At the same time, customers can be compensated with tokens for their commitments, creating incentives for generating reliable user content, particularly for influencers in digital marketing [36].

Online marketing is a vital stakeholder of the Internet's economy [7]. Digital marketing can be influenced by blockchain in many ways and also enables companies to perceive and utilize accurate information to deliver their services [46]. The distributed ledger of blockchain appends all transactions conducted on a public platform that makes it cybersecure and makes fraud detection easy [46]. In digital marketing, the adoption of blockchain technology is lagging as compared to other fields [18]. Several aspects like brand communication, transparency of brand to consumers, marketing performance, and design of online marketing campaigns are impacted by adopting blockchain technology. However, the transactions on the blockchain are entirely private and transparent because verification of transactions is done by consensus algorithms and peers in the blockchain network [4]. Contributors and participants of digital marketing networks at various levels must participate in the blockchain network for it to be successful. Furthermore, the authentication process in a private blockchain is performed by the shared/private keys, so therein is a threat to security and authentication if the shared/private key is lost or hacked.

Enhancing Trust and Transparency

In todays' marketing environment, trust issue is the biggest challenge [22]. The immutability of each transaction is a key feature of blockchain technology [49]. Trust and transparency are the most valuable features of blockchain technology because the level of transparency can be adjusted through cryptographic algorithms [44]. It empowers trust and transparency in digital marketing [49]. No single entity has full control over transactions. The decentralized nature of blockchain means that the user can trust the data—a feature that is crucial for some business models and marketing activities [44]. In using blockchain technology in industry 4.0, truth and trust become embedded in businesses, as every interaction with the customer can be traced and will be transparent in blockchain-enabled digital marketing [49]. Dishonest marketing and fraud can be controlled as the entire blockchain system is transparent [22]. It will force digital marketing companies to tell the complete truth and maintain honesty with customers and partners in the business [22]. The malicious marketing of counterfeit products and violation of copyright law and intellectual property (IP) rights can be avoided with the help of blockchain technology, as the technology facilitates the traceability of end-to-end products.

By integrating blockchain technology in financial and marketing services, consumers can gain better access and information about products and services, the visibility of production process is increased, and supply chain activities and delivery route can be tracked [44]. That leads to a transparent interaction with the consumer and supplier/brand, enabling companies to gain consumer trust in the brand [17]. The transparency and verification of each process made by the marketers can boost the confidence of consumers and will help maintain long-term relationships [17]. A customer can sign the smart contract with the marketer/service provider and agree on a specific date and terms. If the marketer does not meet the conditions, the smart contract will automatically refund the customer [17]. Blockchain has improved transparency with reduced tracking time. Walmart cooperated with IBM to trace all steps of products back to the supplier [17]. Similarly, Starbucks collaborated with Microsoft to trace their coffee from bean to cup using Azure blockchain [52]. The next section will explain how Hyperledger techniques enable monitoring, searching, and maintaining blockchain developments and related data.

Reinforcing Blockchain with Hyperledger Technology

Research into blockchain technology enabled its implementation across different areas. Some common protocols of blockchain technology include Bitcoin, consensus network, Ethereum, Corda, and Hyperledger. It is widely used in the software industry because of its potential benefits. Hyperledger is an open-source system for implementing permitted blockchains in a modular and extensible form [50]. The Linux Foundation initiated it in 2015 as an open blockchain platform [50]. The Hyperledger (Fig. 3) deploys the distributed applications without relying on native cryptocurrency, and it does not support Bitcoin or any other cryptocurrency. Still, it works on blockchain technology and tokens via chain code [35]. The Hyperledger is an incubation project under which open-source blockchain applications and tools are developed. The key purpose of the Hyperledger protocol is to increase the reliability and performance of the ledger [50]. To create a collaborative environment for supporting a wide array of components of different uses, the Linux Foundation has made efforts to provide a modular framework. The goal of Linux is to develop an environment in which different developers and companies meet and coordinate to build a blockchain framework [15]. Hyperledger has gone through major upgradation with the collaboration of other technology players like Microsoft's COCO, Enterprise Ethereum Alliance (EAA), and Cisco [40].

Business blockchain solutions that achieve customization, trust, and transparency in company operations were enabled by Hyperledger, which was a secure and decentralized secret platform for enterprise blockchain solutions [15]. Hyperledger provides a decentralized, immutable blockchain framework coupled with chain

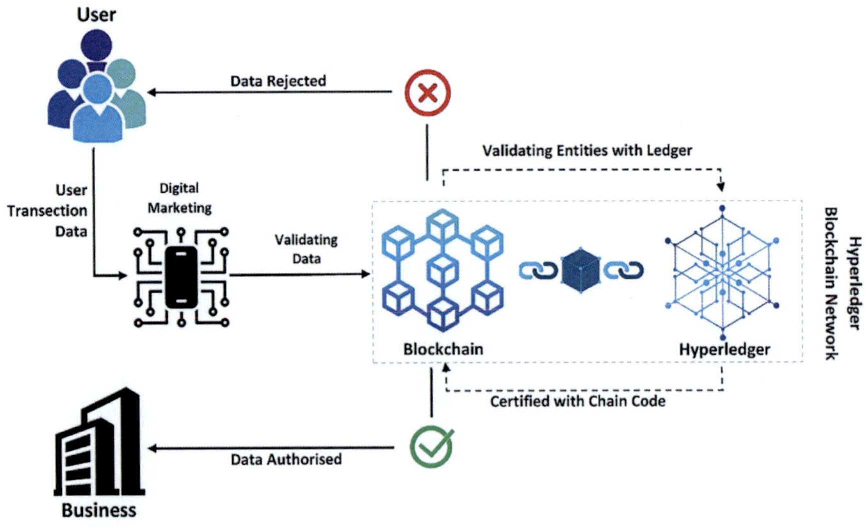

Fig. 3 Hyperledger in blockchain technology

codes (smart contracts) that guarantee data privacy and facilitate cross-industry coordination and engagement [50]. By allowing for multiple transactions to be executed simultaneously, Hyperledger significantly improved data analysis and prediction [38]. Furthermore, in using Hyperledger technology, it became possible to create separate channels for total secrecy, and a personal data feature besides facilitated authorizations within digital marketing platforms [38].

4 Findings and Discussion

Consumers and businesses will benefit from blockchain technology, which will play a critical part in industry 4.0 and simplify digital marketing procedures by creating a safe, trustworthy environment [21]. Blockchain technology uses a cryptographic hashing technique to improve security and reliability [43]. The members can trust in blockchain networks instead of third parties, as the evidence of each transaction is recorded in the hash value of a chain [37]. Blockchain enhances the collection and processing of data and makes data more valuable [44]. Blockchain in industry 4.0 will influence big data and ultimately digital marketing [44]. At the same time, the implementation of blockchain for industry 4.0 will reduce data, as the consumer has control of their data, and marketers are not allowed to access the data without consumer consent [17]. This reduction in data will provide a more holistic understanding of the consumer. The data in the blockchain network will be better than the traditional network as the users are authenticated and verified [17]. In Fig. 4, the application of blockchain in digital marketing and incorporation of big data with blockchain technology have several advantages. The blockchain will increase the accuracy and security of data. In addition, the threat of cyberattacks is eliminated due to the distributed nature of blockchain technology [2]. Data stored in the blockchain is cryptographically signed and immutable, influencing companies to operate and communicate in the context of digital marketing and create customer-relevant content [37]. Blockchain technology in industry 4.0 offers data transparency and data integrity to all participants in the network and allows verified nodes to access and operate the blockchain [44]. When companies do not provide the relevant information, consumers will block companies from processing or accessing their data [18].

Hyperledger provides the essential infrastructure for building diverse blockchain applications for modular, comprehensive, interoperability, and security features [15]. Hyperledger's data are protected from the underlying algorithm due to asymmetric cryptography and zero-knowledge verification; and the digital certificate management service ensures the company's blockchain authenticity [15]. Hyperledger's multichannel design isolates data across channels where private data gathering allows for personal information separation between businesses within the same network [38]. The prominent built-in features of blockchain technology are trust, security, and transparency, which resolve marketing issues in an effective way [23]. The integration of blockchain technology in digital marketing is still in

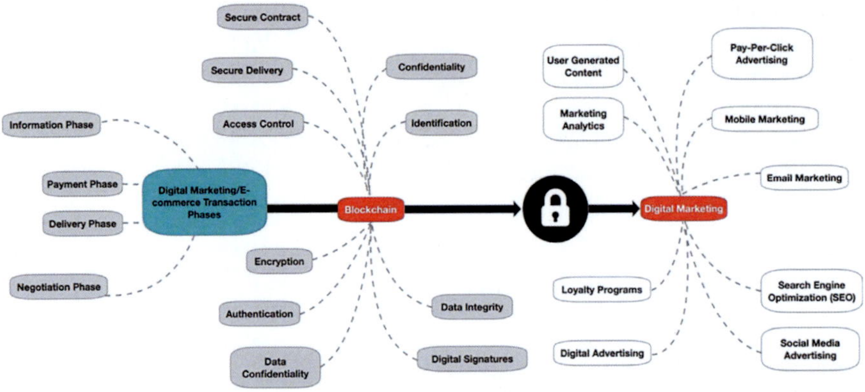

Fig. 4 Application of blockchain in digital marketing

development, but this review of the literature proved that it has enormous potential to improve or even disrupt multiple areas in marketing. In conclusion, blockchain technology ensures the following things:

- A solid technology foundation is required to boost consumer confidence in digital marketing [41].
- Blockchain technology can help corporations and consumers through secure decentralized data storage [43].
- Asymmetric encryption, digital signatures, and access control can protect large amounts of customer data [42].
- The technology can decentralize and self-organize the brand business ecosystem by coordinating and integrating marketing data [33].
- Blockchain technology can preserve Internet shoppers' privacy [35].

Users can trust blockchain systems because transactions are directed to a random network node [36]. Limiting network members' access to block data helps safeguard consumer privacy online, and the network can confirm private transactions [55]. Since blockchain encrypts user credentials (e.g., user IDs, passwords, electronic ID cards), customers have more control over their personal information [35]. Blockchain transaction history can assist consumers in understanding their preferences. Trust is vital in digital marketing, and without trust, consumer faith in business has dwindled [11]. According to Moşteanu and Faccia [41], technical infrastructure greatly influences trust. Digital marketing now facilitates transactions without personal contact, and a business' success hinges on trust and transparency [28]. With this high degree of transparency, marketers can indicate various beneficial characteristics and highlight their altruistic purpose to look out for the consumer's best interests [48]. The use of blockchain technologies for Industry 4.0 promotes a transparent market environment and ensures that the future quality of marketing is traded fairly under guaranteed contracts [48]. Table 1 summarizes the findings of a systematic literature review, which concludes that blockchain technology has

Table 1 Applications of blockchain in digital marketing in the industry 4.0 era

Author	Title	Findings/applications
Abou-Nassar, EM; Iliyasu, AM; El-Kafrawy, PM; Song, OY; Bashir, AK; and Abd El-Latif, AA 2020	DITrust chain: towards blockchain-based trust models for sustainable healthcare IoT systems	Blockchain technology offers a solid foundation to protect and increase the effectiveness of sustainable healthcare IoT systems
Abd El-Latif, AA; Abd-El-Atty, B; Mehmood, I; Muhammad, K; Venegas-Andraca, SE; and Peng, J 2021	Quantum-inspired blockchain-based cybersecurity: securing smart edge utilities in IoT-based smart cities	Blockchain technologies based on quantum-inspired models can protect against message attacks and ensure safe data transmission between IoT devices
Aggarwal, S; Chaudhary, R; Aujla, GS; Kumar, N; Choo, KKR; and Zomaya, AY 2019	Blockchain for smart communities: applications, challenges and opportunities	Blockchain can operate as a shared platform to enhance network and systems integration
Almasoud, AS; Hussain, FK; and Hussain, OK 2020	Smart contracts for blockchain-based reputation systems: a systematic literature review	Blockchain technologies provide a platform in which users may digitally evaluate the trust or confidence of those who provide products or services
Bettiol, M; Capestro, M; and Di Maria, E 2017	Industry 4.0: the strategic role of marketing	A more positive client experience due to the incorporation of industry 4.0 technology
Bezovski, Z; Jovanov, T; and Temjanovski, R 2021	The impact and the potential disruption of the blockchain technology on marketing	The built-in characteristics of blockchain technology, including openness, trust, and security, prove beneficial for tackling key marketing challenges
Bhuvana, R, and Aithal, PS 2020	Blockchain-based service: a case study on IBM blockchain services and hyperledger fabric	Hyperledger is a framework that helps blockchain overcome current technological restrictions surrounding security, authenticity, integrity, efficiency, and accuracy
Brauer, J, and Linnala Eriksson, B 2020	Blockchain's influence on digital marketing: Aan exploratory study examining blockchain in relation to big data and digital marketing	Blockchain is capable of removing mediators and eliminating fraudulent digital marketing activities. Because both big data and digital marketing are heavily data driven, blockchain will influence both

(continued)

Table 1 (continued)

Author	Title	Findings/applications
Clim, A 2019	Cyber security beyond the Industry 4.0 era. A short review on a few technological promises	In preventing cyberattacks and intrusion on a particular network or computer, industry 4.0 measures will be crucial for any business process
Ertemel, AV 2018	Implications of blockchain technology on marketing	Blockchain eliminates the need for trust management mechanisms and also offers complete brand transparency and traceability
Kumar, V; Tripathi, AK; Chandra, N; and Goyal, AK 2020	Blockchain-enabled a transparent and secure framework using smart contract for online advertisements	With blockchain, advertisers, publishers, and end users all have access to greater transparency
Lee, JY 2019	A decentralized token economy: how blockchain and cryptocurrency can revolutionize business	The future could see the emergence of a token economy based on new business models using blockchain technology
Moin, S; Karim, A; Safdar, Z; Safdar, K; Ahmed, E; and Imran, M 2019	Securing IoTs in distributed blockchain: analysis, requirements and open issues	IoT data reliability is lacking due to the absence of data manipulation when data is shared. The emergence of new secure decentralized storage technology, like blockchain, would address these limitations of IoT
Nosalska, K, and Mazurek, G 2019	Marketing principles for Industry 4.0—a conceptual framework	A new approach in industry 4.0 to forming marketing strategies and marketing mix helps identify the major areas according to industry 4.0 concepts
Pal, A; Tiwari, CK; and Haldar, N 2021	Blockchain for business management: Aapplications, challenges and potentials	Blockchain integration can be used to protect financial transactions, decrease errors, facilitate operations, and prevent fraud
Lu, N; Zhang, Y; Shi, W; Kumari, S; and Choo, KKR 2020	A secure and scalable data integrity auditing scheme based on hyperledger fabric	Hyperledger Fabric is a communication platform, and it improves data efficiency when it comes to integrity verification
Rahman, KT 2021	Applications of blockchain technology for digital marketing: a systematic review	Blockchain technology shapes digital marketing by limiting businesses' ability to gather customer data and offering to return the value to customers

(continued)

Table 1 (continued)

Author	Title	Findings/applications
Rejeb, A; Keogh, JG; and Treiblmaier, H 2020	How blockchain technology can benefit marketing: Ssix pending research areas	Blockchain promotes disintermediation, helps fight against online fraud, strengthens confidence and transparency, allows for greater privacy protection, and improves security
Ungerman, O; Dedkova, J; and Gurinova, K 2018	The impact of marketing innovation on the competitiveness of enterprises in the context of industry 4.0	Increased business competitiveness is seen as the most significant impact of innovative marketing in the context of industry 4.0

potential applications in digital marketing. The review also identifies significant benefits of implementing blockchain technology in the industry 4.0 era.

5 Conclusions

Industry 4.0 is driven by emerging and foundational technologies such as the IoT, cloud, AI, automation, and blockchain. In using blockchain technology to secure public confidence and manage data, businesses can automate processes and reduce the need for physical work. In the coming years, blockchain technology will become a critical component of industry 4.0. Blockchain technology can increase security, trust, transparency, confidence, and performance in the digital marketing environment, where data is the most important factor in determining the effectiveness of digital marketing efforts. Blockchain technology allows the collection of high-quality data without putting the data's integrity at risk. Digital marketing and advertising systems powered by blockchain technology establish reliable and secure transactions while maintaining confidentiality. The Hyperledger framework supports blockchain in overcoming the technological limitations in security, authenticity, integrity, efficiency, and accuracy. Blockchain technology in industry 4.0 will keep growing with wider applications in business and marketing as it can give a solution for existing and emerging marketing problems. This review found that blockchain in industry 4.0 can change marketing operations through cryptocurrency, supply chain management, and loyalty programs. It is recommended that digital marketing performed on the blockchain network reflects the clarity, security, and access to accurate information. Blockchain improves digital marketing methods and allows consumers to share their information directly with the right seller. The enhanced security, traceability, and transparency of blockchain in industry 4.0 can open a new horizon in business practices for consumers [14]. Advertisement strategies can also be improved with the blockchain encryption mechanism, as the data is protected and privacy rights are maintained, boosting the confidence level of consumers. Despite its time and transparency efficiencies, implementing blockchain and managing a marketing channel remain challenging. Various areas of digital marketing have been explored in this study that can benefit from the implementation of blockchain technology. There is no question that blockchain technology should be incorporated into digital marketing––the benefits are clear. Although blockchain technology has clear benefits and prospects to improve efficiency and save costs, it also has significant challenges and constraints that cannot be ignored [1]. Future research should investigate precisely how blockchain technology can be implemented in digital marketing in a way that mitigates challenges and limitations.

References

1. K. Abbas, L.A.A. Tawalbeh, A. Rafiq, A. Muthanna, I.A. Elgendy, A. El-Latif, A. Ahmed, Convergence of blockchain and IoT for secure transportation systems in smart cities. Secur. Comm. Netw., 1–13 (2021). https://doi.org/10.1155/2021/5597679
2. A.A. Abd El-Latif, B. Abd-El-Atty, I. Mehmood, K. Muhammad, S.E. Venegas-Andraca, J. Peng, Quantum-inspired blockchain-based cybersecurity: Securing smart edge utilities in IoT-based smart cities. Inf. Process. Manag. **58**, 102549 (2021). https://doi.org/10.1016/j.ipm.2021.102549
3. E.M. Abou-Nassar, A.M. Iliyasu, P.M. El-Kafrawy, O.Y. Song, A.K. Bashir, A.A. Abd El-Latif, DITrust chain: Towards blockchain-based trust models for sustainable healthcare IoT systems. IEEE Access **8**, 111223–111238 (2020). https://doi.org/10.1109/ACCESS.2020.2999468
4. S. Aggarwal, R. Chaudhary, G.S. Aujla, N. Kumar, K.K. Choo, A.Y. Zomaya, Blockchain for smart communities: Applications, challenges and opportunities. J. Netw. Comput. Appl. **144**, 13–48 (2019). https://doi.org/10.1016/j.jnca.2019.06.018
5. I.M. Al-Jabri, I.E. Mustafa, A. Abed, The willingness to disclose personal information: Trade-off between privacy concerns and benefit. Inform. Comput. Secur. **28**, 161–181 (2020). https://doi.org/10.1108/ICS-01-2018-0012
6. A.S. Almasoud, F.K. Hussain, O.K. Hussain, Smart contracts for blockchain-based reputation systems: A systematic literature review. J. Netw. Comput. Appl. **170**, 102814 (2020). https://doi.org/10.1016/j.jnca.2020.102814
7. G. Appel, L. Grewal, R. Hadi, A.T. Stephen, The future of social media in marketing. J. Acad. Mark. Sci. **48**, 79–95 (2020). https://doi.org/10.1007/s11747-019-00695-1
8. N.I. Arkhipova, M.T. Gurieva, Internet of things in digital marketing and data security concerns, in *3rd International Conference on Judicial, Administrative and Humanitarian Problems of State Structures and Economic Subjects, Advances in Social Science, Education and Humanities Research (ASSEHR), Moscow*, vol. 252, (2018), pp. 262–265. https://doi.org/10.2991/jahp-18.2018.54
9. M. Attaran, S. Attaran, Digital transformation and economic contributions of 5G networks. Int. J. Enterp. Inf. Syst. **16**, 58–79 (2020). https://doi.org/10.4018/IJEIS.2020100104
10. M. Bala, D. Verma, A critical review of digital marketing. Int. J. Manag. IT Eng. **8**, 321–339 (2018)
11. G. Bansal, F.M. Zahedi, D. Gefen, Do context and personality matter? Trust and privacy concerns in disclosing private information online. Inf. Manag. **53**, 1–21 (2016). https://doi.org/10.1016/j.im.2015.08.001
12. M. Benz, D. Chatterjee, Calculated risk? A cybersecurity evaluation tool for SMEs. Bus. Horiz. **63**, 531–540 (2020). https://doi.org/10.1016/j.bushor.2020.03.010

13. M. Bettiol, M. Capestro, E. Di Maria, Industry 4.0: the strategic role of marketing, in *XIV Convegno Annuale SIM, Dipartimento di Scienze Economiche, Bergamo*, (2017), pp. 26–27
14. Z. Bezovski, T. Jovanov, R. Temjanovski, The impact and the potential disruption of the blockchain technology on marketing. J. Econ. **6**, 13–23 (2021). https://doi.org/10.46763/JOE216.10013b
15. R. Bhuvana, P.S. Aithal, Blockchain based service: A case study on IBM blockchain services & hyperledger fabric. Int. J. Case Stud. Bus. IT Educ. **4**, 94–102 (2020). https://doi.org/10.5281/zenodo.3822411
16. R. Bohnsack, M.M. Liesner, What the hack? A growth hacking taxonomy and practical applications for firms. Bus. Horiz. **62**, 799–818 (2019). https://doi.org/10.1016/j.bushor.2019.09.001
17. A. Boukis, Exploring the implications of blockchain technology for brand–consumer relationships: A future research agenda. J. Prod. Brand Manag. **29**, 307–320 (2019). https://doi.org/10.1108/JPBM-03-2018-1780
18. J. Brauer, B. Linnala Eriksson, *Blockchain's Influence on Digital Marketing: An Exploratory Study Examining Blockchain in Relation to Big Data and Digital Marketing* (Umeå University, Sweden, 2020)
19. C. Campbell, S. Sands, C. Ferraro, H.Y.J. Tsao, A. Mavrommatis, From data to action: How marketers can leverage AI. Bus. Horiz. **63**, 227–243 (2020). https://doi.org/10.1016/j.bushor.2019.12.002
20. H.S. Chen, T.M. Jai, Trust fall: Data breach perceptions from loyalty and non-loyalty customers. Serv. Ind. J. **1-17** (2019). https://doi.org/10.1080/02642069.2019.1603296
21. A. Clim, Cyber security beyond the industry 4.0 era. A short review on a few technological promises. Informatica Economica **23**, 34–44 (2019)
22. H. Conick, What marketers need to know about blockchain. Market. News **51**, 12–14 (2017)
23. M. El Ghazouani, M.A. El Kiram, L. Er-Rajy, Blockchain & multi-agent system: A new promising approach for cloud data integrity auditing with deduplication. Int. J. Commun. Netw. Inform. Secur. **11**, 175–184 (2019)
24. A.V. Ertemel, Implications of blockchain technology on marketing. J. Int Trade Log Law **4**, 35–44 (2018)
25. A. Grubor, O. Jakša, Internet marketing as a business necessity. Interdiscip. Descr. Complex Syst. **16**, 265–274 (2018). https://doi.org/10.7906/indecs.16.2.6
26. T. Hariguna, Y. Durachman, M. Yusup, S. Millah, Blockchain technology transformation in advancing future change. Blockchain Front. Technol. **1**, 13–20 (2021)
27. M. Hussain, Social media-related cybercrimes and techniques for their prevention. Appl. Comput. Syst. **24**, 9–17 (2019). https://doi.org/10.2478/acss-2019-0002
28. A.K. Jain, S.R. Sahoo, J. Kaubiyal, Online social networks security and privacy: Comprehensive review and analysis. Complex Intell. Syst., 1–21 (2021). https://doi.org/10.1007/s40747-021-00409-7
29. P.K. Kannan, Digital marketing: A framework, review and research agenda. Int. J. Res. Mark. **34**, 22–45 (2017). https://doi.org/10.1016/j.ijresmar.2016.11.006
30. N. Kasambu, R. Sritharan, A study on problems and prospects of e-marketing. Stud. Indian Place Names **40**, 3447–3456 (2020)
31. S. Konyeha, Exploring cybersecurity threats in digital marketing. Mark. Theory **2**, 12–20 (2020). https://doi.org/10.37933/nipes/2.3.2020.2
32. V. Kumar, A.K. Tripathi, N. Chandra, A.K. Goyal, Blockchain-enabled a transparent and secure framework using smart contract for online advertisements. J. Crit. Rev. **7**, 1246–1255 (2020). https://doi.org/10.31838/jcr.07.11.211
33. J.Y. Lee, A decentralized token economy: How blockchain and cryptocurrency can revolutionize business. Bus. Horiz. **62**, 773–784 (2019). https://doi.org/10.1016/j.bushor.2019.08.003
34. W.B. Lee, H.B. Chen, S.S. Chang, T.H. Chen, Secure and efficient protection for HTTP cookies with self-verification. Int. J. Commun. Syst. **32**, e3857 (2019). https://doi.org/10.1002/dac.3857

35. S.Y. Lim, P.T. Fotsing, A. Almasri, O. Musa, M.L.M. Kiah, T.F. Ang, R. Ismail, Blockchain technology the identity management and authentication service disruptor: A survey. Int. J. Adv. Sci. Eng. Inform. Technol. **8**, 1735–1745 (2018)
36. A. Lisi, A. De Salve, P. Mori, L. Ricci, S. Fabrizi, Rewarding reviews with tokens: An Ethereum-based approach. Futur. Gener. Comput. Syst. **120**, 36–54 (2021). https://doi.org/10.1016/j.future.2021.02.003
37. S.K. Lo, X. Xu, Y.K. Chiam, Q. Lu, Evaluating suitability of applying blockchain, in *22nd International Conference on Engineering of Complex Computer Systems (ICECCS), IEEE, Fukuoka, Japan*, (2017), pp. 58–161. https://doi.org/10.1109/ICECCS.2017.26
38. N. Lu, Y. Zhang, W. Shi, S. Kumari, K.K.R. Choo, A secure and scalable data integrity auditing scheme based on hyperledger fabric. Comput. Secur. **92**, 1–16 (2020). https://doi.org/10.1016/j.cose.2020.101741
39. N. Makmor, S.S. Alam, N.A. Aziz, Social support, trust and purchase intention in social commerce era. Int. J. Supply Chain Manag. **7**, 572–581 (2018)
40. S. Moin, A. Karim, Z. Safdar, K. Safdar, E. Ahmed, M. Imran, Securing IoTs in distributed blockchain: Analysis, requirements and open issues. Futur. Gener. Comput. Syst. **100**, 325–343 (2019). https://doi.org/10.1016/j.future.2019.05.023
41. N.R. Moşteanu, A. Faccia, Digital systems and new challenges of financial management–fintech, XBRL, blockchain and cryptocurrencies. Qual. Access Success J. **21**, 159–166 (2020)
42. M. Naz, F.A. Al-zahrani, R. Khalid, N. Javaid, A.M. Qamar, M.K. Afzal, M. Shafiq, A secure data sharing platform using blockchain and interplanetary file system. Sustainability **11**, 1–24 (2019). https://doi.org/10.3390/su11247054
43. G.N. Nguyen, N.H. Le Viet, M. Elhoseny, K. Shankar, B.B. Gupta, A.A. Abd El-Latif, Secure blockchain enabled cyber–physical systems in healthcare using deep belief network with ResNet model. J. Parallel Distrib. Comp. **153**, 150–160 (2021). https://doi.org/10.1016/j.jpdc.2021.03.011
44. M. Niranjanamurthy, B.N. Nithya, S. Jagannatha, Analysis of Blockchain technology: Pros, cons and SWOT. Clust. Comput. **22**, 14743–14757 (2019). https://doi.org/10.1007/s10586-018-2387-5
45. K. Nosalska, G. Mazurek, Marketing principles for industry 4.0—A conceptual framework. Eng. Manag. Prod. Serv. **11**, 9–20 (2019). https://doi.org/10.2478/emj-2019-0016
46. A. Pal, C.K. Tiwari, N. Haldar, Blockchain for business management: Applications, challenges and potentials. J. High Technol. Managem. Res. **32**, 1–12 (2021). https://doi.org/10.1016/j.hitech.2021.100414
47. T. Purcărea, A. Purcărea, Services marketing in the era of disruption and digital transformation. Rom. Econ. Bus. Rev. **12**, 7–26 (2017)
48. K.T. Rahman, Applications of blockchain technology for digital marketing: A systematic review. Blockchain Technol. Appl. Digit. Market. **16-31** (2021). https://doi.org/10.4018/978-1-7998-8081-3.ch002
49. A. Rejeb, J.G. Keogh, H. Treiblmaier, How blockchain technology can benefit marketing: Six pending research areas. Front. Blockchain **3**, 1–12 (2020). https://doi.org/10.3389/fbloc.2020.00003
50. P. Sajana, M. Sindhu, M. Sethumadhavan, On blockchain applications: Hyperledger fabric and Ethereum. Int. J. Pure Appl. Math. **118**, 2965–2970 (2018)
51. P. Siano, G. De Marco, A. Rolán, V. Loia, A survey and evaluation of the potentials of distributed ledger technology for peer-to-peer transactive energy exchanges in local energy markets. IEEE Syst. J. **13**, 3454–3466 (2019). https://doi.org/10.1109/JSYST.2019.2903172
52. D. Sihi, Impacts of blockchain technology in marketing, in *Springer Proceedings in Business and Economics*, (2020), pp. 25–30. https://doi.org/10.1007/978-3-030-47595-6_4
53. O. Ungerman, J. Dedkova, K. Gurinova, The impact of marketing innovation on the competitiveness of enterprises in the context of industry 4.0. J. Compet. **10**, 132–148 (2018). https://doi.org/10.7441/joc.2018.02.09

54. J. Weking, M. Mandalenakis, A. Hein, S. Hermes, M. Böhm, H. Krcmar, The impact of blockchain technology on business models–a taxonomy and archetypal patterns. Electron. Mark. **30**, 285–305 (2020). https://doi.org/10.1007/s12525-019-00386-3
55. X. Zhu, J. Zheng, B. Ren, X. Dong, Y. Shen, MicrothingsChain: Blockchain-based controlled data sharing platform in multi-domain IoT. J. Netw. Netw. Appl **1**, 19–27 (2021)

Applying Advanced Wireless Network Cluster-Tree Topology to Optimize Covid-19 Sanitary Passport Blockchain-Based Security in a Constrained IoT Platform

Sanaa El Aidi, Fatima Zahra Hamza, Siham Beloualid, Abderrahim Bajit, Habiba Chaoui, and Ahmed Tamtaoui

1 Introduction

IoT is a group of infrastructures linking several connected objects that perform one or more functions and communicate through the Internet. IoT allows managing the communication between objects and the transport and the access of data [1]. IoT is a network that must guarantee communication between objects while respecting security and confidentiality requirements.

To ensure this security and fight against cyberattacks, blockchain is one of several solutions proposed to protect data. Blockchain is a storage technology that contains all the information communicated between several users. It ensures the three main requirements of security: (1) confidentiality, i.e., only the user can read the message; (2) integrity, ensuring that the message is received as it is; and (3) availability, services are available to the user.

BC uses a public key as the user's identity. It allows a sender to designate a recipient. Each exchange is accompanied by a key that cannot be modified or deleted to prevent any external disruption of the system. BC then offers a high level of confidentiality [2].

The particularity of blockchain is how all transactions form a chain. Each node of the IoT network has a strong cryptography that guarantees a secure exchange with other nodes and allows for regular monitoring and security updates of the system.

S. El Aidi (✉) · F. Z. Hamza · S. Beloualid · A. Bajit · H. Chaoui
Laboratory of Advanced Systems Engineering (ISA), National School of Applied Sciences, Ibn Tofail University, Kenitra, Morocco
e-mail: fatimazahra.hamza@uit.ac.ma

A. Tamtaoui
National Institute of Posts and Telecommunications (INPT-Rabat), SC Department, Mohammed V University, Rabat, Morocco
e-mail: tamtaoui@inpt.ac.ma

Y. Maleh et al. (eds.), *Advances in Blockchain Technology for Cyber Physical Systems*, Internet of Things, https://doi.org/10.1007/978-3-030-93646-4_15

The blockchain transparently records all transactions so that the source of the error can be identified in case of an incident and the system can immediately solve the problem.

The blockchain is then a database that manages the link between several nodes in a peer-to-peer (P2P) network. The P2P function allows for direct communication between several peers. A node can create a transaction signed with a private key to guarantee the identity of the owner. It can execute a transaction, transmit or receive it, and create new blocks and validate transactions [3].

The consensus of all nodes must verify each transaction. Several consensus algorithms ensure that transactions are reliable and that the rules are respected. There are two common algorithms: the proof of work (PoW) that uses the hash function, which is a function that requires great power to calculate the probability of finding a unique signature for the validation of the transaction. And the second algorithm is the proof of participation (PoP) which ensures that distributed nodes validate the solution found by the miners before the latter can add a new block to the blockchain.

Consensus algorithms are notoriously difficult and require a lot of computation and verification time. However, several kinds of research show that consensus is not really necessary to implement a decentralized asset transfer system using asynchronous trustworthy transfers (AT2) [4]. AT2 can be used to validate transactions either by using the quorum in the case of an authorized or small unauthorized network, in which the quorum checks if the action is correct to validate it, or by randomly selecting the viewpoint of a specific number of nodes in the case of a worldwide unauthorized network.

During the covid-19 health crisis, impacts of the crisis were very strong, and several organizational changes were developed: lockdowns and deconfinement telecommuting, travel restrictions, and others.

Many organizations were challenged to adjust very quickly to the emergency. And for this, the use of communication technologies was and will remain important for maintaining "normal" life between people. The virus pushed several researchers and decision-makers to find a solution to keep the social aspect between people, live their daily lives in the same way as before the covid-19, and protect their lives through several rules such as social distancing, wearing the mask, and vaccination.

Several innovations and adaptations are then deployed during the crisis. The blockchain is then used to secure a large part of these innovations. We quote the project [5], which proposed a solution to decrease the propagation of covid-19 by tracing the trajectories of patients and their relatives. Securing the privacy of patients was developed through a blockchain platform that protects the user's spatiotemporal information.

A team from Nirma University in India has proposed a solution [6] to monitor the social distance between people using artificial intelligence. Using static CCTV cameras and lens-equipped drones, the proposed system uses fast convolutional region-based neural networks (RCNN) and "Only Look Once" (YOLO) models to recognize objects (e.g., people) in real time. It is also possible to calculate the distance between two people using an efficient Euclidean-based method. The physical layer and administrative services can exchange data reliably and securely with the advent of blockchain technology. Payment of fines for social distance violations can also be made using a blockchain-powered wallet. Blockchain is also used to secure the platform to pay the fine in case of noncompliance with the distancing rule. The problem of security of patient data of the pandemic was a challenge to governments. A research was done in this direction to clarify benefits of using blockchain to secure information sharing between people and government accurately and validly, as well as contact tracing [7–9].

In this work, we implemented a smart, synchronized, and secure medical IoT platform that monitors a public area using a set of tests. The people in place must present their vaccination QR code. A first test is performed to read this QR code. If the code is validated, we perform a second test to validate the person's identity corresponding to the vaccination code using the facial recognition algorithm. In the positive case, the person will be able to access the public area. Citizens who are still not vaccinated must present a negative PCR test not exceeding 48 h. We then perform two verification tests, one test to read the barcode of the PCR test and a second test for facial recognition of its carrier. In the situation where a person is presented with neither his vaccination certificate nor a PCR test, we develop a strategy of three tests with three IoT nodes.

The first test is to measure the temperature of the citizen. If it exceeds 37 degrees and/or the citizen exceeds the time, access to the public area is denied. We move on to the third node in the case where the temperature is below 38 degrees. This node consists of identifying the client IoT object by radio frequency identification (RFID), and it then determines the identity of the citizen and his health information. If the citizen exceeds the specified time or the information provided is wrong, then access is denied. If the result of the third node is positive, we move on to the last node, which is based on the principle of facial recognition. This node verifies that the system has used the citizen's photo to recognize his face and identity to access the public area.

Our platform utilizes the InterPlanetary File System (IPFS) as its core system to store secured data and distribute it to other nodes. In the IPFS, initializing the repository generated public/private keys and formed a local folder that hosted the IPFS configuration and stored repository objects from a node. The node ID was made by using a hash of its public key. Peers were restricted to a private IPFS network prepared for the process of distributing files between peers automatically. Peers can access a published file on IPFS to keep in the local copy [10, 11] (Fig. 1).

2 Sanitary Passport Blockchain-Based IoT Platform

Our work consists of securing the transfer and storage of data, and for the transfer, we started by securing the TLS communication using the RSA-SHA256 encryption algorithm, and we also added a security layer on the CoAP protocol using the RSA/AES-SHA256 algorithm to analyze the traffic and see the best in

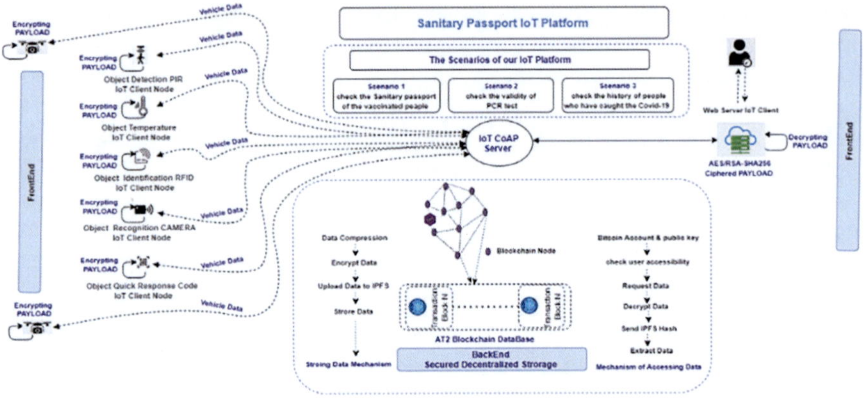

Fig. 1 Intelligent and secure static/dynamic medical IoT platform

terms of execution, memory occupation, and energy consumption, and for data storage security, we used the AT2 (asynchronous trustworthy transfers) blockchain technology which is a class of non-consensus algorithms, and we used IPFS as a central system to store the secured data and distribute it to other nodes in a P2P network.

Medical IoT Applicative Protocols MQTT/CoAP

From the studies already done, it has been found that IoT platforms are vulnerable to attacks from the network system, and especially for healthcare platforms that require better security to ensure confidentiality and data integrity and protect the privacy of patients. For this reason, we thought of adding a security layer on IoT communication protocols such as MQTT and CoAP.

In order to select the best IoT communication protocol according to different criteria, we compared the following two protocols:

The MQTT protocol is defined as a communication protocol based on the publish-subscribe architecture using the TCP/IP protocol. It uses three types of quality of service to ensure a better quality of sending/receiving messages. We define the first type of quality-of-service QoS0 which is used to send the message only once and without acknowledgment of receipt. The second type of QoS is QoS1 which allows to send the message at least once until the MQTT broker acknowledges its reception. The third type of QoS is QoS2 which sends a message while ensuring that the message is received only once, thus avoiding duplicates [12–14].

The CoAP protocol is a lightweight protocol that uses the UDP protocol and is based on the REST architecture that provides *get*, *post*, *put*, and *delete* methods. It uses a 4-byte header, which is appropriate for restricted nodes [15, 16].

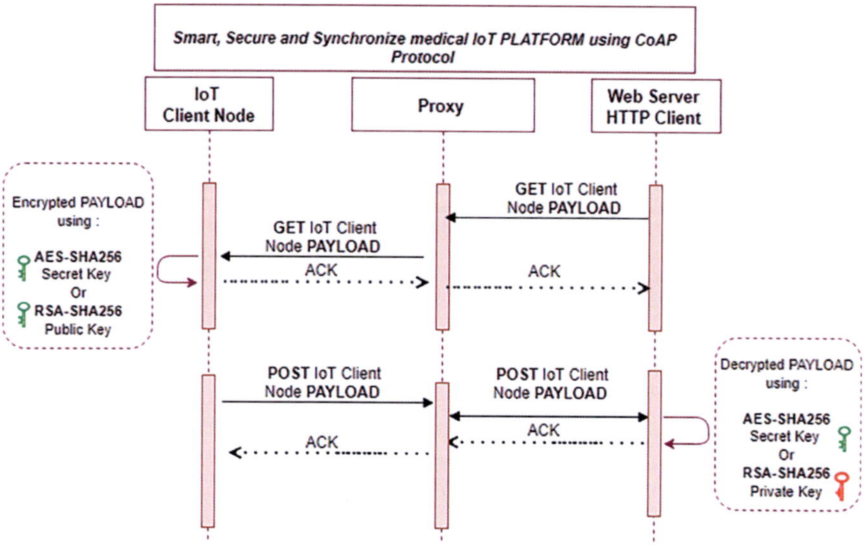

Fig. 2 Diagram sequence of the secured CoAP IoT client's *payload*

In this work, we used CoAP protocol because it consumes less power, energy, resource, bandwidth, and latency comparing to MQTT and HTTP protocols (Fig. 2).

Medical IoT Protocol Payload Security

There are several encryption algorithms to guarantee the five pillars of security using symmetric and asymmetric encryption such as RSA, Digital Signature Algorithm (DSA), AES, Triple Data Encryption Algorithm (3DES), Elliptic-curve cryptography (ECC), Elliptic Curve Integrated Encryption Scheme (ECIES), and SHA. For our work, we have opted for the use of RSA and AES combined with SHA256 [17, 18].

In our platform, we encrypted data with three types of encryption algorithms RSA, RSA, and AES, to see the impact of each algorithm on the platform. The first algorithm, AES, is a symmetric key algorithm in which the sender and receiver share a common key, used to encrypt the message. And we used the RSA algorithm which is based on two keys, public and private key. The public key is used to encrypt messages, and the private key is also derived from these same two prime numbers, and this key is used to decrypt the message, and finally, we have used the AES algorithm [19].

We have also secured the TLS connection by the RSA algorithm with SHA256, TLS [20] is a security protocol of the transport layer, it is used to provide and ensure secure channels for end-to-end communications, and this protocol works in two phases: the first phase is based on the key agreement, the two parties negotiate and agree on a set of encryption algorithms, and then a cryptographic protocol of key exchange is used to perform mutual authentication and to establish a shared session key. In the second phase, the agreed session key performs an authenticated encryption [21].

Medical IoT Network Topologies

Our Medical IoT platform is developed on the cluster topology, which is a mix between advantages of the mesh topology for the reliability of the communication and benefits of star which is fast and provides less memory consumption.

Cluster tree is a case of tree topology where the parent and its child are named cluster. The relationship between nodes is not equal due to the existence of a parent-child relationship. Cluster tree is structured as a tree with the top node, other parent nodes are redirectors, and the end of the tree is end devices. The cluster-tree network has the advantage of easy addition of nodes, making it capable of expansion and bringing together the benefits of star and mesh topologies. The cluster-tree structure allows the network to predict the delay time of data exchanges, and the coordinator has a global monitoring of the system and can have the status of all nodes in the network [22, 23].

Mesh topology is a topology in which each computer and network device is interconnected. The advantage of this topology is that nearby nodes consume less energy and the network is easy to add and remove devices. In case of failure, the structure of this topology allows it to go to other nodes, and the system continues to work. The disadvantage of this topology is that it is more complex and requires a higher overhead, which increases the cost of installation and generally takes longer [24].

Star topology consists of a coordinator and nodes. Nodes communicate with each other through the coordinator. The advantage of this topology is that it is easy to install and the data transmission time is short since only the part between the primary node and the wires can communicate. It is also efficient in fault detection because each device requires only one port and no collision can occur. The connection between nodes is thus easy and does not require a large number of cables. But the major problem is that the exchange between nodes depends on the coordinator, which involves a lot of resources and regular maintenance. In case of a malfunction of the coordinator, the whole network will stop working, so the reliability of communication is lower [25, 26].

Tree topology consists of a coordinator and the routers and end nodes called child. Each child can only communicate with its parent router or core node. This is a combination of the star and mesh topology. The advantage of tree is that if one node goes down, other nodes can continue to operate unaffected. However, the child cannot exchange with the network if the parent is disconnected. The size of the connecting cables between nodes increases with the length of the network, making the system more complex to maintain and very limited and increasing the cost of installing the network [27, 28].

Medical IoT Blockchain-Based Data Storage Security

The development of this platform is based on various topologies (star, tree, mesh, cluster tree) depending on the type of area (urban, tourist, industrial, sports, commercial, etc.). We also used two kinds of security to secure the information: connection security and payload security. For connection security, we used SSL, and for payload security, we used RSA/AES-SHA256 encryption algorithms to provide better security regarding execution time, memory space occupation, and energy consumed. As for data storage, blockchain is used to make information decentralized because it works to gather information in blocks. Our platform relies primarily on CoAP communication protocol, and the application model of IoT CoAP server is implemented in python environment using FastAPI framework. The python application manages the AT2 blockchain and the IPFS P2P network, adds and retrieves the files from the IPFS private network, encrypts the content of the files, and exposes its functionalities through REST services, as seen in Fig. 1. The front end presents IoT client node and IoT Web client, which communicate with the backend using MQTT or CoAP protocol; the back end contains RESTful services that provide interoperability between computer systems on the Internet. RESTful Web services allow the requesting systems to access and manipulate textual representations of Web resources by using a predefined set of stateless operations. Also, the backend includes the repositories, which uses both AT2 bridge and IPFS APIs to communicate with the blockchain and the P2P network, respectively, and it exposes the application functionalities through RESTful Web services; lastly, the P2P network contains AT2 blockchain and IPFS network, which are deployed to Azure cloud properly secured and served with HTTPS using traffic [29].

A peer-to-peer (P2P) network is a collection of nodes that share files with each other. Each node in this network has the same priority level and performs the same tasks as other nodes [7]. This type of network does not contain a central node that manages the sending and receiving of data, so each node can act as both a server and a client depending on other nodes [8].

To ensure data security for our platform, blockchain technology was used, which is considered an open and distributed ledger that can record transactions between two parties in an efficient, transparent, verifiable, and permanent manner, and this technology serves to create trust between different entities where trust is nonexistent or unproven. As a result, these entities are willing to engage in transactions involving the sharing of data that they might not otherwise have done or that they would have required an intermediary to do so, so the elimination of a third party in data exchange transactions promotes the security and privacy of the participants. This principle of data decentralization allows data to be shared within an ecosystem of companies where no single entity is exclusively responsible [7], and blockchain offers better security because it creates an unalterable record of transactions with end-to-end encryption, which excludes fraud and unauthorized activities. In addition, blockchain data is stored on a network of computers, making it virtually impossible to hack. Furthermore, blockchain can address privacy concerns better than traditional computer systems by anonymizing data and requiring permissions to limit access.

Blockchain technology will allow us to save file sharing transactions and their details securely, authentically, fastly, and inexpensively. In our proposed secured medical IoT platform, we have used the AT2 (asynchronous trustworthy transfers) blockchain, a class of consensus-less algorithms.

3 Discussion

The objective of this work is to apply a security layer on our platform by using encryption algorithms such as RSA/AES/SHA256 as well as see the impact each topology has on our platform, in addition to implementing the best communication protocol that matches best the needs of the platform. Therefore, figures show comparative results of implementing CoAP protocol in our platform regarding execution time and memory occupation. We adopted the cluster-*tree* topology for the proper functioning of our platform and also apply RSA and AES encryption methods to finally adopt the one that conserves as much energy as it can and is very hard to crack.

According to the previous results, we noticed that MQTT does not support message labeling, which forces clients to know the type and format of the message sent in advance to establish communication. Unlike the CoAP protocol, which provides default support and offers built-in message handling, nodes can connect. MQTT is also known for its faster execution time than CoAP. MQTT runs on TCP, based on connection establishment and closure, which increases overhead and causes high memory usage. On the other hand, CoAP uses the UDP protocol, which works based on fire and forget, reducing the size of the message and, therefore, the memory used. The results also showed that CoAP requires less energy and thus consumes less resources compared to MQTT.

For security, we opted for CoAP because MQTT is a protocol that supports the lowest level of security based on a username and password. At the same time, CoAP uses multiple encryptions and authentication methods.

From these results (Figs. 3 and 4; Tables 1, 2, and 3), we can observe that this consumption is because of RSA being an asymmetric algorithm that uses two large keys for encryption and decryption (public and private key), which allows it to offer great security and makes it very robust and strong against any attack. However, this

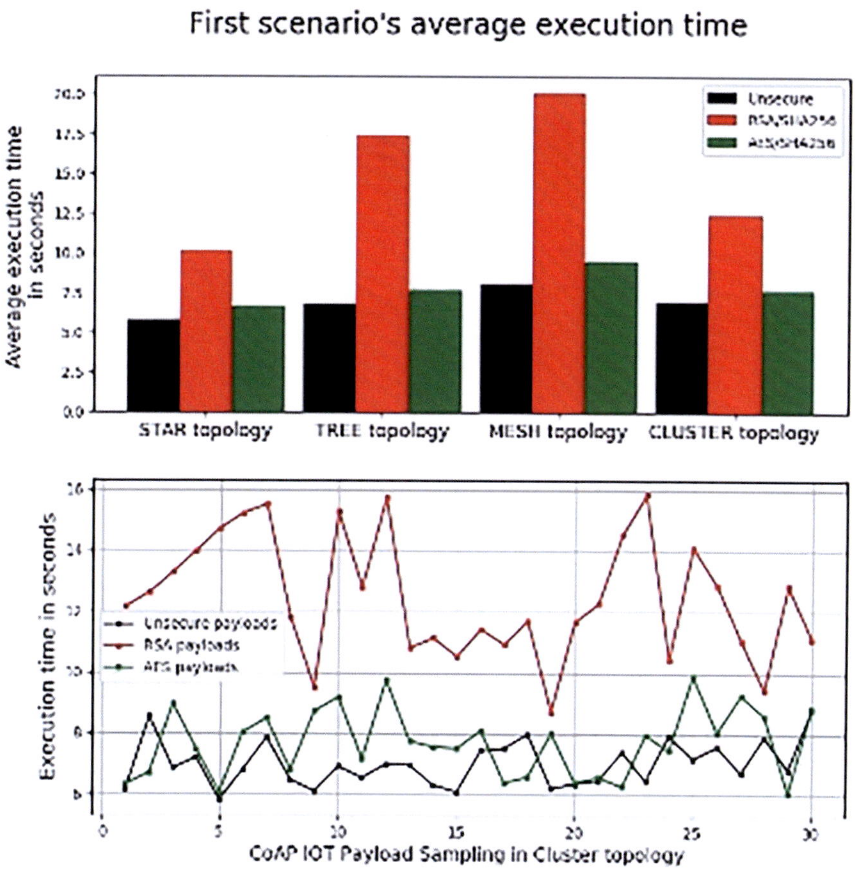

Fig. 3 Average execution time for all scenarios

does not make it suitable for our platform due to its high resource consumption. As for AES, it is a symmetrical algorithm that uses only one key for encryption and decryption, which is why it consumes much fewer resources than RSA (Tables 1, 2, and 3) and is the most recommended for IoT platforms because it is reliable in terms of communication, rapidity, and data security.

When it comes to security algorithms RSA and AES, AES is considered secure. The size of the AES key is nearly 256 bits, while RSA has only 112 bits of security, which makes AES stronger against attacks. The advantage of AES is that it is faster in processing and more powerful. RSA is an asymmetric algorithm, so it uses both encryption and decryption keys, while AES is a symmetric algorithm and uses only one key.

AES is a symmetric algorithm that uses only one key for encryption and decryption, and that is exactly why it is greatly less consuming than RSA. We also

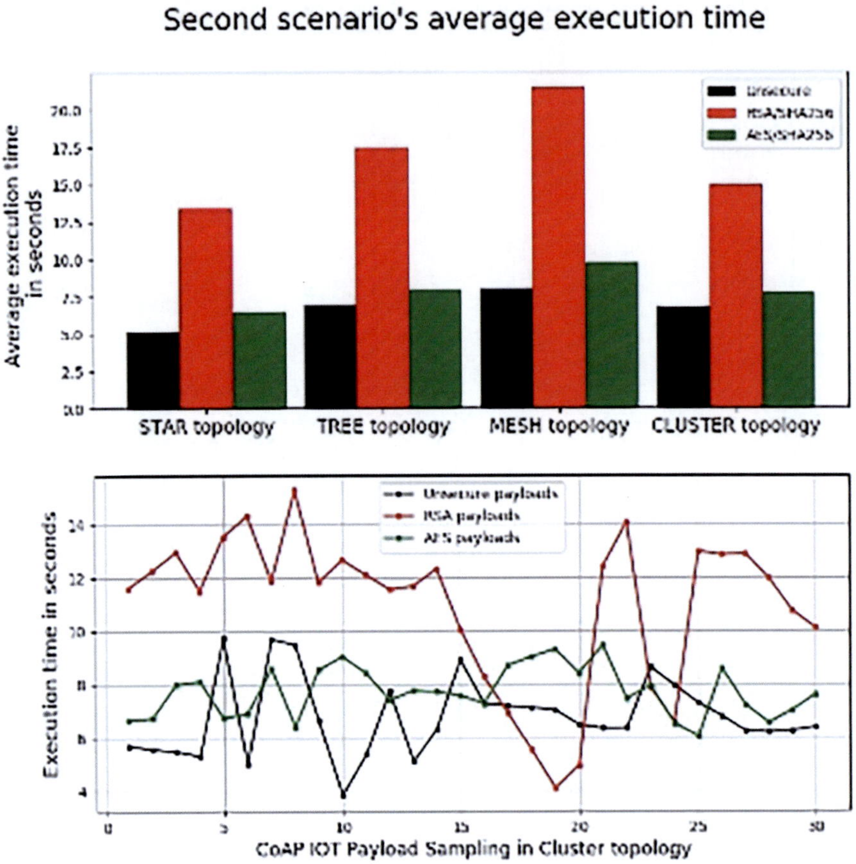

Fig. 3 (continued)

found that AES can be implemented in both hardware and software. It uses higher length key sizes such as 128, 192, and 256 bits for encryption. Hence, it makes the AES algorithm more robust against hacking, and for 128 bit, about 2128 attempts are needed to break. This makes it very difficult to hack it; as a result, it is a very safe protocol.

4 Conclusion and Perspectives

This developed platform will enable us to minimize the spread of covid-19 by controlling citizens' access to public areas and keeping in mind that there are three types of citizens: the vaccinated ones, the ones with the PCR test, and the ones

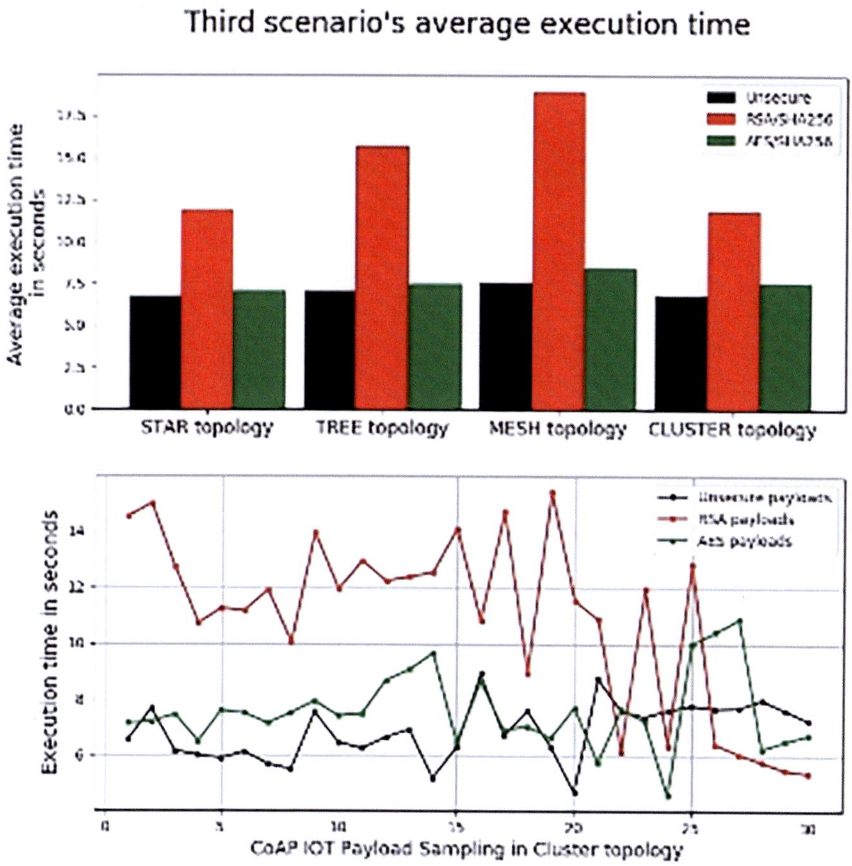

Fig. 3 (continued)

with only an RFID tag. The platform relies on the Constrained Application Protocol to ensure reliable communication between nodes, and it also relies on cluster-tree topology as the main network topology. To further secure the communication between nodes, we opted for the AES-SHA256 cryptography algorithm to encrypt the data and information exchanged.

In future work, we will implement other security methods such as ECIES and discuss radio frequency protocols such as 6LoWPAN, Zigbee, and LoRa, which are known for their efficiency in IoT networks.

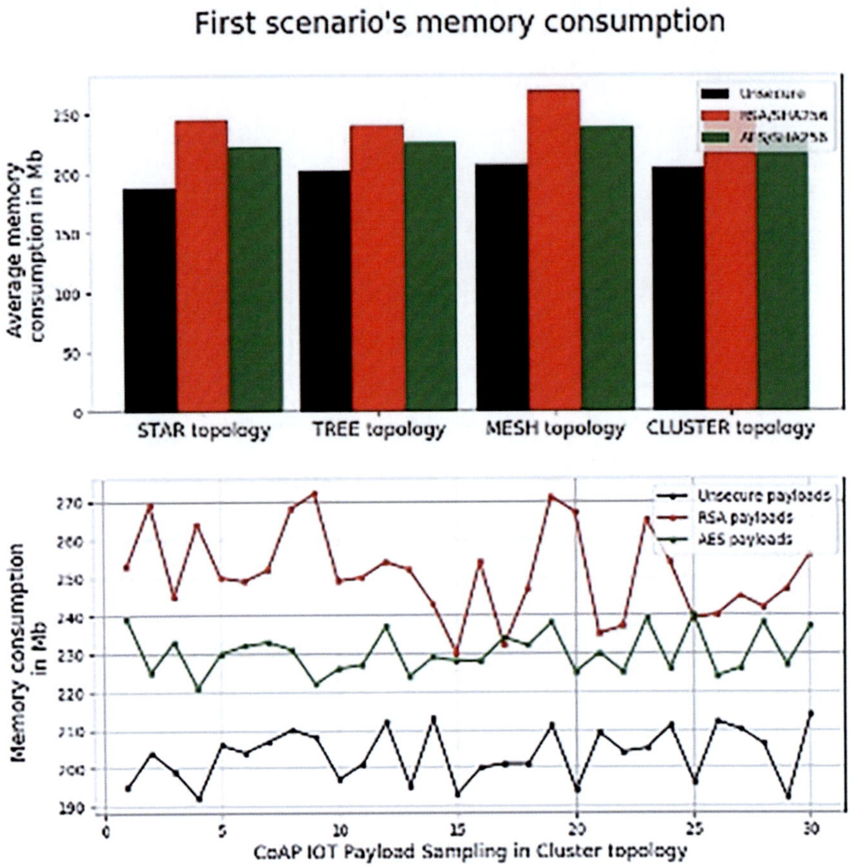

Fig. 4 Average memory occupation for all scenarios

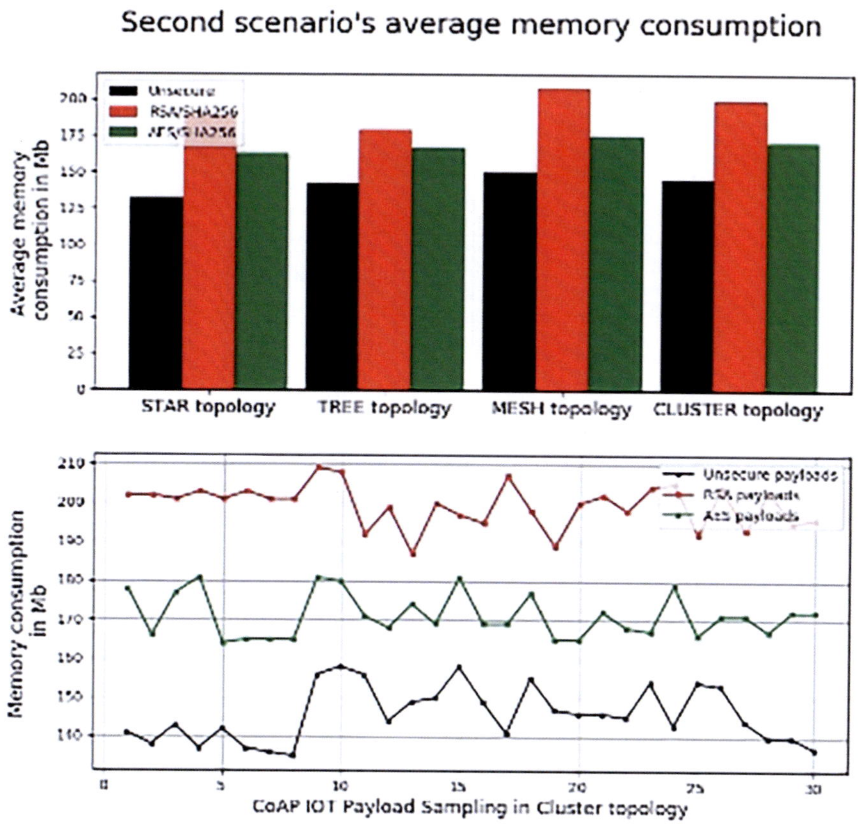

Fig. 4 (continued)

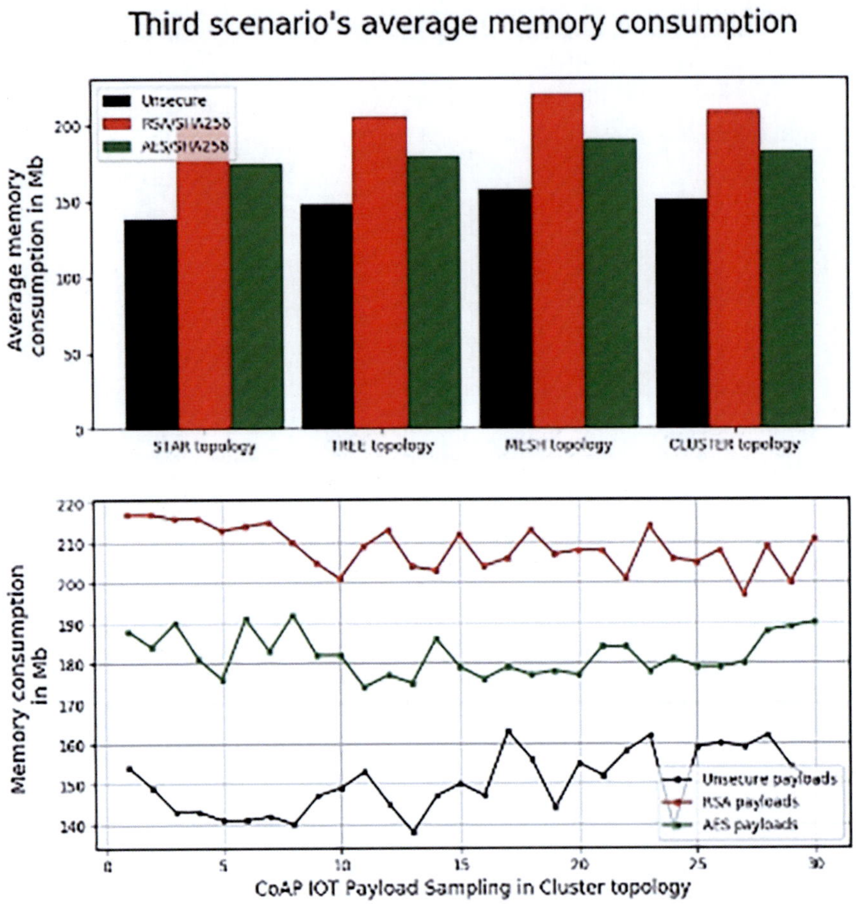

Fig. 4 (continued)

Table 1 Time consumption and RAM occupation of the first scenario

		Star	Tree	Mesh	Cluster
Time (s)	RSA	10.13	17.46	20.12	12.47
	AES	6.71	7.72	9.56	7.70
RAM (Mb)	RSA	244.6	249.4	268.0	251.0
	AES	222.3	225.6	237.4	230.2

Table 2 Time consumption and RAM occupation of the second scenario

		Star	Tree	Mesh	Cluster
Time (s)	RSA	13.38	17.46	21.38	14.91
	AES	6.42	7.92	9.76	7.71
RAM (Mb)	RSA	189.3	179.5	207.5	199.5
	AES	162.9	167.3	175.1	171.1

Table 3 Time consumption and RAM occupation of the third scenario

		Star	Tree	Mesh	Cluster
Time (s)	RSA	11.93	15.72	18.95	11.89
	AES	7.07	7.52	8.51	7.61
RAM (Mb)	RSA	200.7	204.8	219.0	208.7
	AES	173.9	179.4	190.0	181.9

References

1. B. Dorsemaine, J. Gaulier, J. Wary, N. Kheir, P. Urien, Internet of things: A definition & taxonomy, in *2015 9th International Conference on Next Generation Mobile Applications, Services and Technologies*, (2015), pp. 72–77. https://doi.org/10.1109/NGMAST.2015.71
2. Towards an optimized blockchain for IoT, in 2017 IEEE/ACM Second International Conference on Internet-of-Things Design and Implementation (IoTDI), pp. 173–178
3. M. Daniel, O. Benedict, Blockchain mechanisms for IoT security. Internet Things **1–2**, 1–13 (2018). https://doi.org/10.1016/j.iot.2018.05.002
4. R. Guerraoui, P. Kuznetsov, M. Monti, M. Pavlovic, D. Seredinschi, *AT2: Asynchronous Trustworthy Transfers* (2019)
5. Z. Wen et al., Blockchain-empowered contact tracing for COVID-19 using crypto-spatiotemporal information, in *2020 IEEE International Conference on E-health Networking, Application & Services (HEALTHCOM)*, (2021), pp. 1–6. https://doi.org/10.1109/HEALTHCOM49281.2021.9398978
6. S. Tanwar, R. Gupta, M.M. Patel, A. Shukla, G. Sharma, I.E. Davidson, Blockchain and AI-empowered social distancing scheme to combat COVID-19 situations. IEEE Access **9**, 129830–129840 (2021). https://doi.org/10.1109/ACCESS.2021.3114098
7. M. Kaur, M. Murtaza, M. Habbal, Post study of Blockchain in smart health environment, in *2020 5th International Conference on Innovative Technologies in Intelligent Systems and Industrial Applications (CITISIA)*, (2020), pp. 1–4. https://doi.org/10.1109/CITISIA50690.200.9371819
8. L.-Y. Hou, T.-Y. Tang, T.-Y Liang, IOTA-BT: A P2P file-sharing system based on IOTA. J. MDPI Electronics (2020)
9. L. Wang, J. Kangasharju, Real-world Sybil attacks in BitTorrent mainline DHT, in *Proceedings of the 2012 IEEE Global Communications Conference (GLOBECOM) Anaheim, CA, USA*, p. 2012
10. M.S. Ali, K. Dolui, F. Antonelli, IoT data privacy via blockchains and IPFS, in *Proceedings of the 7th International Conference on the Internet of Things, Linz, Austria*, (2017), p. 14
11. L. Balduf, S. Henningsen, M. Florian, Monitoring Data Requests in Decentralized Data Storage Systems: A Case Study of IPFS, arXiv:2104.09202 (2021)
12. S.S. Prayogo, Y. Mukhlis, B.K. Yakti, The use and performance of MQTT and CoAP as internet of things application protocol using NodeMCU ESP8266, in *2019 Fourth International Conference on Informatics and Computing (ICIC)*, (2019). https://doi.org/10.1109/icic47613.2019.8985850
13. S. El Aidi, A. Bajit, A. Barodi, H. Chaoui, A. Tamtaoui, An advanced encryption cryptographically-based securing applicative protocols MQTT and CoAP to optimize medical-IOT supervising platforms. Lect. Notes Data Eng. Commun. Technol. **72**, 111–121 (2021)
14. E.M. Fdil, M. El Haidi, A. Bajit, S. El Aidi, A. Barodi, A. Tamtaoui, An new constrained protocol S-CoAP applied to optimize COVID-19 medical IOT intelligent and security-based DATA supervising platform, in *2020 International Symposium on Advanced Electrical and Communication Technologies (ISAECT)*, (2020), pp. 1–6. https://doi.org/10.1109/ISAECT50560.2020.9523711

15. A. Rhbech, H. Lotfi, A. Bajit, A. Barodi, S. El Aidi, A. Tamtaoui, An optimized and intelligent security-based message queuing protocol S-MQTT applied to medical IOT COVID-19 DATA monitoring platforms, in *2020 International Symposium on Advanced Electrical and Communication Technologies (ISAECT)*, (2020), pp. 1–6. https://doi.org/10.1109/ISAECT50560.2020.9523678
16. A. Cilardo, L. Coppolino, N. Mazzocca, L. Romano, Elliptic curve cryptography engineering. Proc. IEEE **94**(2), 395–406 (2006). https://doi.org/10.1109/JPROC.2005.862438
17. D. Mahto, D.K. Yadav, RSA and ECC: A comparative analysis. Int. J. Appl. Eng. Res. **12**(19), 9053–9061 (2017)
18. C. Arun, B. Hedayath, D. Sivakumar, M. Rizwan, M. Kumar, Communication Engineering, *Secured Image Transmission Using Elliptic Curve Cryptography (ECC)* (2020). https://doi.org/10.37896/jxu14.5/400
19. A. Kajal, G. Badolia, Enhanced cloud storage security using ECC-AES a. Hybrid Approach. **4. 5**, 10.18231/2454-9150.2018.0593 (2018)
20. J. Li, R. Chen, J. Su, X. Huang, X. Wang, ME-TLS: Middlebox-enhanced TLS for internet-of-things devices. IEEE Internet Things J. **1–1** (2019). https://doi.org/10.1109/jiot.2019.2953715
21. http://www.steves-internet-guide.com/mosquitto-tls/, Access on July 21, 2021
22. H.M. Kelagadi, P. Kumar, A cluster-tree based topology control for wireless sensor network, in *2018 International Conference on Electrical, Electronics, Communication, Computer, and Optimization Techniques (ICEECCOT)*, (2018), pp. 643–649. https://doi.org/10.1109/ICEECCOT43722.2018.9001417
23. M. Ouadou et al., Improved cluster-tree topology adapted for indoor environment in Zigbee sensor network. Proced. Comput. Sci. **94**, 272–279 (2016)
24. S.R. Lee, J. Back, J. Oh, M. Jeong, A mesh topology formation scheme for IEEE 802.15.4 based wireless sensor networks, in *2015 Seventh International Conference on Ubiquitous and Future Networks*, (2015), pp. 150–152. https://doi.org/10.1109/ICUFN.2015.7182523
25. K. Kosek, M. Natkaniec, L. Vollero, A.R. Pach, An analysis of star topology IEEE 802.11e networks in the presence of hidden nodes, in *2008 International Conference on Information Networking*, (2008), pp. 1–5. https://doi.org/10.1109/ICOIN.2008.4472755
26. K. Keyur, M. Sunil, *Internet of Things-IOT: Definition, Characteristics, Architecture, Enabling Technologies, Application & Future Challenges* (2016)
27. Ovidiu Vermesan SINTEF, Norway, Dr. Peter Friess EU, Belgium, in *Internet of Things: Converging Technologies for Smart Environments and Integrated Ecosystems*, river publishers' series in communications (2013)
28. V. Lampkin et al., Building smarter planet solutions with MQTT and IBM WebSphere MQ telemetry IBM. O (2012)
29. G. Rachid, K. Petr, M. Matteo, P. Matej, S. Dragos-Adrian, AT2: Asynchronous Trustworthy Transfers. arXiv preprint arXiv:1812.10844 (2018)

Index

Y. Maleh et al. (eds.), *Advances in Blockchain Technology for Cyber Physical Systems*, Internet of Things, https://doi.org/10.1007/978-3-030-93646-4

The manufacturer's authorised representative in the EU is Springer Nature Customer Service Centre GmbH, Europaplatz 3, 69115 Heidelberg, Germany. If you have any concerns regarding our products, please contact ProductSafety@springernature.com

Printed and bound by CPI Group (UK) Ltd, Croydon, CR0 4YY
07/10/2026
02229168-0003